寰宇文献 Universal Library | SINOLOGY 系列

SELECTED WORKS OF BERTHOLD LAUFER

劳费尔著作集

第十二卷

[美] 劳费尔 著

黄曙辉 编

ZHONGXI BOOK COMPANY

图书在版编目(CIP)数据

劳费尔著作集 / (美) 劳费尔著；黄曙辉编. —上海：中西书局，2022

(寰宇文献)

ISBN 978-7-5475-2015-4

Ⅰ. ①劳… Ⅱ. ①劳… ②黄… Ⅲ. ①劳费尔－人类学－文集 Ⅳ. ①Q98-53

中国版本图书馆CIP数据核字（2022）第207067号

第 12 卷

174 翟理斯英译《聊斋志异》书评 …………………………………… 1

175 美索不达米亚的鸵鸟蛋壳杯以及古代和现代的鸵鸟 ……………… 9

176 中国、日本、中亚、印度、埃及、巴勒斯坦、希腊和意大利的墨水历史……………………………………………………………… 81

177 驯养研究中的各种方法 ………………………………………… 159

178 卡特《中国印刷术的发明及其西传》书评 ……………………… 167

179 玛瑙——与之相关的考古学与民间传说 ………………………… 175

180 鸣虫与斗虫 ……………………………………………………… 207

181 麒麟传入考 ……………………………………………………… 259

182 中国的斗蟋蟀 …………………………………………………… 381

183 竞技场上的蟋蟀 ………………………………………………… 387

184 飞机制造的历史背景 …………………………………………… 389

185 电视的历史背景 ………………………………………………… 499

186 马伯乐《古代中国》书评 ………………………………………… 507

187 唤起科学家兴趣的海龟化石 …………………………………… 511

188 巴托尔德《蒙古入侵时期的中亚》书评 ………………………… 515

189 李济《中国民族的形成》书评 …………………………………… 521

190 美国植物的迁移 ………………………………………………… 525

191 关于“烈酒”一词可能源自东方 ………………………………… 541

192 中国学生的使命 ………………………………………………… 547

193 麦吉尔大学葛思德中文图书馆致辞 …………………………… 557

194 1928 年北美考古田野作业 …… 567

195 乔治 · 尤摩弗帕勒斯的中国青铜器 …… 573

196 毡的早期历史 …… 581

197 一份中文－希伯来文手卷:中国犹太人史的新史料 …… 601

198 新增玉器收藏 …… 613

199 中国的铃、鼓和镜 …… 615

174

翟理斯英译《聊斋志异》书评

Vol. 39. JANUARY–MARCH, 1926. No. 151

THE JOURNAL OF AMERICAN FOLK-LORE.

CONTENTS.

			PAGE
1.	Serrano Tales	*Ruth Benedict*	1
2.	Literary Aspects of Winnebago Mythology	*Paul Radin*	18
3.	John Kuners	*Dougald Mac Millan*	53
4.	Yuma Dreams and Omens	*Edward Winslow Gifford*	58
5.	Isleta Variants: A Study in Flexibility	*Esther Schiff Goldfrank*	70
6.	Notes and Queries		79
7.	Book Reviews		86

NEW YORK
PUBLISHED BY THE AMERICAN FOLK-LORE SOCIETY
G. E. STECHERT & CO., NEW YORK, AGENTS.

BOOK REVIEWS.

STRANGE STORIES FROM A CHINESE STUDIO. Translated and annotated by Herbert A. Giles. New York, Boni and Liveright, 1925.

This is a new and welcome reprint of a book first published in 1880 in London (again in 1908) in two handsomely printed octavo volumes of 836 pages. In this American re-edition the same matter is compressed into 488 closely printed pages, without changes or additions in the notes. The English edition made reading more pleasant; but since this was out of print, the publishers deserve thanks for rendering this interesting work accessible to the American public. We have but two minor grudges to register: the dizzy monster on the paper wrapper which is not Chinese at all, but the production of some jazz artist, and a lavish use of inverted p's instead of d's scattered throughout the volume, which irritate a reader with a trained eye for typographical accuracy.

The first edition contained a serviceable index to the notes which unfortunately is here omitted, and some of the notes like that on footbinding (p. 47) have been considerably abridged. Many of the footnotes contain valuable information on Chinese customs, manners, and institutions, but they were written at a time when the Manchu dynasty was still in power, and things have changed so swiftly and radically under the so-called republican regime that most statements which are still left in the present tense should be changed into the past.

The Liao chai chi i, as the Chinese title of the work runs, was written toward the middle of the seventeenth century by P'u Sung-ling, and is justly regarded as a masterpiece of Chinese literature. It consists of a collection of three hundred, for the greater part, brief stories, but also anecdotes, terse sketches, events of interest, and moral examples. Written in a brilliant style which in its brevity recalls the classical language, and seasoned with covert literary allusions, it is eagerly devoured by the literary gourmands of China, but has also become very popular among the masses. There is a special class of public story-tellers who make it their business to recite exclusively the stories of the Liao chai chi i.

Part of the stories were also translated into Manchu; this edition with the Chinese text was printed in 1848 and contains 129 selected stories. Four of the fox stories were recently translated into Russian by V. M. Alekseyev (Vostok, No. 1, 1922). Professor Giles' translation embodies 164 stories, the best contained in the original.

In the introduction it is said that the work was completed in 1679, but this date cannot be quite correct (it is merely the date of the preface), as at the end of No. 20 the year 1682 is given as the date of the employment of a certain woman in the household of an official, who had supplied the author with a story. Again, No. 50 refers to a severe drought in 1682.

It is surprising that although the translation of Giles has been available for forty-five years, no study whatever has been made of its contents, that no one has ever taken the trouble to analyze and trace the sources of information used by the author and that no folklorist has ever availed himself of this material for comparative or any other purposes. And yet this collection is a mine of information, it presents an accurate picture of the state and mind of Chinese society of the seventeenth century, its dream life, its belief in magic and witchcraft, transformations, spirits, devils, purgatory, retribution, its love for the wondrous, uncanny, and adventurous. The most striking feature about this remarkable book is that its author appears to have been in close touch with the people and a recorder of live folk-lore himself. In his preface he remarks, "I get people to commit to writing what they know of the supernatural, and subsequently I dress it up in the form of a story; and thus in the lapse of time my friends from all quarters have supplied me with quantities of material, which from my habit of collecting has grown into a vast pile . . . Midnight finds me with an expiring lamp, while the wind whistles mournfully without, as over my cheerless table I piece together my tales."

At the end of one of the most fascinating stories (No. 23) the author observes, "I learned the above when travelling through I-chou, where I was detained at an inn by rain, and read a biography of Mr. Sang (the hero of the story) written by a comrade of his, Wang Tse-chang. It was lent me by a Mr. Liu Tse-ching, a relative of Sang's, and was quite a long account. This is merely an outline of it." At the end of No. 86 he writes, "Mr. Chao told me this story with all its details." In No. 104 be records a reminiscence of his early boyhood, a marvelous juggler's trick which he had witnessed. A brief item relating to a murder and suicide is taken without embellishment from the Peking Gazette (No. 48).

Altogether the impression prevails that with a few exceptions these stories have not been derived from books, but were recorded from oral accounts or somehow came within the author's personal experience (compare No. 60). In No. 128 he gives a vivid description of an earthquake which he witnessed in 1668.

A prominent place is taken by numerous stories of foxes. The fox is believed to be capable of appearing at will in human form and of doing either good or evil to friend or foe. Foxes in the guise of beautiful girls tempt and bewitch men, and male foxes have relations with women. From the story "The Trader's Son" (No. 13) it would follow that a temporary state of mental aberration in women was ascribed to the action of foxes who had cast spells on them; when the foxes were killed with poisoned wine, the women had peace and their reason returned, though one of them shortly afterwards died of consumption. What the fox cannot get rid of, even if he or she assumes human shape, is the tail which often betrays him. There are also good and kind fox girls who help a man in distress, as in No. 32, where Miss Hung-yü, a fox, provides a poor, struggling scholar with money to pay the purchase price of a wife, and later on, after he has lost his father and wife under tragic circumstances, raises his only son and through her industry restores the prosperity of his house. As the fox has the power of knowing the future, certain classes of soothsayers are believed to be possessed by foxes (No. 154). The daughter of a fox man has the gift of foreknowing whether the harvest will be good or bad, and her advice is taken in such matters (No. 35). A fox lady may be reborn as a good spirit and save the man whom she loved on earth (No. 19). Friendly intercourse with a male fox may not harm a man, and his advice may bring him considerable wealth (No. 22).

The ninth story, entitled "Magical Arts", is interesting because it illustrates the wide-spread belief in the magical power of paper men, clay figures and wooden statues which act as live beings at the instigation of a necromancer who tries to ruin a man's life (compare also No. 134).

Some stories consist of incongruous parts derived from different cycles and but loosely woven together. In No. 63 the hero's adventures are told in the country of the Lo-ch'a (Rākshasa) where ugly features are looked upon as beautiful; this story is a splendid satire of court and official life, worthy of a Swift. Immediately joined to it is a fairy-tale, the hero's voyage to the palace of a sea-god who marries him to his daughter. Though an excellent scholar, P'u Sung-ling, like many others of his countrymen, flunked in the civil service examinations for the higher degrees which opened the gate to an official career. Luckily he did, otherwise we should be deprived of his collection of stories which gained for him everlasting fame. But owing to his failure he often strings his harp to a persiflage of the ruling officialdom with its stupidity, greed, and corruption (compare No. 72).

That the spirits of the dead temporarily return to earth and commune with the living does not surprise us, common enough as this is in our own folk-lore; but what is beyond our imagination is that the spirit of a woman returns in the flesh, marries a man, and gives birth to two sons (No. 15). In another instance, the spirit of a man who had died of the poisonous shui-mang plant marries a girl in the same condition, and both come back to the world to care for the man's distressed mother (No. 171). A dead man appears to his childless wife in a clay image fashioned by her after his likeness, and she gives birth to a son (No. 134).

Of all the expedients of the Chinese story teller the dream-motive is most prolifically used. The reality of dreams is always strongly emphasized. "Looking upon dreams as realities and mistaking realities for dreams," as one of the characters in a story puts it, furnishes the clue to many plots. A dream, in one case, is caused by bees who have settled on a man's pillow and who have derserted their hive because a large snake had settled in it. This snake brought about the man's dream of a huge monster which threatens to destroy a kingdom (No. 70). A dream must not be divulged to any one for fear of destroying its reality. Faith in dreams is also ridiculed, and the lesson is inculcated that dreams should not be believed (No. 122). In a recent book "Märchen und Traum" by G. Jacob, written with special reference to the Orient, not one example is quoted from Chinese folk-lore, while the subject of dreams in Chinese records would be well deserving of a doctor's dissertation.

In some cases it is possible to trace a story or a motif to an earlier source, and through this attempt it becomes more intelligible. To cite one example, — No. 82, entitled The Sea-serpent, reads as follows: "A trader named Chia was voyaging on the south seas when one night it suddenly became as light as day on board his ship. Jumping up to see what was the matter, he beheld a huge creature with its body half out of the water, towering up like a hill. Its eyes resembled two suns, and threw a light far and wide; and when the trader asked the boatmen what it was, there was not one who could say. They all crouched down and watched it; and by and by the monster gradually disappeared in the water again, leaving everything in darkness as before. And when they reached port, they found all the people talking about a strange phenomenon of a great light that had appeared in the night, the time of which coincided exactly with the strange scene they had witnessed."

Professor Giles' comment is that the "sea-serpent" in this case was probably nothing more or less than some meteoric phenomenon. I do not believe that this interpretation hits the mark: rationalistic explanations of myths and tales are usually wrong, because the folk mind is imaginative, not rationalistic. We must first recall Hüan Tsang's story of the merchant prince from Jāguda (cf. S. Beal, Buddhist Records, II, p. 125), who with some other merchants embarked in a ship on the *southern sea* (as above) and lost his way in a tempest. The mariners finally sight a great mountain with steep crags (cf. above, "lowering up like a hill") and a double sun (as above, "its eyes resembled two suns") radiating from afar. The merchants are overjoyed at the prospect of finding rest and refreshment on this mountain, until the merchant-master exclaims, "It is no mountain, it is the fish *makara* (whale); the high crags and precipices are but its fins and mane; the double sun is its eyes as they shine." This is the same sea-monster as in the "Strange Stories," and belongs to the well-known cycle of marine legends in which sailors take the back of a whale or a turtle for an island. This sailor's yarn was known to the Chinese in a perfect form as early as the sixth century A. D. when it was recorded by the emperor Hiao Yüan in his book Kin lou tse ("The Golden Tower"). The same motif appears in the Greek Romance of Alexander the Physiologus, and finally in Sindbad's adventures in the Arabian Nights. I hope to treat the migration of this story in detail on another occasion.

A curious variant of Dido's ruse occurs in No. 100. "Formerly when the Hollanders were permitted to trade with China, the officer in command of the coast defences would not allow them, on account ot their great numbers, to come ashore. The Hollanders begged very hard for the grant of a piece of land, such as a carpet would cover; and the officer above mentioned, thinking that this could not be very large, acceded to their request. A carpet was accordingly laid down, big enough for about two people to stand on; but by dint of stretching, it was soon enough for four or five; and so they went on, stretching and stretching, until at last it covered about an acre, and by and by, with the help of their knives, they had filched a piece of ground several miles in extent." The Chinese have two written versions of the Dido story, formerly discussed in my article "Relations of the Chinese to the Philippine Islands." One fastens the ruse on the Spaniards in connection with the foundation of Manila, and is contained in the Tung si yang k'ao printed in 1618 and repeated in the Annals of the Ming dynasty (chap. 323). The other relates to the settling of the Dutch in the island of Formosa in 1620, and is found in the T'ai wan fu chi ("Gazetteer of the Prefecture of T'ai-wan," i. e. Formosa), the first edition of which was completed in 1694. In both these versions the trick is performed by means of a cowhide, exactly as in the western prototype, and the story is well and clearly told. The substitution of the carpet for the hide in P'u Sung-ling's account is a rather unfortunate idea, but surely it is not the author's invention. The Ming Annals had not yet appeared during his lifetime, and were in fact completed only in 1724; the Gazetteer of Formosa also came out two decades after the conclusion of his work. He might have known the Tung si yang k'ao, but the fact is evident that he did not. The point I wish to make is that P'u Sung-ling's story is independent of the two literary official versions adopted into the official histories and that an oral popular variant must have been afloat at the same time; it is the latter which was recorded by P'u Sung-ling.

No one should fail to read tale 50, "The Flower Nymphs," a love story of great poetic beauty and artistic quality, one of the best ever produced by a Chinese writer. There are many humorous stories like Nos. 4, 5, 62, 122, and each harbors an interesting bit of Chinese thought and belief. This is a veritable source-book of documents for social and psychological research. So many approach me with questions as to the "best" books on China. Of foreign writers I recommend only Marco Polo and Archdeacon Gray, for the rest I warn against the books with the pompous general title "China and the Chinese" and recommend study from within, reading of history, stories, dramas, poetry translated from the Chinese in which the Chinese speak themselves without a chance for the blatant foreigner to interfere. As the "Strange Stories" are now accessible again, it will be the first on my list of recommendation. The translation is excellent and elegant, and if Professor Giles had given us nothing but this fascinating book, this alone would assure to him our lasting gratitude.

B. LAUFER.

Field Museum, Chicago.

THE FOLK MUSIC OF THE WESTERN HEMISPHERE, Julius Mattfeld. A list of references in the New York Public Library, reprinted with additions from the Bulletin of the New York Public Library, November and December, 1924. New York, 1925.

In the reviewer's opinion this bibliography is one of the most valuable and important contributions in the field of folk music that has come out in recent years. The compilation of the list is the result of years of work on the part of Mr. Julius Mattfeld, formerly associated with the music department of the New York Public Library, now with the Radio Broadcasting station WEAF, New York. It covers every reference, obscure or otherwise, to folk music of the western hemisphere contained in the library up to December 1924, although before the bibliography was off the press the accessions to the library had been greatly augmented, so that the bibliography is not quite up to date, much to the compiler's regret. It is conveniently divided into a number of sections so that desired material may be easily located. In addition, the library catalogue numbers have been given for each reference, which saves considerable time to one using the library copies. There are sections on Canadian, Cowboy, Creole, Eskimo, Indian (North American, not including Mexican) Indian (Central and South American, including Mexican), Latin American, Negro (North American), Negro (Central and South American), United States songs, and an appendix on musical instruments, besides an excellent index covering authors as well as important titles. This little book should be in the library of every student of folk song as a

175

美索不达米亚的鸵鸟蛋壳杯以及古代和现代的鸵鸟

Ostrich Egg-shell Cups of Mesopotamia and the Ostrich in Ancient and Modern Times

BY

BERTHOLD LAUFER

CURATOR OF ANTHROPOLOGY

9 Plates and 10 Text-figures

ANTHROPOLOGY

LEAFLET 23

FIELD MUSEUM OF NATURAL HISTORY

CHICAGO

1926

The Anthropological Leaflets of Field Museum are designed to give brief, non-technical accounts of some of the more interesting beliefs, habits and customs of the races whose life is illustrated in the Museum's exhibits.

LIST OF ANTHROPOLOGY LEAFLETS ISSUED TO DATE

No.	Title	Price
1.	The Chinese Gateway	$ .10
2.	The Philippine Forge Group	.10
3.	The Japanese Collections	.25
4.	New Guinea Masks	.25
5.	The Thunder Ceremony of the Pawnee	.25
6.	The Sacrifice to the Morning Star by the Skidi Pawnee	.10
7.	Purification of the Sacred Bundles, a Ceremony of the Pawnee	.10
8.	Annual Ceremony of the Pawnee Medicine Men	.10
9.	The Use of Sago in New Guinea	.10
10.	Use of Human Skulls and Bones in Tibet	.10
11.	The Japanese New Year's Festival, Games and Pastimes	.25
12.	Japanese Costume	.25
13.	Gods and Heroes of Japan	.25
14.	Japanese Temples and Houses	.25
15.	Use of Tobacco among North American Indians	.25
16.	Use of Tobacco in Mexico and South America	.25
17.	Use of Tobacco in New Guinea	.10
18.	Tobacco and Its Use in Asia	.25
19.	Introduction of Tobacco into Europe	.25
20.	The Japanese Sword and Its Decoration	.25
21.	Ivory in China	.75
22.	Insect Musician and Cricket Champions of China (in press)	
23.	Ostrich Egg-shell Cups of Mesopotamia and the Ostrich in Ancient and Modern Times	.50

D. C. DAVIES
DIRECTOR

FIELD MUSEUM OF NATURAL HISTORY
CHICAGO, U. S. A.

OSTRICH EGG-SHELL CUP FROM GRAVE AT KISH, MESOPOTAMIA (p. 2).
ABOUT 3000 B.C. IN FIELD MUSEUM.
About one-third actual size.

FIELD MUSEUM OF NATURAL HISTORY
DEPARTMENT OF ANTHROPOLOGY
CHICAGO, 1926

LEAFLET NUMBER 23

Ostrich Egg-shell Cups of Mesopotamia and the Ostrich in Ancient and Modern Times

CONTENTS

Page
The Ostrich in Mesopotamia 2
The Ostrich in Palestine, Syria, and Arabia 9
The Ostrich in Ancient Egypt 16
The Ostrich in the Traditions of the Ancients 21
The Ostrich in the Records and Monuments of the Chinese 29
The Ostrich in Africa 34
The Domestication of the Ostrich 41
The Ostrich in America 47
Bibliographical References 51

THE OSTRICH IN MESOPOTAMIA

In his "Report on the Excavation of the 'A' Cemetery at Kish, Mesopotamia" published by Field Museum (Memoirs, Vol. I, No. 1), Ernest Mackay writes as follows: "A rare object found in grave 2 was a cup which had been made from an ostrich shell by cutting about one-third of the top of the shell away and roughly smoothing the edge. It was the only one of its kind found in the cemetery, and it was in such a very bad condition with so many pieces missing that it could neither be restored nor drawn. The remains of a similar cup were found in one of the chambers of a large building of plano-convex bricks, about a mile from the 'A' cemetery, which appears to be of the same date. The ostrich is still found in the Arabian desert, and was doubtless plentiful in early times. Its feathers as well as its eggs were utilized by the ancients."

In the course of further excavations on the ancient sites of Kish great quantities of fragments of ostrich egg-shell were brought to light and, together with other collections, mainly pottery, stone, and metal, were recently received in the Museum. Having read in Chinese records of ostrich eggs anciently sent as gifts from Persia to the emperors of China and being aware of the importance of this subject in the history of ancient trade, I took especial interest in these egg-shell fragments and induced T. Ito, a Japanese expert at treating and repairing antiquities, to restore three of these cups completely. The result of his patient and painstaking labor is shown in Plates I and II illustrating two of the cups. These restorations are true and perfect; that is, they consist of some eighty shards each, accurately and perfectly joined, without the use of other substances or recourse to filling-in. Thanks to the admirable skill of Mr. Ito we now have these beautiful cups before us, exactly in the shape, as they were anciently used by the Sumerians. These cups, almost porcelain-like in appearance, have the distinction of representing the oldest bird-eggs of historical times in existence, and may claim an age of at least five thousand years. Being the eggs of the majestic winged camel of the desert, the largest living bird, the fleetest and most graceful of all running animals that "scorneth the horse and his rider," they are the only eggs of archæological and historical interest. But they are more than mere eggs; they are ingeniously shaped into water-vessels or drinking goblets by human hand, a small portion at the top having been cut off and the edge smoothed. They were closed by pottery lids overlaid with bitumen, one of the oldest pigments used by mankind. They are thus precious remains of the earliest civilization of which we have any knowledge. In Plate III single fragments of egg-shell are shown, as they came out of the graves, and some patched together from several pieces. These are decorated with banded zones of brown color brought out by means of bitumen. The shell is extremely hard and on an average 2 mm thick.

The trade in ostrich eggs was of considerable extent and importance in the ancient world. They have been discovered in prehistoric tombs of Greece and Italy, in Mycenæ (Fig. 3), Etruria (Fig. 9), Latium, and even in Spain, in the Punic tombs of Carthage as well as in prehistoric Egypt. We find them in ancient Persia and from Persia sent as tribute to the emperors of China. The Spartans showed the actual egg of Leda from which the Dioscuri, Castor and Pollux, were said to have issued; there is no doubt that the egg of an ostrich rendered good services for this pious fraud. In 1833, Peter Mundy, an energetic English traveler, saw ostrich (or, as he spells, estridges) eggs hung in a mosque in India. In 1771, General Sir Eyre Coote found the cupola of a Mohammedan tomb fifty miles north-east of Palmyra adorned with ostrich eggs, and at present also, devout Moslems of the Near East are fond of honoring the sepulchre of a beloved dead with such an egg which is suspended from a tree or shrub on the burial place. Even in the Christian churches of the Copts they are reserved for the decoration of the cords from which the lamps are suspended.

Pliny writes that the eggs of the ostrich were prized on account of their large size, and were employed as vessels for certain purposes. The eggs were also eaten and found their way to the table of the Pharaohs. The Garamantes, a group of Berber tribes in the oases of the Sahara south of Tripolis, anciently had a reputation for being fond of the eggs. Peter Mundy (1634) found ostrich eggs, whose acquaintance he made at the Cape of Good Hope, "a good meate." The egg is still regarded as a rare delicacy in Africa. The contents of one egg amounts to forty fluid ounces, and in taste it does not differ from a hen's egg. An omelet prepared from one egg is sufficient for eight persons. Cuvier, the French naturalist, remarks that an ostrich egg is equal to twenty-four to twenty-eight fowl's eggs, and that he had frequently eaten of them and found them very delicate.

Arabic poetry is full of praise for the beauty of ostrich eggs, and the delicate complexion of a lovely woman is compared with the smooth and brilliant surface of an ostrich egg. The Koran, in extolling the bliss and joys of Paradise, speaks of "virgins with chaste glances and large, black eyes which resemble the hidden eggs of the ostrich."

The thickness of the egg-shell in the African species (*Struthio camelus*) varies from 1.91 to 1.98 mm; the length of the eggs from 140.01 to 156.75 mm, the width from 121.02 to 138 mm. In *Struthio molybdophanes* (so called from the leaden color of its naked parts) of the Somali country, the egg-shell is even 2.02 mm thick; the length varies from 145 to 159.95 mm, the width from 119.50 to 125.4 mm. The weight of the full eggs is from one to two thousand grams, that of the empty ones varies from 225 to 340 grams.

The eggs of birds living in captivity differ considerably from those of wild birds, both in size, coloration, and structure. The former are frequently larger and more oblong, and have a thin shell; the colors are more lively, and the enamel layer is flat, sometimes entirely obliterated.

The egg of a domesticated ostrich from a Californian farm, 163 mm in length, is shown for comparison in Plate IV. As the Californians are all descendants of birds imported from South Africa, their eggs exhibit to a marked degree the pitting which is characteristic of the South African species and which is associated with the respiratory pores of the shell. In the egg-shell of the North African bird, according to J. E. Duerden, the pores are so small and open so close to the surface, as to be scarcely visible to the naked eye, and are mostly scattered singly, with but little grouping, hence the surface appears almost uniformly smooth. In the southern egg, the shell pores are larger, sunken below the general surface, and mostly in small groups, varying from about six to twelve in a group. It is the close grouping of the sunken pores which give rise to the pitted surface. In both types the outer enamel layer shows differences in thickness, and with it the polished character of the surface. All the eggs are a cream or yellow color when freshly laid, but fade considerably on exposure and harden in course of time.

The egg of the North African bird is larger than that of the southern, the shell is almost free from pores or pittings, and presents an ivory-like smooth surface.

The northern egg is usually rounded in shape and less oval. The egg of the southern bird is deeply pitted all over the surface, the pits often larger and more plentiful at the air-chamber end, hence the shell does not present the ivory smoothness of the northern egg. According to J. E. Duerden, who has devoted a special investigation to the two varieties, no mistake is possible in discriminating the one type from the other in a mixed lot of eggs from northern and southern birds.

In cases where the North African hen was mated with the South African cock, a peculiar feature was noted, namely, that the egg-shells of this cross-breed were only pitted in certain patches, while other patches were quite smooth.

In our Mesopotamian eggs the pores are exceedingly fine, and for this reason it may be concluded that the species represented by them is identical with, or closely allied to the present Syrian and North African ostriches. The latter extends right across the Sahara from the Sudan and Nigeria to Tunis and Algeria and from Senegal eastwards. The egg of the Syrian species, if a distinct species it is, is said to be of smaller size and higher polish than the North African one.

In ancient Elam rows of ostriches are found depicted on early pottery, closely resembling the ostriches on the pre-dynastic pottery of ancient Egypt.

In 1849 Austen H. Layard (Nineveh and Its Remains) wrote, "The only birds represented on the Assyrian monuments hitherto discovered are the eagle or vulture, the ostrich and the partridge, and a few smaller birds at Khorsabad, whose forms are too conventional to permit of any conjecture as to their species. The ostrich was only found as an ornament on the robes of figures in the most ancient edifice at Nimrud. As it is accompanied by the emblematical flower, and is frequently introduced on Babylonian and Assyrian cylinders, we may infer that it was a sacred

OSTRICH EGG-SHELL CUP FROM GRAVE AT KISH, MESOPOTAMIA (p. -).
ABOUT 3000 B.C. IN FIELD MUSEUM.
About one-third actual size.

bird." The statement that the ostrich is represented on an Assyrian king's robe is repeated by Perrot and Chipiez, Handcock, and Meissner; but this bird, in my

FIG. 1.
Assur Strangling Two Ostriches. Engraved on an Assyrian Seal-cylinder. After Dorow.

opinion, is not an ostrich; it has a short neck, and its head is entirely different from that of an ostrich. The fact, however, remains that the latter is clearly represented on seals and cylinders.

FIG. 2.
The God Marduk Executing an Ostrich. Engraved on an Assyrian Seal-cylinder. After W. Houghton.

One of these seals is shown in Fig. 1. It was the seal of Urzana, king of Musasir, a contemporary of King Sargon (eighth century B.C.), and represents

Assur, king of the great Assyrian gods, with four wings, in the act of strangling two ostriches. On another seal (Fig. 2) the god Marduk is shown in the act of executing vengeance on an ostrich. With his left hand he firmly grasps the bird's long neck, and in his right he holds a scimitar which will apparently be used to sever the bird's head. These illustrations apparently hint at a ritual act and seem to indicate that the ostrich was also a sacrificial bird and that its flesh was solemnly offered to the gods. Perrot and Chipiez (History of Art in Chaldea and Assyria, II, p. 153) figure a scene from a chalcedony cylinder in Paris, which represents an ostrich about to attack a man with outspread wings and raised left foot; the man tries to lure the bird with a fruit which he holds in his right hand, while behind his back he hides a deadly scimitar in his left.

In the language of the Sumerians the ostrich was known under the names *gir-gid-da,* which is explained as "the long-legged bird" and *gam-gam,* which means as much as "benefactor" or "well disposed." The latter name was borrowed by the Assyrians in the form *gam-gam-mu.* Other Assyrian designations of the bird are *sha-ka-tuv* and *se-ip-a-rik,* the latter also meaning "long-legged."

THE OSTRICH IN PALESTINE, SYRIA, AND ARABIA

The ostrich was well known to the Hebrews, and as attested by several allusions to the bird in the Old Testament, must in ancient times have been frequent in Palestine. It is included among unclean birds in the Mosaic code (Leviticus XI, 16; Deuteronomy XIV, 15), and its flesh was prohibited. This may hint at the fact that the ostrich had occasionally served as food to the Hebrews, although we have no positive information on this point. The reason for the interdiction is not revealed. The ancient apostolic fathers explain that it was forbidden, because the ostrich cannot rise from the earth; modern commentators, because it is a voracious animal and hunting it is cruel. Those who assert that it was abhorred as an exotic animal in Palestine err in a point of zoogeography. The simplest interpretation seems to be that, like other unclean animals of the Mosaic legislation, it was tabooed by Moses, because the surrounding pagan nations availed themselves of its flesh both as a sacrifice to their gods (see above, p. 8) and for their own use. The Arabs of ancient and modern times feast on the bird, and as related by Leo Africanus of the sixteenth century, its flesh was consumed to a large extent in Numidia, where young birds were captured and fattened for this purpose. There are other tribes like the Shilluks of the Sudan who for superstitious reasons abstain from ostrich flesh. Those who have tasted it state unanimously that it is both wholesome and palatable, although in the wild bird, as might be expected, it is somewhat lean and tough. The meat of domesticated birds, however, especially those fed on alfalfa and grain, becomes juicy and tender. Dr. Duncan of the Department of Agriculture recommends it as a New Year or Easter bird.

Job (XXX, 29) laments, "A brother I have become to the jackals, and a companion to the young ostriches." And the prophet Micah (I, 8) exclaims in a similar vein, "Like jackals will I mourn, like ostriches make lamentation." The comparison alludes to the plaintive voices of these animals. The jackal and ostrich are again combined in a passage of Isaiah (XXXIV, 13): "And it shall be an habitation of jackals, and a court for ostriches." The cry of the ostrich has been described variously by observers: some define it as a loud, mournful kind of bellowing roar, very like that of a lion; others define the common sounds of the cock as a dull lowing which consists of two shorter tones followed by a longer note; in a state of excitement he will give a hissing sound, and his warning cry is an abrupt, shrill note. The Hebrew word *renanim* used for the female ostrich means literally "cries, calls," and refers to the twanging cry of the female. Another designation of the ostrich, *bath haya'anah*, signifies literally "daughter of the desert"; that is to say, a desert-dweller, a very appropriate name for the bird. A parallel term occurs in Arabic with the meaning "father of the desert." Isaiah (XIII, 21), in his prediction of the fate of Babylon, says, "But wild beasts of the desert shall lie there; and their houses shall be full of doleful creatures; and ostriches shall dwell there, and satyrs shall dance there."

The famous passage in Job (XXXIX, 13-18) is thus rendered in the Revised Version: "The wing of the ostrich rejoiceth; but are her pinions and feathers kindly (or, as the stork's)? which leaveth her eggs in the earth, and warmeth them in dust, and forgetteth that the foot may crush them, or that the wild beast may break them. She is hardened against her young ones, as though they were not hers: her labour is in vain without fear; because God hath deprived her of wisdom, neither hath He imparted to her understand-

OSTRICH EGG-SHELL FRAGMENTS PAINTED WITH BITUMEN, FROM GRAVE AT KISH, MESOPOTAMIA (p. 3). IN FIELD MUSEUM.

ing. What time she lifteth up herself on high, she scorneth the horse and his rider."

The text is difficult, especially in the opening paragraph, and various translations have been proposed, thus, for instance: "The wing of the ostriches is raised joyfully; but is it a pinion and feather as kindly as that of the stork? No, the ostrich hen leaves her eggs to the earth," etc.

Professor J. M. Powis Smith of the University of Chicago has been good enough to communicate to me his own translation prepared for his coming version of the Old Testament. It runs thus:

> Is the wing of the ostrich joyful,
> Or has she a kindly pinion and feathers,
> That she leaves her eggs on the ground,
> And warms them on the dust,
> And forgets that the foot may crush them,
> Or the beast of the field trample them?
> She is hard to her young, as though not her own;
> For nothing is her labour; she has no anxiety.
> For God has made her oblivious of wisdom,
> And has not given her a share in understanding.
> When she flaps her wings aloft,
> She laughs at the horse and his rider.

According to those scholars who translate the word *chasidah* by "stork," the Hebrew poet contrasts the ostrich with the stork. The stork, as indicated by its name *chasidah* ("the pious one"), was the symbol of kindness and piety, and was regarded as a model of filial love; for this reason it is venerated by all Oriental peoples. The ostrich may resemble the stork in some respects, but differs from it in the care for its young. The description that follows is based on the widely spread, but erroneous assumption that the ostrich is in the habit of leaving its eggs in the sand to be hatched by the sun. The Hebrew poet is intent on making the point that in spite of the careless treatment of the eggs the bird is propagated as a striking evidence of God's constant solicitude for his creatures. To make amends for the lack of wisdom, fleetness of foot has been granted the ostrich. In case its life is endangered, it leashes the air with its wings which assist in running, and derides horse and rider who are in pursuit of it,—a sign that the ostrich, after all, is not so stupid.

The alleged cruelty of the ostrich to its young is also referred to in the passage, "Even the sea monsters draw out the breast, they give suck to their young ones: the daughter of my people is become cruel, like the ostriches in the wilderness" (Lamentations IV, 3).

The observation made in the book of Job that the ostrich treats her offspring harshly does not conform with the real facts. The birds, on the contrary, are tender parents and feed and watch their young ones very carefully. The eggs are laid in a shallow pit or depression of the soil scraped out by the feet of the old birds with the earth heaped around to form a wall or rampart. The female incubates the eggs during the day, while the male takes her place at night. As eggs are sometimes dropped in the neighborhood of the nest or scattered around, the popular belief in the carelessness of the birds and in the hatching of the eggs by the heat of the sun may have arisen. Any eggs not hatched are broken by the parents and fed to the young for whom they display great solicitude, and whom they defend in case of danger.

As to Palestine, the ostrich still occurs in the farther parts of the Belka, the eastern plains of Moab, and is still obtained near Damascus. It is no doubt now but a straggler from central Arabia, though formerly far more abundant (Tristram, Fauna and Flora of Palestine). The portion of the Syrian desert lying east of Damascus denotes the northernmost limit of the range of the ostrich. According to Burckhardt, it inhabits the great Syrian Desert, some being found in Hauran, and a few being taken almost every year, even within two days' journey from Damascus.

As regards ancient Syria, the ostrich is attested by relief-pictures in the theatre at Hierapolis of Roman times, one of these depicting a lioness seizing an ostrich by the neck, and by its introduction into the Syriac version of the Physiologus.

In the Physiologus, a Greek allegorical natural history, which originated at Alexandria in the second century of our era, the following story is told: "The ostrich looks up to heaven in order to see when her time has come to lay her eggs. She does not lay before the Pleiades rise, at the time of the greatest heat. She lays her eggs in the sand and covers them with sand; thereupon she goes away and forgets them, and the heat of the sun hatches them in the sand. Since the ostrich knows her time, man ought to know his to a still higher degree: we have to look up toward heaven, forget worldly existence, and follow Christ."

This story has doubtless been formed by combining Job XXXIX, 14, with Jeremiah VIII, 7 ("the stork in the heaven knoweth her appointed time"). From the Hebrew name of the stork, *chasidah,* the Greek text of the Physiologus has derived the word *asida* in the sense of ostrich. In mediæval Europe the notion still prevailed that the ostrich hatches her eggs merely by glancing at them or by the steadfast gaze of maternal affection. In consequence of this imaginary exploit the bird was chosen as an emblem of faith.

The great outlets from Syria for the ostrich plumes are Aleppo, Damascus, and Smyrna, where the bazars always contain a good supply.

The Janizaries of Turkey who had excelled in battle had the privilege of adorning their turbans with an ostrich feather. At the time of the Ottoman empire there was an imperial ostrich-park in Beylerbey Serai on the Bosporus.

From times immemorial the ostrich has been an inhabitant of Arabia. Heraclides and Xenophon, subsequently Agatharchides and Diodorus mention it as a native of the peninsula. The valuable white plumes of the wings and tail are in great demand among the Arabs for their own wants in the decoration of tents and spears of the sheikhs. Ostrich hunting is alluded to in early Arabic poetry, and has always been a popular sport with the Arabs, who rely on the speed of their horses and run the birds down. As these are in the habit of circling their favorite haunts, the horsemen hunt in relays, and are apt to overtake the birds by pursuing in a straight line.

Kazwini (1203-83), the Arabic author of a Cosmography in which a section is devoted to animals, tells this story: "When the ostrich has laid her eggs, twenty in number or more, she buries them under the sand, leaving one third in one place, exposing another third to the sun, and hatching another third. When the chicks have come out, she breaks the hidden eggs and feeds her young with them. And when the chicks have grown strong, she breaks the last third on which vermin will collect, and this serves as food for the young until they are able to graze." There is a germ of truth underlying this story, and this is that the old birds feed their young on the contents of eggs which they trample down for them. When the eggs are left during the heat of the day, they are covered up with sand, and the occasional finding of such eggs may have given rise to Kazwini's story. He relates another anecdote to the effect that the ostrich, when it has withdrawn from its own eggs and spies other birds' eggs, will hatch the latter and desert its own.

There is a Moslem legend in explanation of the bird's inability to fly. "Once upon a time the ostrich was winged, and like other birds, was capable of flight. He once laid a wager with the bustard, but relying on his strength he forgot before rising to invoke Allah's assistance. He flew in the direction of the sun which scorched his pinions, so that he pitifully plunged down to earth. His progeny has since suffered from the curse which befell its ancestor, and restlessly roves about in the desert."

The Arabs have many names for the ostrich like camel-bird, father of the desert, the magician, the strong one, the fugitive one, the stupid one, and the gray one (for a young bird). Ostrich fat is regarded as a powerful remedy for both external and internal use.

FIG. 3.
Engraving on an Ostrich Egg from Mycenae, Greece.
After Perrot and Chipiez.

THE OSTRICH IN ANCIENT EGYPT

The ancient Eygptians received the ostrich and its products from Nubia, Ethiopia, and the country Punt on the east coast of Africa. An expedition to Punt, probably of a peaceful nature, is recorded on the wall connecting the two Karnak pylons of King Harmhab of the nineteenth dynasty. A relief shows the king at the right, holding audience, receiving the chiefs of Punt approaching from the left, bearing sacks of gold dust, ostrich feathers, etc. (Breasted, Ancient Records of Egypt, III, 37).

In the rock temple of Abu Simbel are represented scenes depicting a war of Ramses II against the Libyans and the Nubian war. In one of these scenes Ramses sits enthroned on the right side; approaching from the left are two long lines of Negroes, bringing furniture of ebony and ivory, panther hides, gold in large rings, bows, myrrh, shields, elephants' tusks, billets of ebony, ostrich feathers, ostrich eggs, live animals, including monkeys, panthers, a giraffe, ibexes, a dog, oxen with carved horns, and an ostrich (Breasted, op. cit., III, 475).

Fig. 4 illustrates a very instructive Egyptian scene. The man on the left leads a captured ostrich, grasping its neck with his right hand, while his left holds a rope slung around the bird's neck; this double precaution hints well at the strength of the powerful avian giant. The man on the right carries three ostrich feathers and a basket filled with three ostrich eggs. The ostrich was sometimes used as a riding-beast, as may be seen from the scene in Fig. 5. Oppianus remarks that it can easily carry a boy on its back. Heuglin says that the possibility of riding the ostrich has often been doubted, but he assures us that the animal is able to carry a heavy man, but not for a long time, and after a brief run will throw itself on the ground. A prehistoric serpentine figure of a seated ostrich is illustrated by Flinders Petrie (Amulets, No. 246).

Ostrich eggs showing traces of painting and engraving have been found in prehistoric tombs of Egypt, and are figured by Jean Capart (Primitive Art in Egypt, p. 40). They were also imitated in clay and decorated with black zigzag lines in imitation of cords or simply painted with white spots. In the Egyptian department of the British Museum is shown an enor-

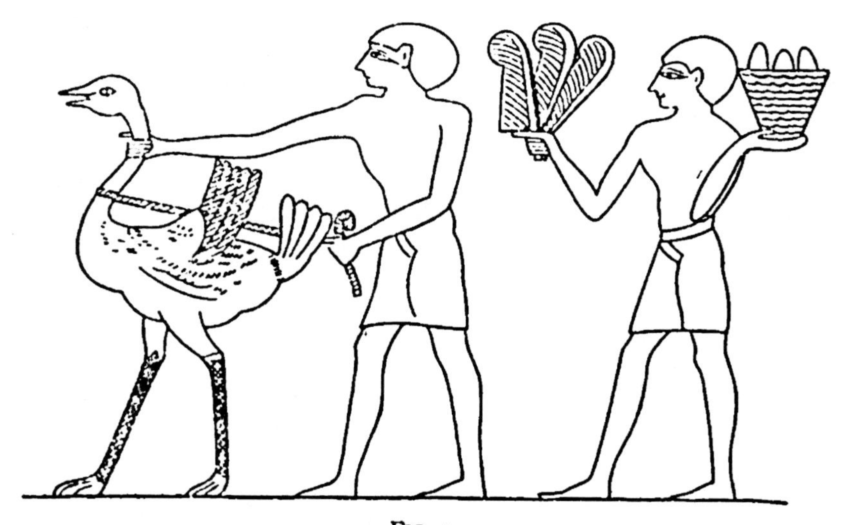

FIG. 4.
Egyptian Scene Showing a Captured Ostrich and Man with Ostrich Feathers and Eggs. After O. Keller.

mous marble egg which is apparently intended for an enlarged ostrich egg, and which was once deposited in a sacred place. During the historic period, ostrich eggs and feathers were imported from the land of Punt and probably also from Asia.

Imitations of ostrich eggs in terracotta have been found in the tombs of Vulci in Italy, which, according to G. Dennis (Cities and Cemeteries of Etruria), seems to indicate that the demand was greater than the supply.

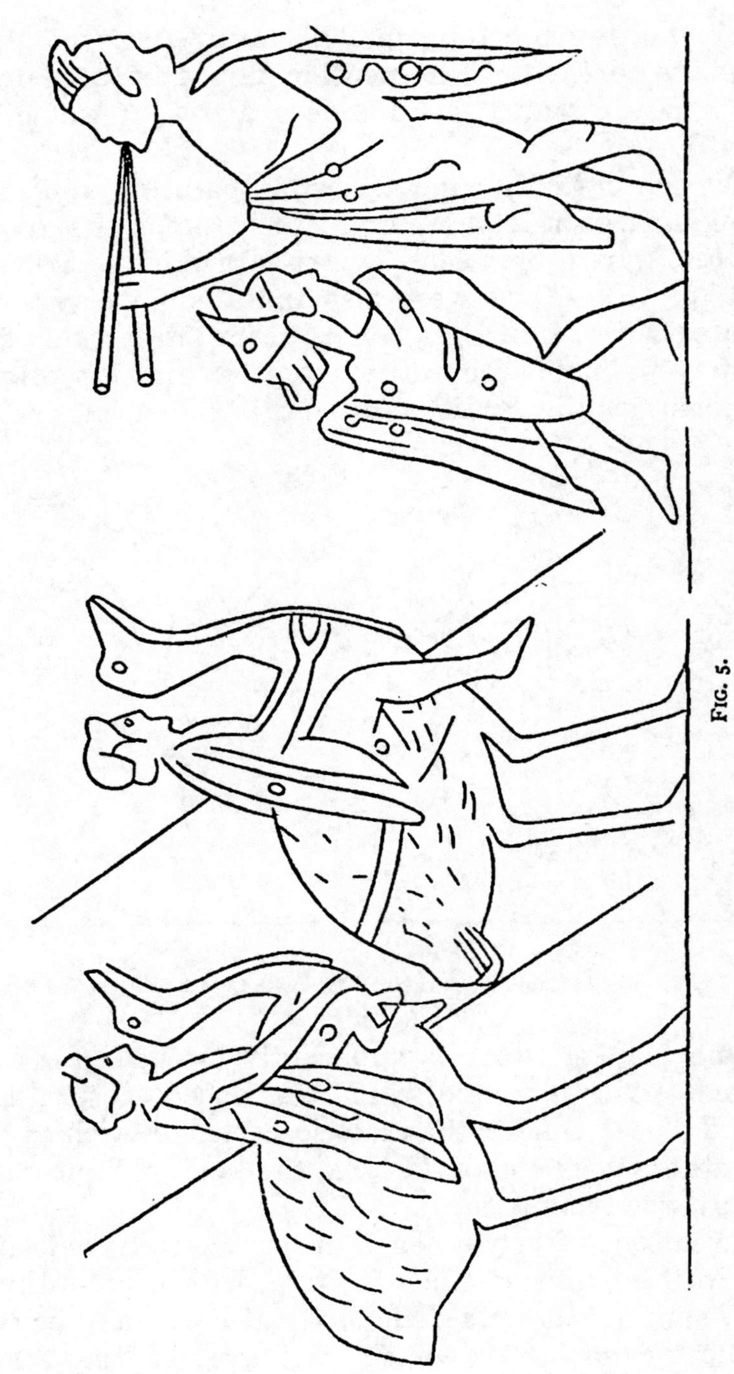

FIG. 5.
Chorus of a Comedy with Spear-men astride Ostriches. Painting on a Greek Vase.
After Daremberg and Saglio.

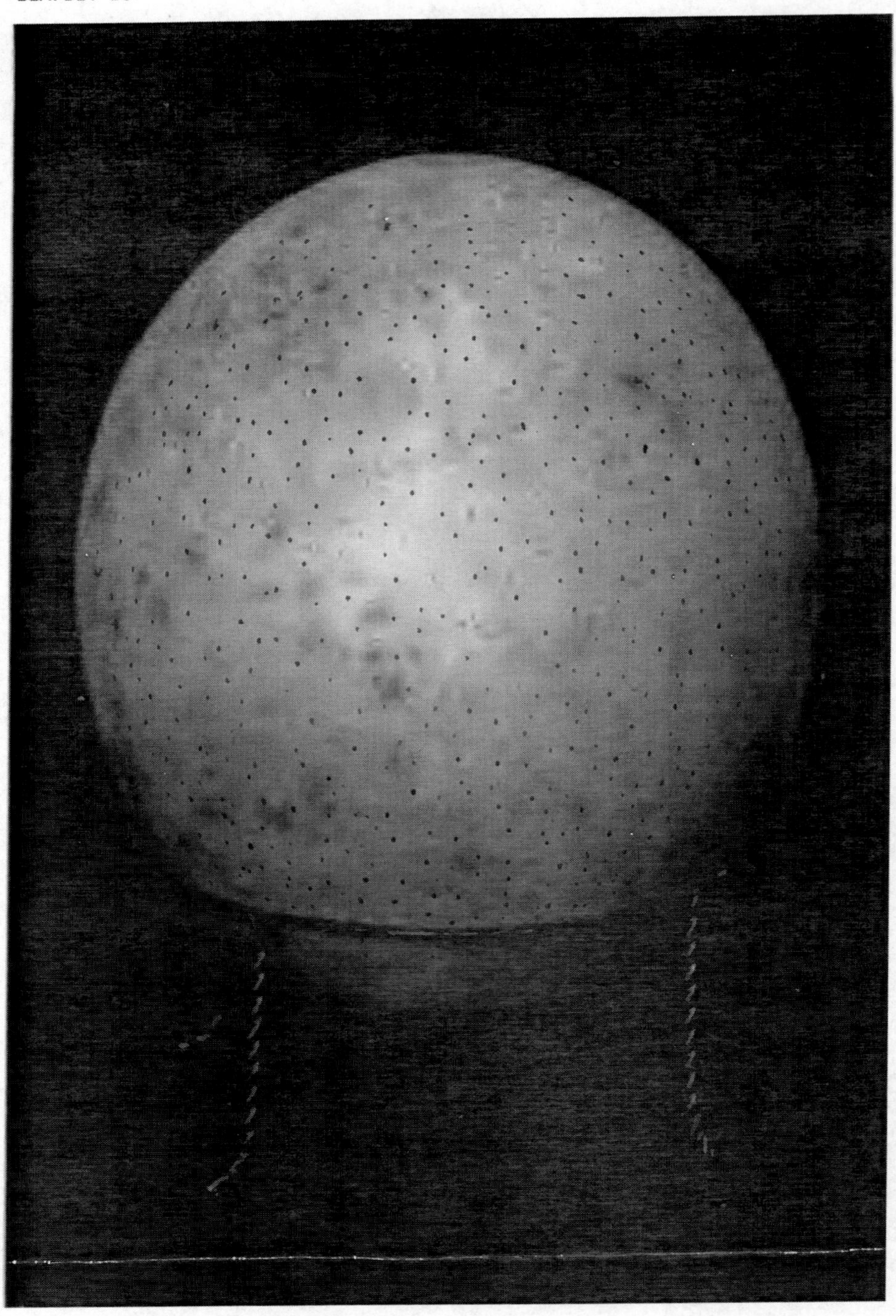

EGG OF DOMESTICATED OSTRICH FROM OSTRICH FARM IN CALIFORNIA (p. 5).
IN FIELD MUSEUM.
About two-thirds actual size.

Flinders Petrie (Naukratis, part 1, p. 14) recovered from the temple of Apollo at Naukratis a piece of an ostrich egg-shell with a pattern of wreath etched out of the inside, and the upper part stained red; the etching was probably done by drawing the wreath with wax on the shell and then eating out the background with vinegar; and the higher surface of the wreath was polished, like the rest of the inside, before etching.

Ostrich feathers were worn by men in ancient Egypt, being stuck in their hair, and a religious significance was possibly connected with this custom. Such feathers are invariably found in the hair of lightly-equipped soldiers of ancient times, and there is a hieroglyph showing a warrior thus adorned. An ostrich plume symbolized truth and justice, and was the emblem of the goddess Ma'at who personified these virtues, and who was the patron-saint of the judges. Her head is adorned with an ostrich feather, her eyes are closed, similarly as Justice is blindfolded. The image of this goddess was the most precious offering for the gods, and was attached to the necklace of the chief judge as a badge of office.

Subsequently when the insignia of the various ranks in the court ceremonial were regulated, the ostrich feather became the exclusive prerogative of the kings, and these and the princes of royal blood exclusively were permitted to wear it. Those decorated with the ostrich feather are designated as "fan-carriers on the left of the king" in the inscriptions of the monuments.

The princesses had fans made from ostrich feathers. In the tomb of the queen Aa Hotep, mother of Amasis I (about 1703 B.C.) was discovered a semicircular fan decorated all over with gold plates and provided along its edge with perforations for receiving the feathers. When the Pharaoh showed himself to the people, high dignitaries carried ostrich-feather fans attached to long poles alongside the royal palanquin.

Among the amulets of power conferred upon the dead were two ostrich plumes supposed to fly away in the wind, bearing the king's soul, and the pair of plumes therefore were provided as a vehicle for the soul of the deceased (Flinders Petrie, Amulets).

The British Museum has a terracotta from Naucratis representing a goddess on horseback with a lyre, wearing a head-dress surmounted by the solar disk, horns, and ostrich feathers (Walters, Cat. of the Terracottas in the British Museum, p. 256, with illustration).

In the eighteenth and first part of the nineteenth century the ostrich still lived in the plains of northern Egypt and along the Arabic coast of the Red Sea (Heuglin). Near the oases of middle Egypt it still occurs at present, likewise along the south-eastern frontier of the country.

THE OSTRICH IN THE TRADITIONS OF THE ANCIENTS

The ancients knew the bird as an inhabitant of northern Africa, upper Egypt, and Arabia.

The first Greek author who mentions the ostrich is Herodotus (IV, 175, 192). With reference to the Macæ who inhabited the coast of Libya, he states that they wore the skins of ostriches as a protection in war. He terms the ostrich "the bird remaining on the ground."

The skin of the ostrich is very thick, and still serves as a cuirass to Arabic tribes. Pierre Belon, a famous French naturalist of the sixteenth century, saw large numbers of ostrich skins with the feathers on in the shops of Alexandria, where they had arrived from Ethiopia. In northern Africa an ostrich skin is valued at about $75.

Xenophon (Anabasis I, 5), when he accompanied the army of Cyrus through the desert along the Euphrates, in northern Arabia, noticed numerous wild asses and many ostriches which he calls "large sparrows," as well as bustards and antelopes; and these animals were sometimes hunted by the horsemen of the army. While they succeeded in catching some asses, no one succeeded in capturing an ostrich. The horsemen who hunted that bird soon desisted from the pursuit; for it far outstripped them in its flight, using its feet for running and raising its wings like a sail. This description is quite to the point. Macaulay said of John Dryden, "His imagination resembled the wings of an ostrich. It enabled him to run, though not to soar." The wings serve the ostrich, while running, as poy and rudder, and it has been observed that with favorable wind they are even used as sails. Xenophon confirms the fact that in ancient times the ostrich ranged right up to the Euphrates. The last record of ostriches in the region of this river was in 1797 when Oliver mentioned them in the desert west of Rehaba, about twenty-three miles due south of Deir-ez-Zor.

Strabo (XVI, 4, 11), the Greek geographer (63 B.C.-A.D. 19), speaks of a tribe of Elephant-eaters near the city Darada in Ethiopia. Above this nation, he continues, is a small tribe, the Struthophagi ("Bird-eaters"), in whose territory there are birds of the size of a deer, which are unable to fly, but run with the swiftness of an ostrich. Some of the people hunt these birds with bows and arrows, others by putting on the skins of the birds. They hide their right arm in the neck of the skin and move the neck as the birds use to do. With their left hand they scatter grain from a bag suspended to the side. They thus lure the birds, driving them into ravines where they are slain with cudgels. Their skins are used both as clothes and as coverings for beds.

This method of hunting by means of a decoy-bird is perfectly credible and universally employed. In South Africa the native hunters hide in a hole which they dig close to the nest of the birds. Having accounted for one bird, they stick up its skin on a pole near the nest, and in this way decoy another ostrich. Other tribesmen who keep tame ostriches avail themselves of the latter to approach wild ones and shoot them with poisoned arrows.

George W. Stow (Native Races of South Africa) gives the following graphic account of the Bushmen's method of hunting (compare Plate V): "In stalking the quagga (*Equus quagga*), the Bushmen generally disguised themselves in skins of the ostrich, with a long pliant stick run through the neck to keep the head erect, and which also enabled them to give it its natural movement as they walked along. Most of them were very expert in imitating the actions of the living bird. When they sighted a herd of quaggas which they

BUSHMAN PAINTING IN A CAVE OF THE HERSCHEL DISTRICT, CAPE COLONY, SHOWING BUSHMAN DISGUISED AS OSTRICH HUNTING OSTRICHES (p. 22).

After G. W. Stow.

wished to attack, they did not move directly toward them, but leisurely made a circuit about them, gradually approaching nearer and nearer. While doing so, the mock-bird would appear to feed and pick at the various bushes as it went along, or rub its head ever and anon upon its feathers, now standing to gaze, now moving stealthily toward the game, until at length the apparently friendly ostrich appeared, as was its wont in its natural state, to be feeding among them. Singling out his victim, the hunter let fly his fatal shaft, and immediately continued feeding; the wounded animal sprang forward for a short distance, the others made a few startled paces, but seeing nothing to alarm them, and only the apparently friendly ostrich quietly feeding, they also resumed their tranquillity, thus enabling the dexterous huntsman to mark a second head, if he felt so inclined. But as these primitive hunters never wantonly slaughtered for the mere sake of killing the game, like those who boast a higher degree of civilization, they generally rested satisfied with securing such a sufficiency as would afford a grand feast for themselves and their families, quite content with knowing that as long as the supply lasted, their feasting, dancing, and rejoicing would continue also."

According to Pliny, the feathers of the wings and tail were used as ornaments for the crests and helmets of warriors. The Athenian general Lamarchus wore two fine, white ostrich feathers on his helmet. The British Museum has a bronze statuette of Harpocrates wearing the Egyptian head-dress known as *atef*, resting on goat's horns; it is composed of three ostrich feathers, flanked by two *uraei* and surmounted by disks. Likewise a bronze statuette of Fortune has on her head a stephane surmounted by a disk, on each side of which is an ostrich plume, resting between a pair of wings (Walters, Catalogue of the Bronzes in the British Museum, Nos. 1494, 1540).

The ostrich was known to Aristotle as the bird who lays the largest number of eggs. He describes it as an animal which has the feathers in common with birds, but shares with the quadrupeds hairs, eyelashes, and the inability to fly; like the birds, it has two feet, but like many quadrupeds, cloven feet, and also resembles them in size; for this reason it has no toes, but claws, for a bird must be small in size, as it is not easy that a large bodily mass moves soaring in the air. Aristotle, accordingly, conceives the ostrich as a connecting link between birds and mammals.

In a similar manner Pliny opens his book on birds with a tolerably exact description of the ostrich which he terms *struthiocamelus* ("sparrow camel"), and which he calls the largest of birds almost approaching the nature of quadrupeds. He assigns it to Africa or Ethiopia and writes, "It exceeds in height a man sitting on horseback, and can surpass him in swiftness, as it is provided with wings to aid it in running. In other respects ostriches cannot be considered as birds, and do not rise from the ground. They have cloven talons, very similar to the hoof of a stag; with these they fight and also use them in seizing stones for the purpose of throwing them at their pursuers. They have the marvellous property of being able to digest every substance without distinction."

The ostrich stands about seven or eight feet high when full-grown, weighs upward of two hundred pounds, and in the wild state defies the horse and rider. At full speed it is said to make about twenty-six miles an hour. The family to which the ostrich proper, the rhea of South America, the emu of Australia, and the cassowary of the Malay Archipelago, South Pacific, and northern Australia belong, differs from other birds in having only small and rudimentary wings unadapted to flight, though they assist greatly in running; the barbs of the feathers are of equal length on each side of the quill and of such a nature as to deprive the animal of the power of flight. The breast is rounded instead of being like a keel as in birds of flight. Aristotle and Pliny are right in attributing to it cloven feet; indeed, it has only two toes, the third and fourth in the pentadactyl system, but unequal in size and not covered with hoofs: the outer toe is much smaller and has no claw. The other members of the family have three toes.

The ostrich, as Pliny points out, is a good fighter. When wounded and hard pressed, it will attack a man, raise its leg to the height of his head, and kick him with its feet, which are hard like steel and yet elastic. A blow from this foot may rip open any animal on which it may fall. The notion that the bird hurls stones at its pursuers (according to Burton, prevailing throughout Arabia) may have been prompted by the observation that when it runs at great speed, it kicks up the stones behind with such violence that they would almost seem to be flung at the hunters in pursuit.

Although the ostrich will swallow almost anything, it is by no means able to digest everything, as Pliny thought. It demands stones instead of bread and swallows them in the same manner as other birds do gravel. They act as mill-stones and assist the gizzard in its function. In the South-African ostrich farms a certain amount of bone and grit is supplied to the birds. White quartz has been found to give excellent results. Grit is so essential that in some parts of the country it is carted by wagon or by rail for many miles, as it was found that without it the birds could not thrive—in fact, could not exist (Thornton).

The fondness for metals has obtained for the bird the name of the "iron-eating ostrich." In 1579 Lyly wrote in his Euphues that "the ostrich disgesteth harde yron to preserve his health." In Shakespeare's Henry VI Jack Cade thus threatens Iden: "I'll make thee eat iron like an ostrich, and swallow my sword like a great pin, ere thou and I part."

The ancients entertained no high idea of the intelligence of ostriches. Pliny comments that their stupidity is remarkable; for, although their body is so large, they imagine that when they have thrust their heads into a bush, the whole of their body is concealed. Diodorus from Sicily, a Greek historian of the first century B.C., was much wiser in remarking that so far from displaying stupidity in thus acting, it adopts a prudent precaution, its head being its weakest part. The same author regarded the ostrich as a missing link between a bird and a camel. The ancient legend is still reflected in our phrases "ostrich policy" and "ostrichism."

In *The Birds* of Aristophanes (415 B.C.), the great Greek writer of comedy, the chorus sings thus:

> To the bird of awful stature,
> Mother of gods, mother of man;
> Great Cybele! Nurse of nature!
> Glorious ostrich, hear our cry!
> Fearful and enormous creature,
> Hugest of all things that fly,
> Oh preserve and prosper us,
> Thou mother of Cleocritus!

Nothing is known about Cleocritus, except that he was unfortunate in his figure, which was supposed to resemble that of an ostrich, while his mother had feet as large as those of an ostrich.

Cornelius Fido, a son-in-law of the poet Naso, is said to have burst into tears when Corbulo called him a bare-skinned or plucked ostrich (struthocamelum depilatum). Seneca who relates this anecdote thinks it funny that a man should lose his temper over so absurd a phrase.

The Arabs have a saying "more stupid than an ostrich," and the French use their *autruche* in the sense of a tall, idiotic fellow. The sweeping judgment of the ancients, however, is based on crude and limited observation of the animal, and on erroneous interpretation of what they did not understand.

In his Report on Ostrich Farming in America Dr. T. C. Duncan (1888) writes that despite its proportionately small brain the bird is anything but stupid, as every one must own who has seen it breaking open the shell to let out a chick that is fast inside, or has seen it managing its chicks.

The Romans indulged in roast-ostrich, and especially enjoyed the wings as a delicacy. Paulus of Aegina, a celebrated physician of the seventh century A.D., writes that they are as juicy and savory as those of other birds. Caelius Apicius, a renowned gormandizer at the time of Augustus and Tiberius, who committed suicide when he saw his fortune shrunk to two million and a half sestertii, has handed down several culinary recipes as to how to prepare good ostrich meat. The emperor Heliogabalus (A.D. 218-222) once served at a banquet six hundred ostrich heads, the brains of which were to be eaten, and was extremely fond of roast-ostrich. The usurper Firmus, who rebelled in Egypt against Aurelianus, performed the *tour de force* to do away with an entire ostrich in the course of a day. Ostrich fat was recommended by physicians as a remedy for all sorts of pain, and the stones found in its gizzard were believed to be a powerful medicine in eye diseases.

Toward the end of the Roman empire ostriches were sometimes shown in the arena of the circus. Three hundred birds are mentioned on one occasion, and a thousand on another as participants in a circus game. A Roman mosaic shows an ostrich acting in the amphitheatre. Eight ostrich teams figured in the triumphal procession of Ptolemy Philadelphus at Alexandria. The Roman emperor Firmus rode with an ostrich team and conveyed the impression as if he were flying. Amor is represented on an engraved gem as being drawn by two ostriches (Fig. 6). A picture on a Greek vase illustrates a comical chorus of spearmen astride on ostriches (Fig. 5). A cut gem of Oriental, perhaps Gnostic origin, shows a running ostrich surrounded by symbolic designs, among these a star with apples (Fig. 7).

FIG. 6.
Amor with a Team of Two Ostriches. Cut Gem of Green Jasper.
After Imhoof-Blumer.

FIG. 7.
Cut Gem of Serpentine Representing a Running Ostrich, Surrounded by Symbolic Designs, among these Star with Apples. Oriental, perhaps Gnostic.
After Imhoof-Blumer.

THE OSTRICH IN THE RECORDS AND MONUMENTS OF THE CHINESE

The ostrich was first discovered for the Chinese by the renowned general Chang K'ien during his memorable mission to the nations of the west (138-126 B.C.). He returned to China with the report that in the countries west of Parthia there were "great birds with eggs of the size of a pottery jar." The "great bird" is the common name of the ostrich among all early Greek writers, while the name "camel-sparrow" or "camel-bird" is found at a later time in Diodorus and Strabo. When Chang K'ien had negotiated his treaties with the Iranian countries in the west, the king of Parthia (called Arsak by the Chinese after the ruling dynasty, the Arsacides) sent an embassy to the Chinese court, and offered as tribute eggs of the Great Bird. In A.D. 101 live specimens of ostriches, together with lions, were despatched from Parthia to China, and at that time were styled "Arsak (that is, Parthian) birds," also "great horse birds." On becoming acquainted with the Persia of the Sasanian dynasty, the Chinese Annals mention ostrich eggs as products of Persia, and describe the bird as being shaped like a camel, equipped with two wings, able to fly, but incapable of rising high, subsisting on grass and flesh, also able to swallow fire. Another account says quite correctly that the birds eat barley. When an attempt was made in Algeria to domesticate them, it was found that they thrive well on barley, fresh grass, cabbage, leaves of the cactus or Barbary leaves chopped fine; and three pounds of barley a day was recommended for each bird, green food according to circumstances.

To the north of Persia, the Annals of the Wei dynasty mention a country Fu-lu-ni, where there is a great river flowing southward; this territory harbors a bird resembling a man, but also like a camel. Again, under the T'ang dynasty, in A.D. 650, the country Tokhara offered to China "large birds seven feet in height, black in color, with feet resembling those of a camel, marching with outstretched wings and able to run three hundred (Chinese) miles a day and to swallow iron." They were then called "camel birds," in accordance with the Greek, Arabic, and Persian designations. Again, in the first part of the eighth century, ostrich eggs were sent to China from Sogdiana. We have to assume that the live birds transported from Persia to the capital of China over a route of several thousand miles must have been extraordinarily tame, and it was a remarkable feat at that. These birds must have been kept in the parks of the Chinese emperors who were always fond of strange animals and plants. What is still more astounding is the fact that in the mausolea of the T'ang emperors near Li-t'süan in Shen-si Province there are beautiful, naturalistic representations of ostriches carved in high relief in stone (Plates VI-VII and Fig. 8). The two sculptured slabs shown in the Plates were erected on the tomb of the emperor Kao Tsung, who died in A.D. 683; the one in Fig. 8 was placed on the tomb of the emperor Jui Tsung, who died in A.D. 712. The artists of the period doubtless received an imperial command to portray the ostriches of the imperial park in commemoration of the vast expansion of the empire over Central Asia during that epoch. As shown by their results, they did not copy any foreign artistic models, but they witnessed and carefully observed and studied live specimens. Their ostriches, in fact, belong to the best ever executed and known in the history of art, and are far superior to any representations of the bird in Assyria, Egypt, and Greece, which are conventional and stiff. The Chinese ostriches are correct in their accentuation of motion and action. The formation and length of the neck allow the bird

CHINESE STONE SCULPTURE OF OSTRICH ON THE TOMB OF THE EMPEROR KAO TSUNG (p. 30). T'ANG PERIOD, SEVENTH CENTURY A.D.

After E. Chavannes.

to turn its head completely around, a characteristic skilfully brought to life in stone by the unknown Chinese sculptor (Fig. 8).

For comparison the sketch of an ostrich by Albrecht Dürer is reproduced in Plate VIII. It is dated 1508 with the addition of the monogram A. D. It is supposed that during his stay in Venice the artist may

FIG. 8.
Chinese Stone Sculpture of Ostrich from the Tomb of the Emperor Jui Tsung.
T'ang Period, Eighth Century.
After E. Chavannes.

have had occasion to view a live ostrich. His sketch is better than that of his contemporary, the naturalist C. Gesner, who had evidently never seen the bird. In the museum of Nuremberg there is a painting of Wohlgemut representing the adoration of the Three Magi; the Moor offers an ostrich egg filled with spices and bordered with gold or silver. The initials A.D. on the egg possibly refer to Dürer, and may hint at his collaboration.

Under the T'ang, the Chinese were also informed of the fact that the ostrich was a native of Arabia. It is on record that "the camel-bird who inhabits Arabia is four feet and more in height, its feet resembling those of a camel; its neck is very strong, and men are able to ride on its back (compare p. 16); the birds thus walk for five or six miles. Its eggs have the capacity of two pints."

When, during the middle ages, the Chinese became slightly acquainted with the east coast of Africa, they learned also that the ostrich was at home in the Somali country. Then they styled it "camel crane," and compared its eggs not unfittingly with a coconut. They even report that the natives of Africa heat copper or iron red and give it to the birds to eat; if the eggs are broken, they give a ring like pottery vessels. In the fifteenth century, under the Ming, ostriches are reported also from Aden and Hormuz.

While in general the Chinese accounts are sensible and make an interesting contribution to the geographical distribution of the species in ancient times, it is noteworthy that they never allude to the bird's plumage; and it seems that its feathers were never utilized in China. We are well acquainted with all articles of trade imported into China by the enterprising Arabs from the early middle ages down to more recent times, but ostrich plumes are never mentioned. Feathers of pheasants were used by the Chinese for the decoration of head-dresses, peacock and eagle feathers for fans, the blue feather of the kingfisher for inlaying in ornaments and screens.

The area of the habitat of the ostrich was formerly much more extended than at present; continued persecution of the bird for the sake of its precious plumage has exterminated it in certain districts or decreased its numbers. It may still occur in some parts of southern Persia, and still lingers in the wastes of Kirwan in eastern Persia, whence individuals may occasionally stray northward to those of Turkestan, even as far as the lower course of the Oxus. No representation of the bird has as yet been found in Persian art.

THE OSTRICH IN AFRICA

The ostrich has reached its greatest extension in the vast grass steppes of Africa, especially those covered with brushwood, where it finds the best conditions for living and satisfying its nomadic habits. There it migrates from pasture to pasture, and may appear and disappear in a certain locality, particularly when forced by lack of food or droughts to move. A gregarious animal, it wanders about in flocks of twenty and thirty and keeps strictly to the steppe, but avoids altogether forests, high tablelands, and damp and swampy tracts. It associates also with other species, like giraffes, zebras, and antelopes (Plate IX), who look upon the ostrich as their guardian. On account of its tallness and far-sightedness the bird is the first to give them a danger signal. The ostrich is extraordinarily keen-sighted, and on its native plains is extremely wary. The whole tribe is characterized by excessive shyness and timidity without which in the struggle for self-preservation it would ere this have ceased to exist.

The ostrich lives on cereals, seeds of grasses, vegetables, leaves, buds, berries, dates, fruits of the tamarind and palms, young birds, lizards, beetles, grasshoppers, etc. In general frugal and capable of withstanding hunger and thirst for days, it becomes greedy and indiscriminately voracious at times. The birds perform strange dances in the sunshine, run around in a circle, flap their wings, and endeavor to rise into the air. During the mating season the male always wooes the female with wild and eccentric dances. He is the original inventor of the Charleston. Ostriches are fond of bathing and swimming, and have been observed to take to the brine. In more than one respect they are almost human, and in their mobility and peregrinations

CHINESE STONE SCULPTURE OF OSTRICH ON THE TOMB OF THE EMPEROR KAO TSUNG (p. 30). T'ANG PERIOD, SEVENTH CENTURY A.D.

After E. Chavannes.

over vast stretches of land they are typical, restless nomads given to a life of hustle and bustle.

The natives of Africa have at all times appreciated both its feathers and flesh. The former are used as fly-whisks and as ornaments on lances, in the kingdom of Congo as war standards, by the Somali as head-ornaments. The empty egg-shells serve as water-vessels, and are suspended in tents and mosques. The roofs of straw huts in the Sudan are adorned with the eggs. Many Negro tribes cut the shell into small button-like

FIG. 9.
Painted Ostrich Egg from Etruscan Tomb of Isis.
After G. Dennis.

pieces which they perforate and string, wearing such strands as necklaces. Perforated and decorated ostrich egg-shells, together with implements of the stone age, were unearthed by Foureau in the Sahara.

Painted ostrich eggs were discovered in the Punic tombs of Carthage and even in the tomb of the valley of Betis in Spain. In the tomb of Isis opened at Vulci, Italy, in 1839, and so called on account of the Egyptian articles found in it, but in fact the sepulchre of two Etruscan ladies of rank, were found six ostrich eggs, one of these being painted with a winged camel or, more probably, a fabulous creature (Fig. 9). G. Dennis (Cities and Cemeteries of Etruria) is inclined to think that these eggs were imported from Egypt, but others assume that they testify to the ancient commercial relations between Etruria and Carthage. There is a fragmentary cup of plated silver, presumably imported from Carthage into Etruria, which is adorned with rows of fantastic and real animals, among these unmistakable ostriches.

The trans-Saharan trade in ostrich eggs has persisted to the present day. The eggs are sent along with the consignment of feathers and emerge at the towns of the Mediterranean coasts of Tunis and Tripolis, where they are in request as pendant ornaments in the mosques.

Richard F. Burton (Lake Regions of Central Africa) wrote in 1860, "The ostrich extends through Unyamwezi and Usukuma to Ujiji. The eggs are sold, sometimes fresh, but more generally stale. Emptied and dried, they form the principal circulating-medium between the Arab merchants and the coffee-growing races near the Nyanza Lake, who cut them up and grind them into ornamental disks and crescents. The young birds are caught, but are rarely tamed. In Usukuma the bright and glossy feathers of the old male are much esteemed for adorning the hair; yet, curious to say, the bird is seldom hunted. Moreover, these East Africans have never attempted to export the feathers, which, when white and uninjured, are sold, even by the Somal, for eight dollars a pound. The birds are at once wild and stupid, timid, and headstrong; their lengthened strides and backward glances announce terror at the sight of man, and it is impossible to stalk them in the open grounds, which they prefer."

The Nandi in eastern equatorial Africa wear a cockade of ostrich feather in times of war. It is with them also an emblem of peace; when after war peace is desired, an ostrich feather is placed in a high-road in a prominent position. The Nandi have the following riddle: "What is the thing which, though so weak that it is blown about by the wind, is able to herd oxen?" Answer: "The ostrich-feather head-dress." The grass in the Nandi country is so high that only a warrior's head-dress can be seen above it, and at first sight it often appears as if a herd of oxen were being guarded by the ostrich feathers, which are the plaything of every gust of wind (A. C. Hollis).

Among the Somalis the ostrich feather is universally used as a sign and symbol of victory. Every man hangs to his saddle-bow an ostrich feather, and generally the white feather only is stuck in the hair. All the clans wear it in the back hair, but each has its own rules. Some make it a standard decoration, others discard it after the first few days. The learned have an aversion to the custom, stigmatizing it as pagan and idolatrous; the vulgar look upon it as the highest mark of honor (R. F. Burton).

In the Lango country the white ostrich feathers are dyed red with iron ochre and worn as head-ornaments. The greater part of the feathers now exported from the Sudan are furnished by ostriches taken young and reared by the Shilluks and Bagaras. These birds become as tame as chickens; in the morning they go to the fields with the cattle, and return home in the evening.

The ostrich-hunters par excellence were the Bushmen of South Africa, now extinct. Their legends prove that they had an intimate knowledge of the life of the ostrich. Two of these are recorded in the "Specimens of Bushman Folklore" by Bleek and Lloyd (1911). One of these, entitled the Resurrection of the Ostrich, gives a good picture of the bird's mating habits and winds up thus: "He will drive away the jackal, when he thinks that the jackal is coming to the eggs, the jackal will push the eggs. Therefore he takes care of the eggs, because they are indeed his children. Therefore, he also takes care of them, that he may drive away the jackal, that the jackal may not kill his children, that he may kick the jackal with his feet."

Fig. 10 represents two sketches of ostriches made by the Bushmen living in the west corner of the Orange

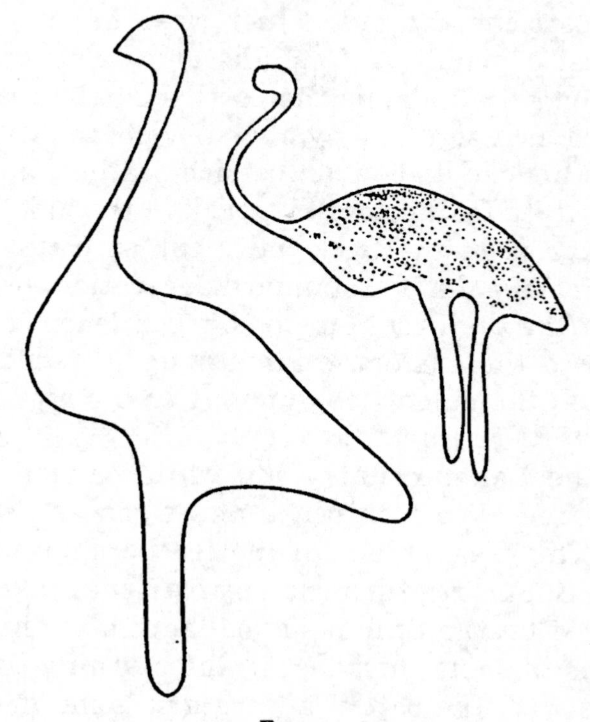

FIG. 10.
Bushmen Rock Carvings of Ostriches.
After M. Helen Tongue.

River Colony. They are reproduced from the work of M. Helen Tongue, "Bushman Paintings" (Oxford, 1909). These pictures are carved in rocks by means of a stone. The markings indicating feathers in the sketch on the right are chipped in the scraped-out surface of the flat stone. The curved neck and the outline of the body testify to good observation. The other sketch is more primitive and not so good. A fine example of

Bushman paintings of ostriches is reproduced in Plate V.

In excavating the diamond-bearing deposits at Du Toit's Pan, in Griqualand West, numerous Bushman beads made of ostrich egg-shell were found at various depths ranging from six to eight feet, and in several spots resting on the bed of calcareous tufa. These local accumulations had evidently been very gradual in their formation. Multitudes of minute land-shells were interspersed throughout them, the animals which inhabited them having evidently perished and been entombed while traversing the arid sand. This place had obviously been a great station for the Bushmen, in the midst of the ostrich country, and had in all probability been a locality, where the manufacture of ostrich egg-shell beads had been carried on for generations. Some were found in various stages of manufacture. Some of those dug out from the lowest depths had become perfectly fossilized, and adhered to the tongue.

A belt from three to six or seven inches in width, formerly worn by young Bushman women, consisted of small circular pieces of ostrich egg-shell bored in the centre and strung like buttons with their flat sides together. Necklaces were made in a similar manner. After the stronger peoples came in contact with the Bushmen bead-makers, they used to purchase these pierced disks of egg-shell from the latter for small bits of iron. Ostrich egg-shells also furnished the Bushmen with water-bottles in which to carry water to the place of their haunt. The openings were closed with a bunch of grass. The women carried twenty or thirty of such egg-shells in a bag or net on their backs.

Spears and poles dressed with black ostrich feathers were stuck in the ground around places, where the Bushmen halted during their hunting expeditions, in order to frighten away lions, which, from their experience, it was discovered were not fond of their appearance (G. W. Stow, Native Races of South Africa).

The Bushmen, further, used the breastbone of the ostrich as a dish, made threads of its sinews and wrought these into nets and bags.

From the remains of ostrich egg-shells is prepared a powder which is also regarded by the Boers as an excellent remedy for man and cattle; it is even said to protect one from blindness.

In South Africa the ostrich exists now only in the domesticated state. In the Kalahari desert, however, in the tablelands of South-west Africa, Matabele and northern Rhodesia it still occurs wild.

THE DOMESTICATION OF THE OSTRICH

The ostrich is the most recent of all domesticated animals, and its complete domestication was accomplished by the white farmers of South Africa as late as the latter part of the nineteenth century. It is the only domestication that occured in modern times, and the only one to which, with certain reservations, our own civilization may lay claim; all other domestications of large mammals and birds were brought about ages ago, either in Asia or Africa, and were simply adopted by the nations of Europe. The first step toward domestication is taming and training, but a tamed animal is not necessarily domesticated. In India, for instance, the elephant has been tamed and schooled as a laborer to a high degree of perfection, but does not propagate in captivity; for this reason it is not classed among domesticated animals. In the stage of domestication the animal propagates its species, and in its breeding is to a great extent influenced by human interference; it will gradually lose many of its savage instincts and acquire new useful qualities. In this manner are formed numerous new varieties which exhibit many differences from the original type and greatly vary in size, color, habits, even in anatomical and mental traits.

Long before the advent of the white man, the natives of Africa had kept ostriches in captivity and tamed them to a certain extent, but they did not succeed in domesticating them. A few examples may suffice to illustrate this point.

In the Uganda Protectorate, where the ostrich is a native of the northern and eastern districts, its eggs and recently hatched young are constantly brought in by the natives for sale. Boyd Alexander (From the Niger to the Nile) writes that in Bornu the ostrich is not much hunted, because it is very difficult to get near, but that the natives catch the young which become domesticated, most of the big men in Bornu owning three or four. The word "domesticated" is here used in a loose fashion, not in the strictly scientific sense.

For generations the Arabs and Berbers of North Africa have kept the ostrich in small kraals and ruthlessly plucked its feathers. These are birds captured as chicks from the nest of the wild bird, but chicks were never bred in captivity.

In Kordofan in the Egyptian Sudan young ostriches are frequently reared, fattened, and slaughtered; the flesh is eaten either fresh or dried. In the Arabic villages of Central Africa ostriches are kept for their feathers; they are hatched from eggs accidentally discovered. In the Sudan chicks are caught, raised by hand and kept until the birds become too old to produce feathers of paying quality, when they are killed and eaten. The system of farming these birds is to enclose each in a small circular mud wall or enclosure, about eight feet in diameter. The birds are never given an opportunity to breed; and this practice, being continued for centuries, has led to the belief that the ostrich will not breed in captivity. The method of removing the feathers from these captive birds was atrocious and the crudest possible; whenever the native farmer required money, he pulled as many feathers from the bird as he could remove in order to turn them into cash. The stage of the growth of the feathers was not considered, the feather sockets were damaged, and in the course of two or three years the birds produced only worthless feathers. It was a system of spoliation inspired by the most sordid greed of profit.

In the eighteenth century ostriches were still plentiful in southern Africa. Peter Kolbe, who spent ten years there on scientific research, wrote in 1742, "These birds occur in the Cape Territory in so large a number

SKETCH OF OSTRICH BY ALBRECHT DÜRER, DATED 1508 (p. 31).
After S. Killermann.

that you cannot travel for a quarter of a mile without seeing some of them. They can easily be tamed, and many are kept in the Citadel of the Cape."

As early as 1662, Jan van Riebeek, Dutch commander of the Cape Colony (1652-62), directed his successor's attention to the taming of young ostriches. On several occasions tame ostriches had been sent to the Indies, where they had proved acceptable presents to native potentates. Their feathers were saleable, but it does not seem to have occured to any one in those days that it would pay to tame the bird for the sake of its plumage (G. M. Theal, History of Africa South of the Zambesi).

When in 1865 the domestication was first attempted in South Africa, natives who had some experience in managing the birds were employed as trainers; but when it has been recognized that the domestication is historically connected with the crude efforts of the natives, it must be frankly admitted, on the other hand, that the success of the actual domestication is solely to the merit of the Africanders. They certainly availed themselves, as it could not be expected otherwise, of some of the experiences previously accumulated for many centuries, not as mere imitators, however, but as novel investigators who grasped the situation with open eyes and energetically applied themselves to a minute study of the bird's life-habits. By creating for their favorite its natural surroundings, by reserving to it vast spaces for movement and exercise, and by proper feeding and care-taking, above all, by sympathy and understanding, their success was permanently insured. Just because these simple farmers were simply human and humane, they achieved what was denied to the Egyptians, Romans, or Arabs with their vain conceit. The barbarous treatment which the poor bird had hitherto received from the hands of African savages gave way to a charitable attitude and an enlightened method prompted by truly scientific research.

The domesticated stocks of South Africa were produced from captured wild chicks who, on reaching the age of maturity, were allowed to breed. Due to careful handling and selected breeding, the quality of the feathers has vastly improved.

The following figures may illustrate the rapid progress made by the industry in South Africa since ostriches were first domesticated. In 1865 there were in South Africa 80 domesticated ostriches; the weight of feathers exported in that year was 17,000 lbs., most of which were feathers of wild birds, valued at £65,000. Ten years later, in 1875, there were 32,000 domesticated birds, and the export of feathers amounted to 100,000 lbs., to the value of £300,000. In 1891 the number of domesticated birds had increased to 154,000; weight of feathers exported was 212,000 lbs., probably including a small amount of wild birds' feathers, to the value of £563,000. In 1904 there were 307,000 domesticated ostriches; the export of feathers was 470,000 lbs., valued at £1,058,000. In 1908 a maximum of 700,000 domesticated ostriches was reached; the weight in feathers exported came to 800,000 lbs., valued at £2,098,000. In 1913 a million pounds of feathers were exported, valued at £2,750,000. There is, accordingly, the remarkable result that during a period of forty-eight years the industry has risen from an export value of £65,000 to £2,750,000; that is, an increase of 4130 per cent.

In mediæval Europe ostrich plumes decked the helmets of knights, later the hats of cavaliers, and the fashion came in again for a time at the Restoration. The fashion of the seventeenth century was dominated by a large felt hat decorated with ostrich plumes laid around the brim. Their natural beauty, particularly the graceful curve taken toward the tip, has always had a strange fascination for the human heart.

The feathers are now utilized for the decoration of ladies' hats, as well as for the making of fans and boas. For the latter the flue or soft portion of the feathers only, also damaged and inferior feathers, are used. The flue of inferior feathers serves also for padding clothes and quilts. The market, of course, is subject to fluctuations due to changes of fashion, but it is very unlikely that the demand for ostrich feathers will ever completely die out.

Each bird has twenty-five white plumes in each wing with a row of protectors, floss feathers underneath. Above these are a row of black feathers and still another row of shorter ones which are black in the adult male and drab in the hen. The feathers are removed by clipping; at the age of six months the birds receive their first clipping, and thereafter are clipped at intervals of nine months. The bird will continue to produce good feathers for practically an indefinite period. This method is perfectly humane, the bird does not receive any injury whatever. Feeding was found to have a very marked effect on the feather growth. This led to the pampering of the bird to such an extent that it is now fed on everything it desires. This method of humorizing its appetite has produced the best results. The fact that the highly-fed ostrich gave the greatest financial return was the cause of erecting the majority of the largest irrigation-works undertaken in South Africa. The return was so enormous that many irrigation-works which could not have been undertaken otherwise were carried out as paying propositions, and are at present a source of immense wealth to the country.

The farmer of South Africa, as R. W. Thornton justly says, is under an inestimable debt of gratitude to the ostrich as being the means by which the best areas of arid land have been converted under irrigation into highly productive fodder-producing areas, which, even if the industry were to fail, would be of incalculable value as fodder-producing areas for any class of farming.

Experiments to introduce ostrich domestication into Algeria in 1881 were unsuccessful. Egypt has an ostrich farm near Matarieh north of Cairo.

In view of the similar climatic and soil conditions of South Africa and Australia and considering the fact that the camel introduced in 1846 was rapidly acclimatized in Australia, it was suggested to naturalize there also the South African ostrich. Its breeding was started in 1880 in the southern part of the continent, but has thus far not been very successful. The statistics of the Australian Government for the year 1922 give the number of ostriches in the Commonwealth as 780, which is a small figure as compared, for instance, with 11,738 camels. Good results were attained on an ostrich farm near Christchurch in New Zealand. Near Buenos Ayres, in Montevideo, Argentina, and Patagonia ostrich farms were also founded.

WILD AFRICAN OSTRICHES WITH EGGS AND CHICKS, IN ASSOCIATION WITH ZEBRAS AND ANTELOPES (p. 34).
After O. Schmeil.

THE OSTRICH IN AMERICA

In 1882 Dr. Charles J. Sketchly, one of the greatest ostrich-farmers in South Africa, transported a troop of two hundred picked ostriches from Cape Town via Buenos Ayres to New York. From there the birds were forwarded by railroad via Chicago and Omaha to the Pacific coast, having covered a distance of 23,000 miles. Twenty-two arrived in California in fair condition, and were at once taken to Anaheim. The California Ostrich Company was soon formed with a capital of $30,000, and Dr. Sketchly was made superintendent. The first year these birds resided in America they presented the company from April to October with 270 eggs.

At the same time the American Ostrich Company was organized in Maine with E. J. Johnson as manager. He went to Africa and spent there a year, studying the habits and management of the birds. He started with twenty-three of them and landed at New Orleans in December, 1884, after a voyage of fifty-three days, with all the birds alive,—a remarkable result, as the usual loss at sea is about 25 per cent. He settled in the valley of the San Luis Rey, about seven miles from the town of Fallbrook, north of San Diego, in southern California. The clear, dry air, the excellent water, and the shelter afforded by the Santa Rosa hills furnished suitable conditions for the establishment of an ostrich farm. The birds took kindly to their adopted home, and have thriven well, the old ones maintaining their natural vigor, and the American-born being at two years unusually fine, both in size and quality of feathers. The breeding birds are kept paired in corrals of an acre in extent. Those one and two years old are left a range of some thirty acres on the mesa, while the young chicks are allowed to run with the other barnyard fowls. The ostrich is now perfectly acclimatized in this country, and it is even asserted that the American birds are finer and larger than their African progenitors.

Other farms soon followed, and are now established near Los Angeles, San Diego, and San José, California; Hot Springs, Arkansas; Jacksonville, Florida; Phoenix, Arizona, also in Oklahoma and Texas. Ostrich farming has developed into an industry of great importance.

The scientist is gratified at the domestication of the ostrich, because it supplies the demands of the feather trade and will therefore lessen or ultimately stop the reckless slaughter of wild ostriches and, let us hope, also the killing of song-birds formerly sought for their feathers. It will enable us, further, to obtain an accurate account of the bird's life history and habits and to render it justice and correct the numerous errors to which the ill-founded fables of past centuries have subjected it. Until recent times it was believed, for instance, that the ostrich is polygamous and mates with from two to five or more females. A. Reichenow (1900) is the first who observed that the wild cock pairs only with a single hen, and I am inclined to assume that he is right; for it has been found that the birds reared in captivity are monogamous, and it can hardly be supposed that the ostrich, perhaps under the influence of American environment, should have suddenly repented and changed from harem habits to a state of monogamy. Some years ago F. J. Haskin, after studying an ostrich farm near Los Angeles, reported, "The ostrich is abnormally finicky about mating. Some birds remain determined bachelors all their lives, and every one chooses his mate only with great delay and caution. Usually it takes two or three years of earnest and patient courtship on the part of the hen before she ensnares her prey. But once captured, the male ostrich is her devoted slave for life. He flutters anxiously about her while she sits on the family eggs and takes up an unnecessarily combatant attitude, one deadly toe-nail raised for fight, whenever another bird or the keeper ventures within ten feet of her. If she dies, moreover, he remains a melancholy widower to the end of his days. When one of the males was widowed as a result of his wife getting her head caught in the fence, the keeper picked out the finest female in the flock and offered her as a substitute. She was in the pen just three seconds when the keeper had to risk his life to get her out. As it was, she received such a hard kick that she nearly died and had to be removed to the hospital pen. The hen has no such scruples when it comes to remarrying, and is polite, if not enthusiastic, to every suitor introduced to her. Once in a long while, also, a male bird is found who is not so sternly monogamous. There is one of this type at the farm who has condescended to espouse two wives. They call him 'Brigham Young'."

Captivity has brought about a remarkable change in the attitude of the old birds toward their young. Whereas in the wild state they are good and tender parents, they apparently do not recognize the young bred at the farms under the incubator system. They cherish no affection for their offspring which has thus not been hatched or raised by them, and their impulse usually is to kill the young on sight. What is said in the book of Job about the ostrich's want of regard for its young now sounds like a true prophecy. Pliny, however, if he could come back to life and would visit one of our ostrich farms, would doubtless offer an apology for his somewhat hasty verdict. Civilization, after all, advances: from a mercilessly persecuted and tormented creature we have transformed the ostrich into a happy and contented bird and an eminently useful denizen of our soil. The domestication of the ostrich is a positive contribution to the progress of humanity and humaneness, and may be designated one of the great achievements of modern civilization of which the Africander may justly be proud and for which we have every reason to be grateful to him.

B. LAUFER.

BIBLIOGRAPHICAL REFERENCES

Only articles which might prove of interest to the general reader are listed here.

CARUTHERS, D.—The Arabian Ostrich. Ibis, 1922, pp. 471-474.

DOUGLAS, A.—Ostrich-farming in South Africa. Ibis, 1906, pp. 46-52.

DUERDEN, J. E.—The Domesticated Ostrich in South Africa. Report of the South African Association for the Advancement of Science, Vol. VI, 1909, pp. 155-161.

The Plumages of the Ostrich. Smithsonian Annual Report for 1910, pp. 561-571. 8 plates.

Some Results of Ostrich Investigations. Report of the Sixteenth Annual Meeting of the South African Association for the Advancement of Science, 1918, pp. 247-284.

DUNCAN, T. C.—Ostrich Farming in America. Report of the Commissioner of Agriculture, U. S. Department of Agriculture, 1888, pp. 685-702.

PEARSON, T. G.—The Ostrich as a Protector of Wild Birds. The Craftsman, Vol. XXV, 1913-14, pp. 470-476.

PRATER, S. H.—The Arabian Ostrich. In: A Survey of the Fauna of Iraq. Made by Members of the Mesopotamia Expeditionary Force "D" 1915-19. Bombay, 1923, pp. 43-46.

SCHALOW, H.—Beiträge zur Oologie der recenten Ratiten. Journal für Ornithologie, 1894, pp. 1-28.

THORNTON, R. W.—The Ostrich Feather Industry in South Africa. The South African Journal of Science, Pretoria, Vol. XII, 1916, pp. 272-279.

176

中国、日本、中亚、印度、埃及、巴勒斯坦、希腊和意大利的墨水历史

PRINTING INK

A HISTORY

With a Treatise on Modern Methods of Manufacture and Use

BY FRANK B. WIBORG

NEW YORK AND LONDON
HARPER & BROTHERS, PUBLISHERS

CHAPTER I

THE HISTORY OF INK IN CHINA

THE celebrated calligrapher Wang Hi-chi (A.D. 321-379), whose handwriting is said to have been "light as floating clouds and vigorous as a startled dragon," is credited with the dictum, "Paper represents the troops arrayed for battle; the writing-brush, sword and shield; ink represents the soldier's armor; the ink-stone, a city's wall and moat; while the sentiments of the heart symbolize the chief commander." In this saying the mental attitude of the Chinese toward the arsenal of the learned is well crystallized: paper, brush, ink, and ink-slab are the four great emblems of scholarship and culture, inventions which the Chinese may justly claim as their own, which constitute the fundamentals of their civilization, and which have largely contributed to make them a nation of studious, well-bred, and cultured men.

In extolling the art of printing as one of the great achievements which has remodelled our intellectual life, we must not overlook the fact that the merit of this invention rests to a lesser degree on the basic idea than on its primary conditions,—the existence of an economic material suitable for writing and printing and easy to manufacture in large quantity, and a medium that will permanently fix the written thought to the paper. That rag-paper is a Chinese invention and that the Arabs transmitted the method of its manufacture to Europe is a fact established beyond any doubt, not only through historical records, but also through archæological discoveries and microscopical and chemical analyses of ancient paper remains.

The same cannot be said about ink: Egypt, the ancients, and mediæval Europe were familiar with ink of different kinds; but the Chinese product is so superior to anything accomplished in the West that for centuries it was employed by artists of Europe under the misnomer "India ink," and is still unrivalled.

The date for the first manufacture in China of ink in the proper sense of the word is variously given in current literature. Palladius, a prominent Russian sinologue, writes that ink is said to have first been produced in A.D. 220; and Geerts (*Les produits de la nature japonaise et chinoise,* 1878, p. 197), in accordance with S. Julien, gives the date more specifically as that of the Wei and Tsin dynasties (A.D. 220-419). This indeed is the period commonly fixed in Chinese sources. In consequence of a misprint in M. Jametel's little book *L'Encre de Chine* (1882, p. xi), where the date of the Wei is given as "220 à 260 [instead of 265] avant [instead of après] J.-C.," several authors have adopted the error in assigning the invention to the third century B.C. Thus F. M. Feldhaus (*Technik der Vorzeit,* 1914, col. 1198) and Rein (*Industries of Japan,* p. 417) even turn the figures around, giving 260-220 B.C. as the date for the invention of ink. Giles (*Glossary of Reference on Subjects connected with the Far East,* p. 132) states that ink was used all over the empire since the third century of our era, though, according to one native authority, it was manufactured as early as 140 B.C.

While Chinese records give us a name for the inventor of rag-paper and the writing-brush, there is no name on record for the inventor of ink, simply for the reason that ink is not the invention of an individual. The situation is the same as that with regard to porcelain: Porcelain is not an invention that can be attributed to the efforts of an individual; but it was a slow and gradual process of finding, groping, and experimenting, the outcome of the united exertions of several centuries and generations. The same observation holds good for the history of ink. It took the Chinese several centuries of tests and trials until they eventually discovered an acceptable formula for a good ink, and even after this discovery they made constant improvements and developed the method, as they also enlisted new materials. It is one of the outstanding examples of progress in Chinese technology and an eloquent refutation of the dogma of the stationary character of Chinese culture.

If, in accordance with Chinese conception, ink properly so-called was only the result of the labors of the early middle ages (third to the beginning of the fifth century A.D.), our historical inquiry is mainly concerned with three questions: (1) What were the writing-materials in the times of the earliest antiquity of China? (2) What was the medium of writing in the age of the closing antiquity (period of the two Han dynasties, 209 B.C.-A.D. 220), when the writing-brush and finally paper (from A.D. 105) existed? (3) What did the invention of the Wei and Tsin periods consist of, and how was it further developed?

In a certain class of popular Chinese books of recent date whose main object it is to trace the history of cultural objects and inventions, and which have the undisguised tendency to advance them as far as possible into the dim past, it is boldly asserted that the beginnings of ink and ink-slabs go back into the mythical days of the emperor Huang-ti (alleged date 2698 B.C.), and some even give as the name of the "inventor" Tien Chen, supposed to have lived at that time. No such tradition exists in any ancient book as the *Chu shu ki nien* (*Annals written on Bamboo Tablets*) or the *Shi ki* (*Historical Memoirs*) of Sema Ts'ien. This modern construction is purely fictitious and arbitrary, and is contradictory to all historical facts known in the case.

In a prehistoric age we find knotted cords in use for the conveyance of messages, chiefly in the transaction of government business. Lao-tse, the famed philosopher, in a sentimental yearning for the past, expressed the desire that he might bring his people back to the ancient usage of knotted cords; that is, the simple life of old. The Tibetans have a tradition to the same effect, and certain aboriginal tribes in the south of China availed themselves of this method as late as the twelfth century A.D. In early historic times calendars, calculations, and contracts were made by means of wooden tallies in which notches were carved with a knife; the creditor, for instance, received the left; the debtor, the right half of the tally. Under the Shang dynasty (1783-1123 B.C.) bone and tortoise-shell served as the conveyance of writing, the characters being slightly incised in the surface; such bones were chiefly inscribed for purposes of divination, and many have been unearthed during the last two decades. The earliest form of Chinese script is preserved on them. Further, we have from the early dynasties inscriptions on bronze vases and bells, the writing being produced in the wax mould, and being either incised or raised. Tablets of jade were used for writing by the emperor; tablets of ivory, by the nobles and higher officials. The most common material, however, particularly under the Chou dynasty (1122-247 B.C.) consisted of bamboo slips or square wooden splints which were perforated at their upper ends and fastened together by means of a silk cord or fine leather strip. The main difference between the utilization of bamboo and wood was this, that a message containing upwards of a hundred words was written on bamboo slips; when it contained less than a hundred words, on wooden boards. The bamboo tablets were naturally narrow, and could be piled up in any required number formed into a pack. The wooden documents, being too heavy to allow of a combination of many, served only for brief texts, as official acts and regulations, statistics of the population, and prayers, but they could not be united into books.

The early canonical literature was handed down on bamboo slips of different lengths, each slip as a rule containing a single line of writing varying from eight to twenty-five or thirty words, and inscribed on one side only. Such books, of course, were exposed to many causes of destruction, chiefly from humidity and pernicious insects, so that bamboo books of early antiquity have long since disappeared. Another inconvenience of these books was their heavy weight. A curious incident in allusion to this fact is recorded anent the emperor Ts'in Shi, who was compelled to examine daily state documents to the weight of a hundred and twenty pounds. Neither writing-brush nor ink was invented in those early days, and the bamboo and wooden memoranda were inscribed by means of a pointed bamboo or wooden stylus (*pi*) dipped in a black varnish (*ts'i*). The bamboo or wooden stylus has survived in Tibet and among several other tribes akin to the Chinese. Varnish, according to the unanimous opinion of all competent scholars of China, was the earliest vehicle of committing thoughts to writing. What the composition and preparation of this ancient varnish was, however, is not known; but it is more than probable that it was a product obtained from the sap of the lacquer or varnish tree (*Rhus vernicifera* D. C., family *Anacardiaceæ*), a sumach indigenous to China. In fact, this tree is designated by the same word *ts'i* as applied to the varnish used in writing. The corresponding Tibetan word *r-tsi* denotes any thick vegetable sap, varnish, or paint. The tree was cultivated by the Chinese in ancient times; in the Book of Songs (*Shi king*) it is mentioned three times, in one case as having been planted by Duke Wan. In superior quality its varnish was produced in Yü-chou in the province of Ho-nan and in Yen-chou in the western part of the province of Shan-tung, as may be gleaned from the chapter Yü kung inserted in the *Shu king* (*Book of Historical Records*).

In the *Chou li,* the State Handbook of the Chou dynasty (1122-247 B.C.), varnish is referred to as being applied to bows, to spikes of chariot-wheels, and to hides used for drums. The fact that it was regarded as a precious substance becomes evident from the Book of Rites (*Li ki*), which says that it was employed for coating the covers of the coffins of princes and the highest officials, but that this privilege was not conceded to plain officials. When a prince ascended the throne, his coffin was made and stored away, a coat of varnish being laid on once a year. The Chinese character for the tree consists of the symbols for wood and water written one above the other, alluding to the sap oozing out and dripping down the trunk. The varnish is extracted by making a horizontal slit upon the tree, and this can be done throughout the warm season, from April to the end of October. The varnish released in the spring is least valuable, because it is very watery. The autumn product is much thicker, but also granulous and slow in exudation. Midsummer is the best time for the harvest, and the varnish is then at its best as to quality and quantity. The tapping, as a rule, begins when the tree is from nine to ten years old. The tree was introduced from China to Japan (Japanese *urushi-no-ki*), where it is eagerly cultivated for the same purposes as in its mother country—the preparation of a lacquer from the sap.

There is direct testimony for the fact that books written on bamboo or wooden tablets by means of varnish were actually produced. In A.D. 279 a tomb was opened at Ki in the prefecture of Wei-hui, province of Ho-nan, and yielded several ancient manuscripts of which we have a contemporaneous record: it is stated that they were written with varnish in "tadpole" characters, an ancient form of script. On the other hand, it is reported that one of the manuscripts discovered in the tomb of Ki, and interred there in 299 B.C., the famous *Mu t'ien tse chwan* (the romantic narrative dealing with King Mu's travels to the west) was written on bamboo slips, each containing forty words, with ink (*mo*); and this is the word still used for "ink." This word appears frequently during the Han period (209 B.C.-A.D. 220), and even a few centuries earlier.

Chou Sho, the councillor of Chao Kien-tse, who died in 458 B.C., said to his master, "With my brush [*pi*] soaked in ink [*mo*] and the tablet held in my hand I shall watch over the faults of your highness." In the Annals of the Later Han Dynasty we read of large and small pieces of ink, and even the term "paper and ink" (*chi mo*) occurs. In view of this fact it is curious that the majority of Chinese scholars who have made ink the subject of special research are agreed on the point that ink in the modern sense was made but as late as the age of the Wei and Tsin dynasties; that is, the period from A.D. 220-419. In order to understand and reconcile these anomalies it is necessary to scrutinize the subject at closer range.

The word used throughout the centuries and still at

the present time for the designation of writing and printing ink is *mo* (ancient form *ma-g,* Canton *mök,* Amoy *bat,* Yün-nan *muk,* Korean *mik,* Japanese *moku* or *boku*). The question arises, what notion was conveyed by this word in times prior to the invention of real ink? The etymology of the word gives no direct response to this query. Its primeval and fundamental significance is "black, dark, obscure"; and in this sense it is used in a passage occurring in the work of the philosopher Mong-tse (called Mencius) in the fifth century B.C., where the face of a minister is described as deep black (*shön mo*). From a comparative viewpoint ancient Chinese *mag* corresponds to Tibetan *nag* ("black"), *s-mag* ("dark"), and *s-nag* ("ink"), which goes to show that both in Chinese and Tibetan the word "black" has assumed the meaning "ink," as soon as the invention of ink was made; in fact, the word *nag* or *mag* (thus in some Karen dialects) for "black" is common to all Indo-Chinese or Sinic languages.

In consulting the written symbol or character for the Chinese word *mo,* we find that it is not a simple, spontaneous formation, but presents a composition of two well-known signs, those for "black" (*hei,* anciently *gag*) and "earth" (*t'u,* anciently *du*). The meaning connoted by the character, accordingly, is "black earth" or "black clay," and may hint at the fact that a mineral or clayish substance of black color may be hidden under this term. This opinion is indeed advanced by Chinese writers. Thus Li Shi-chen, author of the famous herbal *Pen ts'ao kang mu* in the latter part of the sixteenth century, infers from the formation of the character for *mo* that the ink of the ancients was made from black earth (*hei t'u*), referring also to the definition in the ancient dictionary *Shwo wen* (about A.D. 100) that *mo* is a kind of earth formed by smoke and glue. It must be borne in mind, however, that any such conclusions as to material, merely deduced from the construction of written symbols, are usually fraught with danger or may be deceptive, that the present explanation is more or less an afterthought, and that the old contemporaneous tradition is lost.

Several passages in ancient texts show us that *mo* was a black pigment. The philosopher Mong-tse, mentioned above, speaks in another chapter of his work of the carpenter's marking-line, expressing it by the compound *sheng mo* ("string and ink," or more cautiously "string and black substance *mo*"). It is clear that the carpenters of his time, in the same manner as those of the present day, must have availed themselves of an instrument for marking lines in black.

Another instance of the ancient application of the word *mo* occurs in the penal code of the Chou dynasty first issued on bamboo slips in 501 B.C. in which five kinds of punishment are laid down, the first of these being branding of a criminal's forehead. This process is denoted by the term *mo,* which in this case is imbued with verbal force ("to blacken, to brand"). We are not informed as to what this black pigment used for the mark of infamy was; it was applied to the forehead by means of an incision with a cutting instrument, and must have been of a rather indelible nature (cf. Couvreur, *Chou king,* p. 386).

Among the ancient Chinese the tortoise was one of the principal vehicles of divination. The carapace of the animal was coated with a layer of a black pigment (*mo*) and exposed to a fire. Thereupon the delineations of the cracks produced by the action of the fire were examined, and the will of Heaven was read from them. This process was styled *ting mo* ("to determine the pigment": Couvreur, *Li Ki,* Vol. 1, p. 682). It was the rule, however, to burn the tortoise-shell without the application of a pigment.

The invention of the writing-brush must have acted as a stimulus to the improvement of all writing materials. Traditionally, the brush is credited to Mung T'ien, who served as general to the first emperor Ts'in Shi, and who died in 209 B.C. This tradition, however, is to be taken *cum grano salis*. It is not recorded by a serious historian like Se-ma T'sien, whose *Shi ki* is the chief source for the history of that emperor's reign; and some Chinese authors are inclined to think that the invention of the brush was merely attributed to Mung T'ien for the glorification of his imperial master, who wished everything to begin from his reign. It would be absurd to assume that the general was the first Chinese who ever invented a brush; so simple an implement must doubtless have existed centuries before his time.

The eminent scholar Yen Shi-ku (A.D. 579-645) probably hits the truth with the following comment: "The tubes of the ancient writing-brushes were made from dried wood with deer's hair backed by sheep wool, but there were no bamboo tubes with hare's hair; this was the work of Mung T'ien." In other words, Mung T'ien may have applied two improvements to the writing-brush, which itself pre-existed; he made it lighter in weight, with finer hair, and perhaps more elegant in form. If he really had anything to do with the whole affair, it is striking that an instrument exalted by the literati and essentially one of the learned owed its perfection to a man of the much despised military class; or, we might rather say, it is surprising that popular imagination, in seeking for the inventor of the writing-brush, has fastened the honor on an old general who is not known as a writer.

Another important innovation took place in the third

century B.C., presumably under the Ts'in dynasty (246-207 B.C.), when a paper made from silk refuse including both raw and woven silk came into being. This silken paper was preceded by bands of silk stuff serving as writing material. The refuse from silkworm cocoons was soaked and beaten in water in order to eliminate coarse particles. This mass was reduced to a paste, and thus purified was spread over a fine bamboo mat mounted on a wooden frame. This served as the mould on which the paste precipitated, and when dried, produced a sheet of paper. Such bamboo mats are still utilized in the modern manufacture of paper from bamboo or tree-bast fibres.

It is clear also that the underlying principle of the subsequent manufacture of rag-paper was forestalled by that of silk paper. Since the writing-brush was perfected in the same period, we can hardly call this coincidence accidental, but must admit that the two inventions were dependent one on the other; hence the further conclusion is justified that the two again may have stimulated the production of a better expedient for writing, which resulted in the black pigment called *mo*. At the outset it is not very likely that varnish continued for silk material and silk paper, nor is it likely that the ancient wooden stylus was applied to silk. Thus there are good technical reasons for the conviction that the varnish of early antiquity was gradually replaced by a more convenient substance in the three or four centuries preceding our era.

A still more powerful impetus to the improvement of writing-ink was received from the invention of bark and rag paper by Ts'ai Lun in A.D. 105. Ts'ai Lun was born at Kwei-yang, a city of Kwei-chou Province in southern China. In A.D. 75 he entered the service of the emperor Ho, and in 89 was appointed director of the imperial arsenals. He was deeply given to study, and whenever he was off duty, he would shut himself up for that purpose. He survived his invention for thirteen years, being ennobled as marquis in 114. In the same year, however, as he could not find favor with the empress, he died a suicide by swallowing a dose of poison. Ts'ai Lun considered that both silk and bamboo tablets were inconvenient writing-materials, the former being too expensive, the latter too cumbersome and perishable. Hence he conceived the idea of using bast-fibre, hemp, and old rags like fishing-nets for making three kinds of paper.

This invention created a profound impression on the contemporaries, who at once turned it to practical use. Although Ts'ai Lun perpetuated and advanced a pre-existing process, and his principal merit consisted in the substitution of little valuable or even valueless substances, which simultaneously yielded better results, for the comparatively costly silk-refuse, he must be honored as the man to whom we are indebted for one of the most far-reaching discoveries ever made in the annals of technology. Without paper there would have been no adequate record of the past, no progress, no science; it marks the dawn of civilization, it sets off civilization from savagery.

Thus the life of the Han dynasty, during the last century of its existence, was signally enriched by the acquisition of paper. Nevertheless, in the outlying colonial possessions, the use of wooden tablets persisted with conservative force. A great number of these were rescued from the sand of Chinese Turkestan by Sir Aurel M. Stein, and have been edited and translated by E. Chavannes (*Les Documents chinois découverts dans les sables du Turkestan oriental,* Oxford, 1913). The wooden slips studied by Chavannes range in date from 98 B.C. to A.D. 153. He observes that a goodly number of these are inscribed with characters of extreme *finesse*, and that their delicate traits could have but resulted from the use of a brush; unfortunately he is silent as to the ink. In the reproductions of the documents the writing appears in black ink, in appearance not different from the Chinese ink to which we are accustomed. This, of course, does not mean that the Han ink might not have been an entirely distinct affair. Microscopical and chemical analysis would be the only means of solving the problem, and it is hoped that the authorities of the British Museum will consent to having this investigation made some day.

Meanwhile, we are thrown back on Chinese literary sources. It is certain that in the Han period an ink-like substance for writing, called *mo*, was utilized; but what its composition was, is not revealed. No recipe has been handed down from that epoch. The fact that it was different from the later ink is manifest from the persistent tradition that this product made its appearance only under the Wei and Tsin. Ch'ao Shwo-chi of the Sung period, who wrote an interesting treatise on the technology of ink, opens it by saying that formerly two kinds of ink were in use,—one prepared from pine-tree lampblack (*sung yen*), another styled "mineral ink" (*shi mo*, literally "stone ink"). The latter, he comments, has disappeared since the Tsin and Wei periods, but he does not explain what it was. He assumes that under the Han period ink was also manufactured from pine charcoal, as *Yü-mi mo* ("ink from Yü-mi") is mentioned at that time, and the Chung-nan mountains at Yü-mi (in Shen-si Province) were covered with pines; but this is a theory which remains in the realm of conjecture.

T'ao Tsung-i, who wrote the *Cho keng lu* in A.D. 1366 (under the Yüan dynasty), states that in times of earliest antiquity there was no ink, documents being written by means of a piece of bamboo dipped in varnish, and that in the middle period of antiquity they utilized the sap of *shi mo,* which is believed by some to be identical with the stone of Yen-ngan in Shen-si (that is, rock-oil or kerosene). This mineral product is described by the Chinese as a stone grease floating on the surface of water, like varnish, and collected to be burnt in lamps or made into torches. From the Sung period onward the lampblack from petroleum was used in the manufacture of ink, called *Yen ngan shi i* ("secretion from the stone of Yen-ngan"), these words being engraved on the ink-cake.

It is on record that Lu Ki or Lu Shi-heng, who lived in the latter part of the third century A.D., one day ascended the T'ung ts'iao t'ai ("Copper Sparrow Terrace"), a tower built by the famed Ts'ao Ts'ao in A.D. 210, and found several jars full of *shi mo* collected and stored by Ts'ao Ts'ao, such as existed no more at any later period. Subsequent authors indulged in speculations as to the material contained in these jars. In this case it is again assumed by some that this substance was identical with the lampblack produced from kerosene (*shi chu yen,* "stone torch smoke") in Shen-si, mentioned by Shen Kwa in his *Mong k'i pi t'an,* written in the middle of the eleventh century. It is supposed also that the cosmetic used for painting the eyebrows by the women of the palace of the Sui dynasty (A.D. 583-617) was a kind of *shi mo.*

Under the term *shi mo* quite a number of different minerals appear to be confounded. *Shi mo* is also a synonym of *shi tan* ("mineral coal"), and Li Shi-chen (*Pen ts'ao kang mu,* Ch. 9, p. 20) affirms that in times of antiquity bituminous coal was utilized for writing. The possibility of this cannot be denied. Incidentally, one Chinese author declares that lead was anciently used for writing.

Several minerals were formerly utilized as substitutes

for ink. The *Yün lin shi p'u,* a treatise on economic mineralogy written by Tu Wan (pseudonym Yün-lin) in A.D. 1133 (Ch. 3, p. 11b), says that in Kwei-chou (prefecture of I-ch'ang, Hu-pei Province) there are black stones appearing in the water of the Yangtse River, of coarse substance, which can be ground and will yield an ink. This stone is called *ta t'o shi* ("stone of the great river"), *t'o* being the local name for the Yangtse among the inhabitants of the gorges near I-ch'ang, who prize this stone very highly. This same work mentions a stone of Ts'ing-chou in Shan-tung, found deep in the soil, being carved and ground, not containing much ink, but used locally. A "dragon-tooth stone" (*lung ya shi*), according to the same author, is found in the district of Ning-hiang in the prefecture of Yo-chou, Hu-nan Province, both in water and in the mountains, of purple color, somewhat glossy, capable of being ground into ink and rather appreciated by the people of the place. Finally, in the river of the district of Fen-i in the prefecture of Yüan-chou, Kiang-si Province, occurs a stone, dark in color, hard and bright, sonorous when struck, gathered by the natives in the water, and ground into ink, which is suitable for the brush; but the material is so coarse that instruments for cutting and grinding it are required. For Kwang-tung Province mineral ink mountains furnishing writing-ink of excellent quality are mentioned as early as the Tsin period in the *Kwang chou ki* of Ku Wei.

There is another kind of *shi mo,* which is identified with *hei shi chi* ("grease of black stone"), described as sticking to the tongue when licked and used for writing, as well as for painting the eyebrows (much practised in ancient China). This is doubtless graphite (Geerts, *Produits,* p. 203). F. de Mély (*Le lapidaire chinois,* p. 256) is inclined to take it for "sulfure d'antimoine." Now the term *shi mo* in the sense of graphite occurs for the first time in the early work *Pie lu,* the foundation of which goes back to at least the Han period, and possibly even to an earlier date. This would well indicate that graphite was one of the substances enlisted as writing material in the epoch of the Han. Besides, products of mineral coal, bitumen, and rock-oil may have been utilized; the earliest ink, accordingly, was of mineral origin, in opposition to the vegetable products laid under contribution in the following period, under the Wei and Tsin (A.D. 220-419).

The oldest recipe for the preparation of an ink that has come down to us is contained in the *Ts'i min yao shu,* a work on practical husbandry, written by Kia Se-hie, who lived in the fifth or sixth century A.D. Unfortunately this important work is handed down in mutilated form. The original was in 92 sections, part of which were lost long ago, and much additional matter has been interpolated by subsequent editors. The recipe for ink, entitled "Method of mixing ink," is apparently incomplete, since the substance from which the lampblack is derived is not even mentioned; in some places the text is enigmatic and evidently corrupt.

Ch'ao Shwo-chi, an author of the Sung period, who wrote a very interesting treatise on the manufacture of ink, quotes three passages from the recipe of Kia, but his text is different from that found in the present editions of the *T'si min yao shu.* The principal points of the formula are as follows: Good and pure lampblack is to be pounded and strained through a sieve of fine pongee, which is placed in a vat of stoneware. The object of this process is to free the lampblack of any adhering vegetable substances so that it becomes like fine sand and dust; but as it is so light in weight, great care must be exercised in preventing it from being scattered around. Five ounces of glue are required for one pound (catty) of ink, and the sap of the bark of the *ts'in-p'i* tree (*Fraxinus pubinervus*) is dissolved in the glue. This bark is green in color like water, and contains a glutinous substance which also improves the color of the ink. The white of five hen's eggs, one ounce of cinnabar, and the same amount of musk may likewise be added, after being well strained. All these ingredients are mixed, and the paste thus obtained must be beaten in an iron mortar with a stick thirty thousand times; the more frequently it is beaten, the better the quality of the ink. In weight a cake of ink should not exceed two or three ounces, and its size should conform to this rule; that is, it must be small, not large.

However imperfect this formula may be, it leaves no doubt that it carries the directions for a real ink, and that in principle it is identical with the process still in vogue. Even though the source of the lampblack is not clearly indicated, it follows from the context that it was of vegetable origin. Li Shi-chen, the great herbalist at the end of the sixteenth century, states expressly that the best ink in his time was made from lampblack of pine-wood with an admixture of the sap of *Fraxinus pubinervus,* boiled together with glue as well as aromatic substances. The green bark of this tree steeped in water is still utilized for obtaining a bluish indelible ink.

It is obvious that the formula divulged by Kia Se-hie did not spring up spontaneously, but that it presents the result of long experiences and experiments conducted during several generations. In fact, he had predecessors, as we see from the *Mo king* of Ch'ao Shwo-chi of the Sung period, who quotes the "ink method" of Wei Tan or Wei Chung-tsiang of the Wei dynasty (A.D. 220-264); this author has also the thirty thousand beatings—doubtless an exaggeration. Coming down to the T'ang dynasty,

Wang Kiün-te, an ink manufacturer, who used a stone mortar, speaks more moderately of from two to three thousand beatings. Under the Ming, toward the end of the sixteenth century, a wooden mortar with a metal pestle was employed, and one was satisfied with several hundred beatings. There is every reason to believe that Wei Tan is the first originator of an ink-formula, as handed down in the *Ts'i min yao shu,* and to a certain degree may be regarded as the real inventor of ink as still manufactured. His name is the first which appears in the *Mo shi,* a history of ink manufacturers by Lu Yu of the Yüan dynasty, who observes that there is no great difference between the formulas of Wei Tan and Kia Se-hie. One of the former's peculiarities was that he mixed one ounce of genuine pearls and a half ounce of musk in his ink.

In the *T'ai p'ing yü lan* published by Li Fang in A.D. 983 (Ch. 605, p. 5) the recipe, as contained in the *Ts'i min yao shu,* is in fact quoted as that of Wei Tan, the text being substantially the same, with the one exception that his "ounce of cinnabar" is replaced by an "ounce of genuine pearls." We must therefore admit that Kia Se-hie merely copied Wei Tan. The latter lived from A.D. 176 to 251 to the age of 75. His ink was still renowned at the end of the fifth century. Siao Tse-liang, a prince of the Southern Ts'i dynasty (about A.D. 484), in one of his letters, speaks admiringly of "Chung-tsiang's ink, every drop like varnish!"

Chinese literature on ink is considerable. It begins to develop under the T'ang dynasty when a treatise on ink (*Mo king*) by Ch'eng Lao-po existed. This work, however, has not survived, but a few brief extracts are preserved in the *Yün sien tsa ki,* written by Fung Chi in the commencement of the tenth century, and in later works.

PRINTING INK

Several technical treatises on ink were written under the Sung, above all, the *Mo king* by Ch'ao Shwo-chi, which is reprinted in the great cyclopædia *T'u shu tsi ch'eng*. This author devotes a systematic discussion to the various kinds of pine-trees suitable for lampblack, with exact enumeration of the mountains and localities where they grow; then he has discourses on charcoal, glue, sifting, the processes of mixing and pounding, chemical ingredients, drying, grinding, color, sound, weight, new and old ink, preserving ink, proper season for making it, workmen and manufacturers.

Another treatise on ink, entitled *Mo p'u*, was published by Su I-kien at the close of the tenth century. The same author wrote a book on paper (*Chi p'u*), and in A.D. 986 summarized his experience in a comprehensive work which he called *Wen fang se p'u* (*The Four Departments of the Study*). This is a repository of information regarding the materials of the study, consisting of four parts which treat of writing-brushes, paper, ink, and ink-pallets, with historical memoranda, essays, and stanzas. Ho Yüan of the Sung wrote a *Mo ki* (*Ink Memoirs*) in which he deals with ink manufacturers; a long list of these from the T'ang to the Yüan dynasty is also inserted in the *Cho keng lu* by T'ao Tsung-i, published in A.D. 1366.

The most complete and interesting work of this class, however, is represented by the *Mo shi* (*Ink History*) of Lu Yu of the Yüan dynasty, who gives a series of brief notices of about a hundred and forty manufacturers whose names had been handed down in connection with their productions from the Wei, Tsin, T'ang and Sung dynasties down to the Kin. He also notes the ink of the Koreans, the Kitan, and Turkestan, with a number of miscellaneous observations respecting ink appended.

In the beginning of the Ming dynasty, in 1398, appeared a manual of the ink manufacturer, by Shen Ki-swan, illustrated by twenty-seven woodcuts showing the various stages in the process of manufacture. The author was a manufacturer himself and professes to divulge only the tricks of his own trade, save some information communicated to him by a monk. The little work teems with technical routine and detail, but in comprehensiveness and clarity does not compare with the *Mo king* of the Sung. It was first translated into Russian by I. Goshkewich (in Works of the Russian Mission of Peking, Vol. I, 1852; cf. W. Schott, Entwurf einer Beschreibung der chines. Literatur, p. 107), then from the Russian into German. A French translation with reproductions of the woodcuts is due to M. Jametel (Paris, 1882).

About 1637 Ma San-heng issued a *Mo chi* in which he treats of ink manufacturers with their productions and the marks that distinguish them. Kao Lien of the Ming wrote an essay, *Lun mo* (*Discourse on Ink*), and T'u Lung published the *Mo tsien,* a short work on ink, during the sixteenth century. The *Süe t'ang mo p'in* is a small treatise on ink, written by Chang Jen-hi in 1671, in which he classifies the productions of various manufacturers and points out the peculiar characteristics of the different kinds. The *Man t'ang mo p'in* is a similar record, supplementary to the preceding, written in 1685 by Sung Lao, who presents notices of thirty-four specimens of ink of the Ming dynasty, with their respective weights. Aside from such monographs there are innumerable references to ink in technological books like the *T'ien kung k'ai wu,* written by Sung Ying-sing in 1637, in the essay literature of the Sung, in cyclopædias, and in the herbals (*Pen ts'ao*).

While the Ming and Manchu dynasties hardly added anything new to the technical side of the subject, great care was bestowed on the artistic embellishment of ink-cakes, and we have several excellent books issued by ink-manufacturers giving reproductions of the designs engraved upon their ink-cakes. These will be considered in the next chapter.

CHAPTER II

THE HISTORY OF INK IN CHINA

(Continued)

AFTER this digression, let us revert to the history of ink. The poet Ts'ao Chi (A.D. 192-232) has left a verse in which he says that ink is produced by the smoke of the green pine. Now he was a contemporary of Wei Tan (A.D. 176-251), and it is therefore a reasonable conclusion that Wei Tan was the first who used the lamp-black from pine in the manufacture of ink. There is another formula handed down from the Wei period and attributed to a certain Ki kung (Mr. Ki) about whom nothing is known otherwise. His ink is said to have been made of two ounces of pine-black mixed with a little clove, musk, and dried varnish; this was compounded with glue, soaked in water, and heated over a fire, the entire process taking a full month; he produced inks of two colors,—a purple ink by adding the root of *Lithospermum officinale* L. var. *erythrorhizon* (the Chinese name of this plant means "purple herb"; it is still cultivated for the purple dye yielded by its root) and a bluish ink by using the bark of *Fraxinus pubinervus* (ts'in-p'i). Ink was formerly put up in various forms. The Chinese language has a large number of numeratives by which objects are counted, and the numeratives vary according to different categories of things and notions. Under the Han ink pieces were usually counted as so and so many *wan* (that is, "pills, pellets, balls"); ink being formerly taken as a medicine. It is very likely that it was first made into pills to be easily swallowed. This practice was continued under the Wei and Tsin, and under the T'ang also we occasionally hear of ink balls; but from that time onward this shape fell into disuse. It seems that in some out-of-the-way places ink is still made into balls; at least E. H. Parker (*Up the Yang-tse,* p. 135, Shanghai, 1899) reports that at An-si Ch'ang in Kwei-chou Province he saw exposed for sale balls of ink made from the soot of the *Aleurites* oil (wood oil or *t'ung* oil).

Further, ink pieces were formerly counted by *liang* (a measure of capacity) and, curiously enough, by conch-shells (*lo*), which may indicate that ink was kept in shells. An early author, T'ao Tsung-i, in A.D. 1366, asserts that in times after the Tsin there was a kind of ink called *lo-tse mo* ("shell ink"), as though shell powdered or burnt had formed an ingredient in its composition; but this notion surely rests on a misunderstanding, for the ancient texts contain nothing concerning such an ink, but what is spoken of is merely that someone, for instance, sent another "two shells of ink," which means ink of a quantity as two shells may hold, whether it was actually transmitted in shells or not.

The prismatic shape of ink seems to have come up under the T'ang: during his last exploration of Turkestan, Sir Aurel Stein (*Serindia,* Vol. 1, p. 316) discovered an octagonal prism of Chinese ink. On his previous journey he found in the stupa of Endere a "cylindrical piece of hard Chinese ink, drilled for a string at one end" (*Ancien-Khotan,* Vol. 1, p. 438). The prismatic or cylindrical shape (so-called sticks) has persisted to this day, and this is the form of ink destined for common use. It is, further, made into small, flat, rectangular cakes, sometimes also cast into circular forms if required by the artistic subject impressed upon the cake.

PRINTING INK

Under the great T'ang dynasty (A.D. 618-906) the manufacture of ink took an unprecedented development. Several events conspired to contribute to this result. Under the T'ang sovereigns Central Asia was annexed to the empire, and this political expansion led to the predominance of Chinese civilization all over Asia. Ink was required in the outlying dominions, whether politically dependent or merely under the influence of the Chinese culture-sphere. It was eagerly demanded in Turkestan as well as in Tibet; in Annam as well as in Japan. The same epoch marks China's Augustan age in literature and painting which then reached their climax, and above all, it was the new invention of printing books by means of wooden blocks which gave a fresh impetus to further progress in the production of ink.

At the end of the sixth century, under the Sui dynasty, when printing first became known, the imperial library contained some 37,000 books; in the early part of the eighth century, being the most flourishing period of the T'ang, the number of works described in the official record of the imperial library, amounted to 53,951 books, besides which there was a collection of recent authors, numbering 28,469 books. These figures will give an idea of the important function which paper and ink must then have performed in the national culture. Fortunately we now have at our disposal both manuscripts and prints of that epoch.

Woodcuts of the T'ang period have been discovered in the Cave of the Thousand Buddhas (Ts'ien Fu Tung) by Sir Aurel Stein (*Serindia,* p. 893) and Paul Pelliot. They illustrate, as Stein writes, the high stage of technique which the art of printing from wooden blocks attained comparatively soon after its first invention, and also the earliest use to which it is likely to have been put. A printed roll, dated A.D. 868 and containing in its 16 feet of length the complete text of a Chinese version of the Vajracchedika, is the oldest specimen of printing at present known to exist, and its fine frontispiece is the earliest datable woodcut. It shows Buddha seated on a lotus throne attended by a host of divine beings and monks and discoursing with his aged disciple Subhuti.

The first novel and significant departure in the age of the T'ang is the large number of ink factories springing up under the guidance of highly trained specialists. No less than twenty-five names of manufacturers of repute have been handed down from this epoch. In ancient times everyone was his own maker of ink, or the ink-maker was a man of no consequence. Under the T'ang, the business was taken out of private hands, and began to be industrialized and commercialized. With the vast expansion of the empire, governmental affairs increased in volume, and state correspondence assumed unparalleled proportions: thus the government was compelled to maintain its own ink establishments, and we hear of an "ink official" (*mo kwan,* or *mo wu kwan,* "official of ink affairs") who was placed in charge of the government works, under instruction to send an annual supply of ink to the metropolis for the feeding of the administrative machine.

The most famous of these ink directors was Tsu Min, whose reputation was widely known throughout the empire, and whose best ink was prepared with a glue concocted from deer's antlers; his fame was so lasting that even in the fourteenth century his name was still forged on ink-cakes. Another ink expert, Wang Kiün-te, worked exclusively for the imperial ateliers, and very little of his products reached the general public, so that any of his inks in private possession were looked upon as veritable family treasures and heirlooms. He availed himself of two sets of chemical accessories, one consisting of pomegranate peels preserved in vinegar, buffalo horn, and sulphate of copper; the other being composed of the bark of *Fraxinus pubinervus*, pods of *Gleditschia chinensis*, sulphate of copper, and *Verbena officinalis* (Chinese *ma-pien-ts'ao*, "horse-whip herb").

It is worthy of note that another species of Verbena is used in India for the manufacture of ink. The T'ang emperors also maintained an ink factory at Jao-chou in Kiang-si Province, where the slopes of the Lu Mountains (Lu shan) were covered with pines. It is on record that the emperor Hüan Tsung (A.D. 713-755) sent every season 336 pellets of ink to two colleges which he had founded, the T'u shu fu and Tsi hien yüan; and the same monarch is said to have himself manufactured an ink with the juice of lotus-flowers (*Nelumbium speciosum*) blended with an aromatic powder, his product being known as "imperial ink" (*yü mo*). This is a very interesting fact in that it demonstrates that the occupation of ink-making was then a perfectly honorable and even dignified and exalted profession, and this is no wonder in a society where learning was so highly esteemed and worshipped.

Most of the ink manufacturers of the T'ang were men of culture, literati and officials, and thanks to their social status, their names have come down to posterity. This is in striking contrast to the fact that, as known to everyone, China has produced a long line of ingenious and clever potters and bronze founders, men of humble standing, and that hardly any of their names have survived. We know and admire their works, but are ignorant of their names; of the ink artisans we have their names, not their works. In China, so far as is known, no ink-cakes of the T'ang and Sung periods are preserved; among the oldest are those which have come down from the Ming period. As regards the esteem in which the ink maker is held even in recent times, Du Halde states in his *Description de la Chine* (1738) that "everything which relates to writing is so reputable among the Chinese that even the workmen employed in making the ink are not looked upon as following a servile and mechanical employment."

In conformity with the specialization of work, the T'ang ink makers signed their pieces, covered them with inscriptions, and even provided them with a date. Thus under Kao Tsung (A.D. 650-683) was issued an ink bearing the inscription, "Ink guarding the treasury [*chen k'u mo*], made in the second year of the period Yung-hui" (A.D. 651). One of these pieces is said to have weighed two catties. There was another ink inscribed, "Made by Li Ts'ao, second secretary of the Board of Water Communication of the T'ang." Li Ts'ao was the ancestor of a family the members of which devoted their lives to the production of ink. They originated from Yi-shwi, but emigrated to Shö-chou which forms the prefectural city of Hui-chou in An-hui Province; extensive pine-tree forests in that locality induced them to choose it as their domicile. This region has remained the principal seat of the ink industry until the present time.

Of the various Li it was Li T'ing-kwei who attained the greatest fame. He was the "ink official" of the Southern T'ang dynasty (Nan T'ang, 937-975, also known as Kingdom of Kiang-nan, with Nanking as capital), and his products were regarded as the best in the empire and the goal which all subsequent manufacturers endeavored to reach; many also borrowed his name and counterfeited his ink-cakes. His originals usually bore an inscription consisting of four characters: *Li T'ing-kwei mo* ("ink of Li T'ing-kwei"); some also were dated, for instance, "In the first year of the period Pao-ta (A.D. 943) offered by Shö-chou and made by the official in charge of ink affairs, Li T'ing-kwei." His ink-cakes were four inches long, one inch wide, and one inch thick, and were frequently adorned with gilded dragons. They were as hard as metal and stone, and so hard and sharp that they could serve for smoothing printing blocks. Even after a hundred years, when rubbed, they still emitted an odor of camphor; and they yielded no sound in being rubbed on the stone, which the ancients always regarded as one of the criteria of good ink. A crackling sound was naturally caused by grittiness adhering in the ink, and good ink had to be free of any sandy matter. Other qualities demanded were that it should be deep black, light in weight, and extremely hard, its hardness being compared with that of jade.

In the palace the ink of Li T'ing-kwei was burnt, and the soot thus obtained was used as a paint for the eyebrows; it was hence styled "eyebrow-paint ink" (*hua mei mo*). His recipe was kept a secret and died with him. Ch'ao Shwo-chi, in his *Mo-king,* states that he availed himself of twelve chemical ingredients, but is only able to name four of them; these are gamboge, the inspissated sap derived from incisions into the bark of *Garcinia morella* or *G. hanburyi* (this beautiful reddish-yellow pigment is still used by Chinese draughtsmen and painters), rhinoceros-horn, genuine pearls, and the seeds of *Croton tiglium*. There is no doubt, of course, that lampblack of pine, as with all manufacturers of the T'ang period, formed the essential of his ink. He is also credited with the production of an ink of blue color.

Next in reputation to the ink of Li T'ing-kwei was that of Chang Yü from Yi-shwi, whose products were known as "tribute ink of Yi-shwi" (*Yi shwi kung mo*). He lived toward the end of the T'ang period, and his inks were made in the shape of copper coins, which was rather inconvenient for rubbing them. Another noted ink artisan was Chu Fung, who worked for Han Hi-tsai, a scholar and minister of state. His atelier at Shö-chou was known as the "hall where pine-trees are transformed" (*hoa sung t'ang*), and his ink was inscribed *Yüan chung-tse* or *Shö hiang yüe hia* ("musk moon box").

In the beginning of the T'ang dynasty (from A.D. 618) Korea (Kao-li) sent to the court of China an annual tribute of ink made from the lampblack of very old pine (*sung yen mo*) mixed with glue obtained from the antlers of the tailed deer (*mi lu, Cervus davidianus*). This Korean ink is described as black as if it were coated with varnish, glossy, and floating in water (*Wei lio,* Ch. 12, p. 1). It is sometimes asserted that the Koreans were the first who manufactured ink from pine lampblack, and that it was the Chinese who subsequently adopted the process; this conception of the matter, however, is not correct. The pine lampblack was utilized in China prior to the T'ang, probably as early as the third and fourth centuries A.D., and the Koreans learned the whole technique from the Chinese; but the Koreans succeeded in perfecting the product to a degree unknown in China by employing a particularly suitable pine-wood well dried in the course of years and a specially fine hart's-horn glue. The latter was known to the Chinese in early times, and is looked upon as the finest kind of glue (those next in rank being derived from the hides of horse, cow, rodents, and rhinoceros); it is called white glue (*pai kiao*) or yellow bright glue (*huang ming kiao*), the latter designation being also applied to the ink thus prepared. It was naturally expensive, as this glue was hard to obtain. The Korean product elicited the admiration of the Chinese, and for some time its method of manufacture remained a secret to them. They made several endeavors to imitate the art of the Koreans, but only attained the desired result toward the close of the T'ang dynasty (about A.D. 900); yet under the Ming the secret of making this ink was lost (*T'ung ya,* by Fang I-chi, Ch. 32, p. 17).

Under the Sung (A.D. 960-1279) we hear of several manufacturers who availed themselves of antlers' glue; thus, for instance, Chang Kü-tsing. Others, like P'an Ku, who lived during the latter part of the eleventh century, and who belongs to the most renowned ink manufacturers of the period, remodeled specimens of Korean ink by breaking them up, pounding the mass, and mixing it again with glue. P'an Ku is also noted for having used in his ink a sort of ivory-black made from bones.

The poet Su Shi, better known as Su Tung-p'o (1036-1101), prepared his own ink by using Korean charcoal and glue of the Kitan. When he lived as an exile on the island of Hai-nan, he caused P'an Heng to make for him an ink bearing the legend, "Hai-nan pine-tree charcoal, ink made after the method of Tung-p'o." This ink was presumably made according to the Korean method.

The Koreans were inventive and ingenious people, who not only advanced the cause of ink and conveyed its manufacture to the Japanese, but also improved on processes of paper-making and printing. They produced a very fine and durable paper from silkworm cocoons which achieved a great reputation in China, and they printed as early as 1403 with movable type cast of copper actual specimens of which may be seen in the American Museum of Natural History, New York. One peculiar custom of Korea, which may not have been practised in China or Japan, is particularly noteworthy: Ink-cakes were formerly offered by the emperor of Korea as a sacrifice to the gods, and there was attached to the Court a special department whose duty it was to manufacture this sacrificial ink. In the collections of Field Museum, Chicago, there is an old ink-cake coming from Korea, of oblong, rectangular shape. The obverse shows two dragons carved in high relief, two large characters being outlined in gold. They read *kwo pao* ("treasure of the country, national treasure"). The reverse contains an inscription in Chinese to the effect that this ink was made in the Yung-lo period (1403-25) of the Ming dynasty; but whether manufactured in Korea or China, is not stated.

Under the Sung a good many innovations as to detail were introduced, but the old principles virtually remained the same. Some improved on the lampblack, others on the glue, fish-glue was then first utilized, while others again devoted much thought and pains to the proper proportions and methods of bonding of the two. Shön Kwei, a native of Kia-ho in Hu-nan, produced a lampblack from charcoal of old pines which he blended with pine-resin and the sediments of varnish; this compound when burnt yielded an extremely fine lampblack which received the name "varnish-smoke" (*ts'i yen*).

The principal innovation that took place under the Sung was the substitution of vegetable, mineral, and animal oils for pine lampblack. The latter method continued in An-hui; the new method sprang up in the central and western provinces, notably Hu-nan and Se-ch'wan. It is clear that pine lampblack, after all, was a comparatively costly matter, that pine trees were not available everywhere, and that the ever-increasing demand for ink must have resulted in a despoliation of forests and contributed its share to that lamentable state of deforestation which has proved so grave a calamity to China. It is not surprising, therefore, that in view of the huge expansion of literature and art, writing, printing, drawing, and painting in the glorious age of the Sung, cheaper materials for the manufacture of ink were sought for.

Hu King-shun, a native of Ch'ang-sha (at that time called T'an-chou) in Hu-nan, obtained lampblack by burning the oil of *Aleurites vernicia* Hassk. (=*A. cordata* Steud., formerly *Elaeococca verrucosa* and *Dryandra cordata*); this was called "t'ung flower smoke" (*t'ung hua yen*) and proved a great success. His ink was much prized by painters for drawing the pupils of the eye. This oil, commercially known as t'ung oil or wood oil, is still accorded preference to any other in the modern manufacture of ink; a hundred catties of it yield eight catties of pure lampblack. Next in appreciation come the seeds of *Sterculia platanifolia* (*wu-t'ung*), which contain a good oil used in making ink. It is a stately, ornamental tree, frequently planted in the courtyards of temples and houses, its large leaves affording an excellent shade. It may be mentioned here that the juice from the crushed seeds is rubbed into gray hair, with the reputed virtue of causing the gray to fall out and the new hair to come in black.

In the period between A.D. 1067 and 1084, Chang Yü manufactured what became known as *yü mo* ("imperial ink"), as he presented his product to the Court. He used lampblack made from oil blended with musk, camphor, and gold-leaf, and is said to have been the first who availed himself of oil. His product was called *lung hiang tsi* ("dragon fragrance compound").

The oil-combustion ink seems to have at first roused the suspicion of some people, for an anecdote has it that when P'u Ta-shao produced his oil ink, people anxiously asked him how such ink could be strong and lasting; he assuaged the skeptics by responding that half of it contained pine lampblack, without which it could not be permanent. Whether this was correct or merely a subterfuge we do not know; it may be that such a "half-and-half" composition was really made, but it is certainly untrue that pine lampblack is necessary to give ink permanency. P'u Ta-shao was a native of Lang-chung, which forms the prefectural city of Pao-ning in Se-ch'wan Province, and his ink was widely used by scholars and offered as present to the throne.

K'ou Tsung-shi, author of the herbal *Pen ts'ao yen i*, written in A.D. 1116 under the Sung, has the following interesting account:—

"Ink is made from the black produced by the smoke of pine-trees. Our contemporaries manufacture a sham product from the ashes of grain stubbles, which should not be used. Only pine-soot ink is serviceable in the materia medica. Solely distant smoke is fine and yields an excellent product; the coarse one should be discarded. At present Korea dispatches ink to China with every mission of tribute, but the ingredients of this ink are not known, nor is it beneficial as a medicine. In Fu-chou and Yen-ngan fu (both in Shen-si Province) there is kerosene (*shi yu*, 'stone oil'); the smoke emanating from it is very thick; the black produced by it can be made into ink. It has a black gloss like varnish, but cannot be employed medicinally."

The use of this petroleum ink inaugurated under the Sung has already been pointed out. It is also mentioned by Shön Kwa, who wrote the *Mong k'i pi t'an* in the middle of the eleventh century. He states that the natives of Yen-ngan in Shen-si burnt petroleum in their lamps, swept the lampblack together, and made it into ink, which had a deep black brilliancy like varnish, and which was superior to ink made from pine-resin.

There seems to have been some good reason for K'ou's

cautioning against the medicinal employment of Korean ink; for it is on record that flies when sucking the juice of it would die, but by dint of what poison remained unknown. The same author alludes to the use of oak-galls as a hair-dye and ink for domestic purposes; but this is an isolated instance, and such ink has never become popular.

Under the art-loving emperor, Hui Tsung (1100-25), attempts were made to mix lampblack with storax, the sweet-scented resin of *Liquidambar orientalis* and *L. altingiana* (cf. *Sino-Iranica,* p. 456). It is said that an ounce of this ink was worth a pound of gold, and that efforts to imitate it failed.

Of oils of animal origin, preference was given to lard. On this point the Jesuit Louis Le Compte, at the end of the seventeenth century, writes that "China ink is not so difficult to make as people imagine; although the Chinese use lampblack drawn from divers matters, yet the best is made of hog's grease burnt in a lamp: they mix a sort of oil with it to make it sweeter, and pleasant odors to suppress the ill smell of the grease and oil. After having reduced it to a consistence, they make of the paste little lozenges, which they cast in a mould; it is at first very heavy, but when it is very hard, it is not so weighty by half, and that which they give for a pound, weighs not above eight or ten ounces."

Fan Ch'eng-ta, in his interesting work *Ling wai tai ta,* which deals with the geography and products of southern China, and which was written in A.D. 1178, has the following note on ink (Ch. 6, p. 2b): "In Jung-chou (prefecture of Wu-chou, Kwang-si Province) there is an abundance of large pine-trees from which the inhabitants manufacture ink. Good qualities are sold by the pound (catty) which is worth two hundred copper coins only.

The merchants raise the money jointly for the purpose and sell the stock as a whole. The ink of Kiao-chi (Tonking), although not very good, is not quite worthless either. The people barter their ink for horn, ink-slabs, and writing-brushes which they carry suspended from their loins."

John Francis Davis (*China,* Vol. II, 1857, p. 180) writes, "Chinese ink has been erroneously supposed to consist of the secretion of a species of sepia, or cuttle-fish. It is, however, all manufactured from lampblack and gluten," etc. Likewise J. Dyer Ball (*Things Chinese,* 4th ed., 1903, p. 105) asserts, "A curious idea was prevalent at one time in the west that the so-called Indian ink was prepared from the coloring matter of the cuttle-fish, instead of being made from lampblack as its principal ingredient." These statements are only partly true. The fact remains that the Chinese formerly made use of sepia also, though to a limited extent.

The cuttle-fish styled by the Chinese "*wu-tse* fish" was already known to T'ao Hung-king (A.D. 452-536), a celebrated physician and alchemist, who says that this creature carries ink in its belly, and that this substance was used as ink in his time. It is therefore called also "ink-fish" (*mo yü*). It is popularly believed to be a transformation of a crow, and another legend has it that it owes its existence to the emperor Ts'in Shi when he dropped his writing outfit into the sea. This cephalopod is met with all along the coast of China and forms an article of trade at Ning-po and Wen-chou in Che-kiang. Large quantities are eaten dried or pickled, or taken as a tonic. The small bag of inky fluid situated near the liver of the cuttle-fish is understood to be its gall. The preparation of sepia as a pigment, however, was never understood; and the employment of the secretion as ink has never attained popularity, as it fades within a few years.

PRINTING INK

Peter Mundy, who visited Canton and Macao in 1637, toward the end of the Ming dynasty, is one of the earliest European travellers who has recorded the use of ink in China. He noted that "they all write with pencills and blacke and red incke made into dry paste which they distemper with water when they will use it." He likewise gives a rough sketch showing a Chinese at his desk engaged in writing and an ink-well "which holds his incke, the one side containing blacke, the other redde; little partitiones with water where he dippes his pensill and so tempers his incke."

A formula for making ink from orpiment is given as follows: "Ochre should be pounded into a very fine powder over which water is swiftly poured to clarify it. The water is poured out. Take the bark of *ts'in-p'i* (*Fraxinus pubinervus*), fruit of Gardenia, and pods of *Gleditschia chinensis,* one-tenth of an ounce of each, one grain of the seeds of *Croton tiglium* after removal of the skin, mix this with a half ounce of bright yellow Kwang-tung glue made from ox-hides, boil this mass, mix it with the ochre and form it into cakes; or let it dry in the shade and use it thus."

I add two recipes from Du Halde's *Description of China* (1738) which he says "are taken from the Chinese, and which perhaps may suffice to make the ink of a good black, which is looked upon as an essential property. Burn, say they, lampblack in a crucible, and hold it over the fire till it has done smoking. In the same manner burn some horse-chestnuts, till there does not arise the least vapor of smoke. Dissolve some gum tragacanth; and when the water in which the gum is dissolved becomes of a proper consistence, add to it the lampblack and horse-chestnuts, and stir all together with a spatula. Then put this paste into moulds; and take care not to put in too much of the horse-chestnut, which would give it a violet black.

"A third receipt, much more simple, and easier to be put in practice, has been communicated to me by P. Contancin, who had it from a Chinese, as skilful in this matter as anyone can be expected to be; for we ought not to suppose that the ingenious workmen discover their secret; on the contrary, they take the greatest care to conceal it, and make a mystery of it, even to those of their own nation.

"They put five or six lighted wicks into a vessel full of oil, and lay upon this vessel an iron cover, made in the shape of a funnel, which must be set at a certain distance, so as to receive all the smoke. When it has received enough, they take it off, and with a goose feather gently brush the bottom, letting the soot fall upon a dry sheet of strong paper. It is this that makes their fine and shining ink. The best oil also gives a lustre to the black, and by consequence makes the ink more esteemed and dearer. The lampblack which is not fetched off with the feather, and which sticks very fast to the cover, is coarser, and they use it to make an ordinary sort of ink, after they have scraped it off into a dish.

"When they have, in this manner, taken off the lampblack, they beat it in a mortar, mixing with it musk, or some odoriferous water, with a thin size to unite the particles. The Chinese commonly make use of a size, which they call *niu kiao* ('size of neat's leather'). When this lampblack is come to the consistence of a sort of paste, they put it into moulds, which are made in the shape they design the sticks of ink to be. They stamp upon the ink, with a seal made for that purpose, the characters or figures they desire, in blue, red, or gold color, drying them in the sun, or in the wind."

An Arabic author, Abu'l Faraj (A.D. 988), writes that the Chinese have an ink composed of a mixture resembling

Chinese grease. He pretends to have seen specimens in the form of tablets representing the image of the emperor, and adds that a piece like this will last for a long time. This notice is of interest in that it shows that ink-cakes were embellished with designs at that early date; yet, the Arabic writer must have been mistaken as to the subject of the picture, for the emperor's portrait was never allowed to be turned to so profane a purpose, nor has it ever been customary in China to circulate an emperor's portrait during his lifetime. No imperial portrait appears on any Chinese coin, the only exception being in recent times that of the emperor Kuang-sü on the Tibetan rupee (coined in Ch'eng-tu for the purpose of counteracting the influence in Tibet of the Anglo-Indian rupee bearing the portrait of Queen Victoria).

In Chinese records we read of representations of dragons on inks of the T'ang and Sung periods, but the Ming dynasty (1368-1643) was the great era when ink manufacturers appealed to noted artists for designs and pictorial representations to be applied to their products. Many specimens of this art have survived, and are eagerly bought by Chinese collectors. Some ink manufacturers published books containing illustrations of all pictures placed on their inks. Two of these works are especially prominent. One is the *Fang shi mo p'u,* published in 1588 in six volumes by Fang Yü-lu, who manufactured ink at Shö hien, forming the prefectural city of Hui-chou in An-hui Province. Most of his illustrations were contributed by an eminent artist, Ting Yün-p'eng, who usually signs his pictures Nan-yü. He was a native of Hiu-ning in An-hui, and thus in close contact with his countryman, Fang Yü-lu. His designs are highly artistic, exceedingly fine, and well drawn, and represent a microcosm of Chinese mythology, as well as a thesaurus of art-motives.

Nan-yü revived several ink designs of which merely a literary reminiscence was preserved; thus an ancient text says that there formerly was an ink of nine children (*kiu tse mo*), which implied the wish that the owner of the ink may be blessed with an abundance of progeny.

Hence Nan-yü introduced the drawing of nine boys engaged in play, flying a kite, riding a hobby, manipulating a movable puppet, beating gongs and a drum. A goodly number of ink pieces are fashioned in the shape of archaic jade ornaments; others, in the shape of a peach-leaf, conch-shell, or bell; others are adorned with designs of palaces, terraces, landscapes, flower-pieces, birds and quadrupeds, the eight famous steeds of King Mu, star-gods and other deities of ancient lore, or Buddhist emblems accompanied by Indian scripts. This, of course, is not the place to give a detailed account of the variety and significance of these designs; suffice it to call attention to this artistic development of ink-cakes which is a unique phenomenon in the history of art.

The other work is the *Ch'eng shi mo yüan* in twelve volumes, issued by Ch'eng Kün-fang (or Ta-yo, but commonly called Kün-fang) between 1594 and 1606, an excellent copy of which, printed on Korean paper, was secured by the writer for the American Museum of New York in 1901, another for the John Crerar Library of Chicago in 1908. This belongs to the finest and most artistic examples of Chinese book-making. It contains 385 cuts accompanied by explanations, essays, and poetry, the handwritings of the authors being reproduced in facsimile. One of the interesting features of Ch'eng's work is that it contains four Christian pictures contributed by the celebrated Jesuit Matteo Ricci, who arrived at Macao in 1582, and who died in 1610; the engravings are accompanied by an essay and explanations written in

Chinese by Ricci himself and reproduced in facsimile. They have been published by the writer in an essay entitled "Christian Art in China."

The Newberry Library of Chicago received from the writer a similar Japanese book, entitled *Ko-bai-en boku-fu* [in Chinese: *Ku mei yüan mo p'u,* (*Collection of the Inks of the Old Plum-tree Garden*)], published in 1773, in five volumes. The first of these contains a number of prefaces, two volumes are filled with eulogies of ink facsimiled in the handwritings of the authors, and two others are occupied by engravings of ink designs. In the main, these follow the models of their Chinese prototypes, but do not quite display their vigor and power; yet, in some cases, they also exhibit original subjects, for instance, ink-cakes in the form of a daimio's suit of armor, bow, and boots. One bears the name of the illustrious Li T'ing-kwei of the T'ang; another is adorned with the picture of the Envoy of the Black Pine, the spirit or genius of ink who, according to a legend, appeared one day to the emperor Hüan Tsung of the T'ang dynasty.

Ink is a favorite gift among scholars and gentlemen, and for this purpose special boxes are made up in a very elegant and tasteful manner. In consideration of some courtesies which had been shown his son, a high Chinese official of Peking once presented the writer with a box containing eighteen ink-cakes, each adorned with the portrait of a famous scholar in low relief, his name being written in gold, the reverse of each cake bearing a stanza accompanied by seals in gold. Nine cakes are arranged in a finely lacquered tray, the inside of which is mounted with yellow silk. Each tray has its separate cover likewise lacquered black, decorated with a border design in gold, and inscribed with four large characters in heavy gold (*fang ku tsang yen,* "in imitation of ancient lampblack"). The two lacquer boxes fit into a card-board case mounted on decorated cloth in the style of bookbindings, and the whole presents the appearance of a book. Other such series are exquisitely adorned with celebrated landscapes or mountain scenery, or show all the stages in the process of tillage and weaving.

For preserving ink, wrapping it up in a leopard-skin is recommended by Fung Chi in his *Yün sien tsa ki* (Ch. I, p. 7), written in the beginning of the tenth century. The remedy may be efficient, but leopard-skins were not within everyone's reach, nor hardly ever available in a quantity sufficient to go round. The average man therefore had to be content with a plain box in which mugwort (*Artemisia vulgaris*) was placed as a means of preserving the ink. The main point is that it should not be exposed to sunlight which would cause it to crack and crumble to pieces. The box therefore must be tight-fitting. As early as the T'ang period special boxes were turned out for keeping ink, and great luxury was displayed in them under the Sung and Ming. They were made of a kind of sandalwood (*Pterocarpus santalinus*), ebony, or nan-mu (*Persea nanmu*), a valuable timber of Se-ch'wan, which does not easily rot, and which for this reason is much used for buildings and furniture. The wood was inlaid with jade plaques derived from the court-girdles of the T'ang period or with designs of hydras, tigers, and genre-scenes carved in jade; it was also coated with red and black lacquer.

The prominent qualities of Chinese ink are well known. It produces, first of all, a deep and true black; and second, it is permanent, unchangeable in color, and almost indestructible. Chinese written documents may be soaked in water for several weeks without washing out. It is safely used to mark linen. In documents written as far back as the Han period in the beginning of our era, and discovered under the sand of Turkestan, the ink is as bright and well preserved as though it had been applied but yesterday. The same holds good of the productions of the printer's art. Books of the Sung, Yüan, and Ming dynasties have come down to us with paper and type in a perfect state of preservation. Above all, we owe to the perfect ink of the Chinese many masterpieces of the brush in black and white and that charming art of monochrome drawing. Li Kung-lin or, as he is better known, Li Lung-mien, was one of the greatest masters of the line who ever lived, and the inspirations of his genius were merely expressed by black ink on white paper; he made the ink live and speak, drawing his lines in hundreds of shades.

Speaking of the qualities of China ink, the Jesuit Louis Le Compte (Memoirs and Observations made in a Late Journey through the Empire of China, p. 192, London, 1697) wrote at the end of the seventeenth century, "This ink is shining, extreme black, and although it sinks when the paper is so fine, yet does it never extend further than the pencil, so that the letters are exactly terminated, how gross soever the strokes be. It has moreover another quality, that makes it admirable good for designing, that is, it admits of all the diminutions one can give it; and there are many things that cannot be represented to the life without using this color."

Naturally Europe endeavored to rival the Chinese competition. Attempts were made in France at an early date to imitate China ink. Father Le Compte writes with reference to this subject, "It is most excellent, and they have hitherto vainly tried in France to imitate it; that of Nanking is most set by: And there are sticks made of it so very curious and of such a sweet scent that one would be tempted to keep some of them though they should be of no use at all." Likewise Du Halde, in his *Description of China* (London, 1738) observes, "The Europeans have endeavored to counterfeit this ink, but without success. Painters and those who delight in drawing know how useful it is for tracing their sketches, because they can give it what degree of shade they please." And still in 1869 P. Champion (*Industries anciennes et modernes de l'empire chinois,* p. 129) wrote that many attempts were made in France to manufacture Chinese ink, but that the results have never been entirely satisfactory, and that the ink of Chinese origin, superior in quality to the French products, has always been preferred by the draughtsmen. Chinese ink can be made only in China, and will never be equalled anywhere else.

Hui-chou in An-hui Province, where ink manufacture was so successfully initiated under the T'ang, is the high seat of the industry also at the present day. It is still of interest to read Du Halde's account of the Hui-chou factories.

"We are assured," he writes, "that in the city of Hui-chou, where the ink is made which is most esteemed, the merchants have great numbers of little rooms, where they keep lighted lamps all day; and that every room is distinguished by the oil which is burnt in it, and consequently by the ink which is made therein. Nevertheless many of the Chinese believed, that the lampblack, which is gathered from the lamps in which they burn oil of gergelin (sesame), is only used in making a particular sort of ink, which bears a great price, but considering the surprising quantities vended at a cheap rate, they must use combustible materials that are more common, and cheaper.

"They say that lampblack is extracted immediately from old pines, and that in the district of Hui-chou where the best ink is made, they have furnaces of a particular structure to burn these pines, and to convey the smoke through long funnels into little cells shut up close, the insides of which are hung with paper. The smoke being conveyed into these cells, sticks to every part of the wall and ceiling, and there condenses itself. After a certain time they open the door, and take off a great quantity of lampblack. At the same time that the smoke of these pines spreads itself in the cells, the resin which comes out of them runs through other pipes, which are laid even with the floor.

"It is certain that the good ink, for which there is a great demand at Nanking comes from the district of Hui-chou, and that none, made elsewhere, is to be compared with it. Perhaps the inhabitants of this district are masters of a secret, which it is hard to get out of them. Perhaps also the soil and mountains of Hui-chou furnish materials more proper for making good lampblack than any other place. There is a great number of pine-trees; and in some parts of China, these trees afford a resin much more pure, and in greater plenty, than our pines in Europe. At Peking may be seen some pieces of pine-wood which came from Tartary, and which have been used for above these sixty years. Nevertheless, in hot weather, they shed a great quantity of big drops of resin, resembling yellow amber. The nature of the wood which is burnt contributes very much to the goodness of the ink. The lampblack which is got from the furnace of glass-houses, and which the painters use, may perhaps be the properest for imitating Chinese ink.

"As the smell of the lampblack would be very disagreeable, if they were to save the expense of musk, which they most commonly mix with it; so by burning such drugs, they perfume the little cells, and the odors mixing with the soot, which hangs on the walls like moss, and in little flakes, the ink they make thereof has no ill scent."

PRINTING INK

Hui-ning in the prefecture of Hui-chou is now the centre of the ink industry of An-hui Province. From there the whole of China is supplied with ink. That of the family Hu K'ai-wen enjoys a special reputation. It sells its product from 300 copper coins up to 48 ounces of silver (taels) for the pound which includes 30-32 cakes of medium size. The ink of the first class is made by varnish and sesame oil; that of the second class, by sesame oil and lard; that of the third class, by the oil of colza; and that of the fourth class by t'ung oil. Each of these substances makes inks of very different quality, according to the number of lamps and the degree of slowness of combustion. Two good workmen can turn out eighty pieces daily, each half a pound in weight (cf. H. Havret, La Province du Ngan-hoei, p. 38).

In 1863, according to S. Wells Williams' *Chinese Commercial Guide,* the finest ink was priced as high as Mexican $5 a catty (about 1⅓ lbs.), common sorts ranging from $0.40 to $1.50. The boxes destined for export to Europe usually contained a hundred cakes.

The usual method of printing books in China and Tibet is that by means of wooden blocks. For this purpose, the manuscript is first written on thin paper by a professional caligraphist. This paper is pasted over the finely planed block with the characters turned face downward, the thinness of the paper displaying the writing perfectly through the back. Then commences the engraver's work, who chisels down the surface of the block around the characters, so that the writing in negative stands out in relief. In this state, the blocks go to the printer, who lightly rubs ink over them with a round brush of coir-palm fibre, places a sheet of paper on them, and takes the impression by passing another brush over.

In regard to printer's ink, old Du Halde (1738) has

supplied the following information from the reports of the Jesuits:—

"The ink which they use for printing is a liquid, and therefore much more convenient than that which is sold in sticks. To make it, you must take lampblack, pound it well, expose it to the sun, and then sift it through a sieve. The finer it is, the better. It must be tempered with Aqua-vitæ till it comes to the consistence of size, or of a thick paste, care being taken that the lampblack may not clot. After this it must be mixed with a proper quantity of water, so that it may be neither too thick, nor too thin. Lastly, to hinder it from sticking to the fingers, they add a little neat's leather glue, probably of that sort which the joiners use. This they dissolve over the fire, and then pour on every ten ounces of ink almost an ounce of glue, which they mix well with the lampblack and Aqua-vitæ, before the water is added to them."

De Guignes (*Voyages à Peking* 1784-1801, Vol. II, p. 229) gives the following note on printer's ink: "For purposes of printing they avail themselves of a particular and rather fluid ink. It is made from lampblack finely ground, which is passed through a very fine sieve. It is then soaked in rice wine, and when it has the consistency of a pap, glue is added at the dose of an ounce for ten ounces of lampblack. The whole is mixed together, with the addition of the necessary quantity of water."

For printing-ink, S. Wells Williams informs us, the lampblack is mixed with strained congee or a vegetable oil; and when the paste is properly dried, it is kneaded on a slab and cut into strips shaped like wrought nails. The printers grind it, or dilute it in oil as they use it, laying it on the wooden blocks with a brush made from the bark of the coir-palm.

At the present time large quantities of printer's ink are

imported into China from Japan, and that of Japan is made according to European methods.

A specimen of powdered printer's ink obtained by the writer in 1910 from a Tibetan monastery, but presumably of Chinese origin, was examined by Henry W. Nichols, associate curator of geology in the Field Museum, Chicago. It was actually used with good success in the Museum's printing office. Mr. Nichols reports as follows:—

"I have made a qualitative chemical examination of the Tibetan ink powder submitted. The material is a nearly black powder of medium coarseness. A sizing test shows 30% retained on a twenty mesh sieve and only 10% fine enough to pass a hundred mesh sieve. The powder is evidently to be reground before used for ink. Under the microscope some of the larger particles take the form of broken fragments, others show spherical and botryoidal surfaces covered with small projecting points. Still others have a rough cellular appearance like that of clinkers and cinders from a coal fire. There is no trace of organic structure apparent.

"The ink is composed principally of carbon. When it is burned, the ash is too great in quantity for charcoal and too little for bone or ivory black. The ash is principally phosphate of calcium with smaller quantities of iron, alumina, silica, and undetermined elements. The ink cannot be a soot like lampblack or carbon black on account of the coarseness. The quantity and character of the ash show that it cannot be a pure charcoal nor a pure bone black.

"The examination indicates that this ink is similar to the older form of drop black except that the process of manufacture has been carried one step further than is the case with bone black. Charred ivory, teeth or the denser parts of bone has been mixed with charcoal and finely ground. The powder has been cemented into a cake with glue, gum or some similar substance. The cake thus formed has been ground to a coarse powder which has been recharred. Heath and Milligan described the old style of drop black thus: Drop Black, as the name implies, was first placed on the market in the form of small lumps or drops and consisted of various mixtures of animal and vegetable blacks ground to a fine powder with water, mixed with a little gum and then moulded into drops and dried. The Chinese material differs from this in that the moulded drops or cakes have been powdered and reburned. The object of adding bone black to the charcoal is to improve the color."

As any substance found in nature and any artifact, ink also has invaded the materia medica of the Chinese. In the *Pen ts'ao kang mu,* written at the end of the sixteenth century and regarded as the standard herbal of the late Ming and Manchu periods, ink is described as astringent, diuretic, emmenagogue and vulnerary in its qualities. It is recommended as an application to the eye when irritated by the presence of foreign bodies. Not so long ago, and perhaps still at present, stale ink was administered as a kind of paint for daubing over tumors and swellings of all kinds, also for treating ulcers and wounds. This does not appear so bizarre if we remember that similar practices prevailed among us. Thus Francesco Carletti (*Ragionamenti sopra le cose da lui vedute ne' suoi viaggi,* Vol. I, p. 84), who visited Peru in 1595, relates that the wound caused by an injurious insect called *higna* when removed from the skin was healed by the application of a little ink.

Du Halde writes, "When the ink has been preserved a long time, it is then never used for writing, but becomes, according to the Chinese, an excellent and refreshing remedy, good in the bloody flux, and in the convulsions of children. They pretend, that by its alkali, which naturally absorbs acid humors, it sweetens the acrimony of the blood. The dose, for grown persons, is two drachms, in a draught of water or wine."

At the end of the eighteenth century de Guignes even recommended old Chinese ink for hemorrhages and the stomach, provided that it is of superior quality. This effect, he comments, is not surprising, as it is combined with *ngo-kiao* or glue from asses' skins, which is a supreme remedy in blood-vomiting. According to de Guignes' description asses' glue would enter the composition of every ink, which, of course, is not true.

Writing-brush and ink have become so essential requisites and attributes of scholarship that *pi mo* ("brush and ink") has developed into a term denoting literature. The phrase "he talks of nothing but pen and ink" means that his hobby is literature. "Eating ink" is a common phrase for studying, from the habit of the Chinese of putting the writing-brush into the mouth in order to give it a fine point. The question addressed to a scholar, "how many years' ink have you eaten?" means as much as "how long have you been studying?" A skilled writer is said "to scatter ink and make pearls."

Under the first emperor of the Liang dynasty (A.D. 502-556) candidates who failed in the examinations for the degree of *siu-ts'ai* were made to drink long draughts of liquid ink.

Ink has naturally entered into proverbial sayings also. "It is not the man who rubs (wears out) the ink, it is the ink which wears out the man,"—by his application to study. "He who is near ink gets black; he who goes near vermilion will make himself red." Written notes are regarded as preferable to memorizing. This is expressed by a proverb which says, "The palest ink is better than a capacious memory"; also quoted in the form, "A clever memory is not equal to a clumsy brush."

Poetical names for ink are "black metal," "dark incense," "black-jade ring."

The Chinese never keep liquid ink in bottles or ink-wells, but prepare only as much as they actually need at a time. For this purpose they avail themselves of a slab of marble or other stone which has a small rounded cavity at one end. A few drops of water are poured over the finely polished surface, and the stick or cake of ink is gently rubbed against it, the ink flowing into the cavity. Many ink-sticks are provided with a rounded notch at the lower end to secure a firmer hold for the finger, while the upper part to be rubbed is rounded; in this manner one avoids confounding the two ends, as the wetted portion will naturally leave black spots on the fingers. The marble, before being used, must be carefully washed, so that no trace of old ink remains upon it; for even a small particle of old ink adhering to it is said to spoil both the marble and the fresh ink. The marble should not be cleaned with hot water or cold water just drawn from a well, but with water that has been boiled, and has grown cool again. The selection of the proper materials for ink-pallets and their preparation and carving has developed into a science in itself, and this subject has called forth a literature as exuberant as that on ink.

White jade makes the finest ink-stones, and there is a specimen in the Field Museum in the form of a well-frame, where the cavity is suggestive of a well,—a veritable ink-well. Another very ancient specimen of cast iron is provided with a lower compartment for heating water with charcoal. A peculiar fad came into vogue during the eighteenth century to convert the ancient roofing-tiles from the palaces of the Han emperors into ink-pallets. The Field Museum has a very fine stone ink-slab of the T'ang period with carved figures of a lion and lioness of realistic style. The Twan-k'i stone has been famed from remote times for its use as ink-pallets; Twan-k'i is an old name for Te-k'ing, a district in the prefecture of Chao-k'ing in Kwang-tung Province, where the quarries are situated. Several monographs have been written on this stone alone.

Glancing back at the preceding sketch which gives a mere outline of the history of ink in China, we recognize a constant ascending development, a gradual improvement and perfection of methods finally culminating in the best and most durable ink produced in the world. In principle, the composition of all Chinese inks is identical: the fundamental substance making the ink is lampblack from whatever source it may be derived; this is compounded with glutinous matter, the glue serving the purpose of uniting the fine particles of carbon and fixing the ink on paper by means of the brush. The perfumes sometimes added, like musk, camphor or patchouly, have the function of hiding the unpleasant odor of the glue, but are unessential. The numerous different varieties and grades of ink depend upon the fineness and quality of the lampblack, the quality of the glue, the proper proportions of the two, the process adopted in mixing them, and general methods of manufacture usually kept very secret. Another differentiation comes in from the addition of accessory vegetable, mineral, or chemical substances, which seem to vary in the hands of every manufacturer. For these reasons it is obvious also that an absolutely correct description of the process which would hold good or be typical of all factories cannot be given. Few foreigners had occasion to obtain access to them, and still fewer possessed the technical knowledge to describe exactly what was going on.

PRINTING INK

A French chemist, Paul Champion, has studied the process at Shanghai and Hankow toward the middle of the nineteenth century, and has given a brief description of it (*Industries anciennes et modernes de l'empire chinois*, p. 136, Paris, 1869).

CHAPTER III

THE HISTORY OF INK IN JAPAN

ACCORDING to the Nihongi (*Annals of Japan*), the king of Korea sent in A.D. 610 two Buddhist priests to Japan, one of whom, the Korean priest Tan-cheng or Tam-ch'i from Kao-li, was skilled in preparing painters' pigments, paper, and ink. He introduced into Japan the technique of manufacturing ink and paper. This industry was ardently advocated and promoted by the celebrated Japanese prince, Shotoku Daishi (A.D. 572-621), the propagator of Buddhism in Japan.

In the beginning the Japanese availed themselves only of the black derived from resinous woods like pine-tree (Japanese *sho-yen* or *matsu no kemuri*). At a later date, however, they learned from the Chinese the method of making a superior ink from the black of oil-lamps (Japanese *yu-yen* or *abura-susu*), and this process has now superseded the pine-soot method. Although Japan itself manufactures the greater part of the ink (*sumi*) required by the country, the Chinese product is looked upon as superior in quality and commands a higher price.

In principle, the process of manufacture is identical with that of China, but deviates somewhat in details. The lamps used for the purpose are small crucibles or dishes of stoneware, with wicks of rush-pitch. A cone-shaped soot catcher or reversed bowl of burnt clay, but unglazed, is placed over each lamp and is replaced with a new one every hour. The rough clay is preferred so that the black matter precipitates in the porous surface. When exposed to the flame too long, it would become too compact. The soot is carefully brushed off and swept together, and is then sifted through a fine hair-sieve. The glue to be added to this substance is made from ox-hides and isinglass, and must be very bright, acting as it does as a cement. To ten catties (13⅓ lbs.) of lampblack from the oil of *Aleurites vernicia,* four catties of old ox-hide glue and one-half catty of old isinglass are reckoned; oil of sesame and colza also are utilized.

These ingredients, after the glue has been boiled in the necessary amount of water, are thoroughly mixed in a porcelain dish or copper basin,—a toilsome process, as the lampblack does not readily combine with water. This being done, the mass may be kneaded and pressed like dough, and is shaped into round balls which are wrapped in cloth. They are placed in a stoneware jar with perforated bottom to be subjected to steam for fifteen minutes. The material is then taken out and wrought with a pestle in a mortar for at least four hours, until it is thoroughly homogeneous and plastic. It is, further, fashioned into large prismatic bars which for a moment are exposed to a temperature of about 50° Celsius in a jar, and then stretched into longer sticks. These are beaten with mallets on an anvil and constantly turned till they have acquired not only the proper form, but also the desired lustre. They are once more kneaded on a smooth table; musk, camphor, or some other odoriferous substance being added, and then shaped by hand and put in a wooden press.

Ashes from rice-straw, carefully sifted and dried in the sun, are used for drying the sticks. For this purpose a layer of ashes about an inch thick is placed in the drying-box to be followed by a layer of ink-sticks; then ashes again, and another layer of sticks covered by ashes on the top. The length of the drying process depends on the quantity of water contained in the ink. When satisfactorily dry, the sticks are removed from the ashes, brushed off, laid in a small sieve, and for a day or two are left in a shady spot, where the drying process is completed. They are then polished by means of a brush, sometimes varnished and gilded, and any required legends as manufacturer's name, name of place, devices or mottoes are impressed on the surface.

This ink should not be used for several years after making, as hardness, blackness, and lustre increase with age. The same rule is observed in China. The quality depends largely on the fineness and lightness of the lampblack, the purity of the glue, and carefulness observed during the several stages of manufacture. Sound and a tinge of brown color are regarded as criteria by which to recognize and judge the best pieces. Ink of the first quality is uniform, without cracks or blemish, and brilliant in its fracture. In rubbing it with water on the inkstone it must not crackle; that is, it must be entirely free from any kind of sandy matter. The odor is required to be that of a pleasing blend of musk and patchouly; the color, that of a black-brownish with a slightly russet tint. The writing when dry must have frigid and glossy tones. The sticks of prime quality, as a rule, have a plain surface without much decoration, and are completely gilded; while those of secondary or inferior quality are usually more highly ornate.

Musa from the province of Omi, Kaibara from the province of Tamba, and Taihei from the province of Yamashiro formerly were renowned brands of ink. At the present time it is the city of Nara, the ancient capital of Japan situated between Kyoto and Osaka, and the manufacturers Matsuda and Matsumura in the province of Kaga who enjoy the greatest reputation for their output of ink.

The pallets for rubbing the ink on (Japanese *sudzuri*) are made in imitation of Chinese and Korean models, usually of a fine-grained dark stone, chiefly old slate, serpentine, or colored marble. In Japan an old, dark blue slate is especially prized for this purpose, and is generally used. It is found in the neighborhood of Amabata, a small town in the province of Kiushiu, and is hence known throughout the country as "stone of Amabata" (*Amabata-ishi*). A cavity is inserted on one side of the stone to serve as a receptacle for water. When ink is required, a few drops of water are poured into the hollow, the stick is dipped in, the water being brought up by it to the surface of the pallet. The ink-cake is rubbed against the stone, and the ink gradually flows back into the well, ready for use.

The Japanese business man always carries with him a portable writing-case (*yatate*), including a holder for fluid ink and a writing-brush enclosed in a metal case. For household purposes is furnished a box with several compartments (*sumi-ire*),—one for brushes, another for ink-cakes, and a third for the ink-stone.

The illustration of a Japanese writing-desk may be seen in Edward S. Morse's *Japanese Homes and Their Surroundings* (p. 317). The usual form consists of a low stool not over a foot in height, with plain legs for support, sometimes having shallow drawers. This is about the only piece of furniture in a Japanese house that would parallel the style of writing-table used in the western part of the world. Paper, paper-weight, ink-stone with ink, water-bottle, brushes, and brush-rest are placed on the desk in the same manner as in China. On page 141 of the same work Morse gives a sketch illustrating the writing-place in a guest room.

PRINTING INK

In preparing ink for wood-block printing, ink-cakes are macerated in water for a few days, until the glue contained in it is dissolved and the mass becomes sufficiently softened. It is then ground by means of pestle and mortar. After it has been mixed with water, glue solution or rice paste, according to the printer's judgment, has to be added. If glue solution is employed, it should be mixed with the lampblack in a basin; but rice paste is mixed with the pigment on the plank by means of the brush.

CHAPTER IV

THE HISTORY OF INK IN CENTRAL ASIA

OWING to the geographical position of the country, the culture of Tibet is of a dualistic character in its absorption of foreign ideas: on the one hand these have filtered in along its southern border from India, and on the other hand along its eastern frontier from China. While the alphabet, literature, and religion were received from India, all practical industries came from China, and so it was with paper and ink.

Under the first powerful Tibetan king, Srong-btsan sgam-po, who died in A.D. 650, writing was introduced from northern India, and soon afterwards the king invited scholars from China to draft his official reports to the emperor whose daughter he had received in marriage in A.D. 641. In 648 he applied to his imperial father-in-law for workmen capable of manufacturing paper and ink, and this request was granted. From this date onward the Tibetans joined the ranks of literary nations, and in a few centuries developed a literature of an astounding extent. They likewise adopted from the Chinese the art of block-printing, and we now have at our disposal Tibetan writings as early as the ninth century. It follows from the preceding account that the Tibetans learned the preparation of ink from the Chinese; but the bulk of their ink is still imported from China.

Colored inks, and especially writing in gold and silver, are mentioned in Tibetan literature at an early date. Copying a religious book means accumulation of religious merit, and the merit is graded in accordance with the color of the ink: gold is regarded as first in rank; silver, as second; vermilion, as third; and black, as fourth. Gold and silver manuscripts are written on a stiff, heavy paper of black glossy background surrounded by a blue border.

As early as the beginning of the fourteenth century the first copies of the Buddhist canon, known as the Kanjur and Tanjur (making about 230-245 large volumes), were produced in gold writing by Sa-skya Pandita, and subsequently we hear of a number of such editions of the sacred scriptures both in gold and silver, also in an alloy of both of these metals. A superb edition of the Kanjur in vermilion was issued in 1700 from the press of the imperial palace of Peking (so-called palace edition) by order of the emperor K'ang-hi, and another of the same character by his successor, K'ien-lung.

In China and Japan also Buddhistic manuscripts in gold are occasionally found; hence we may infer that this practice was propagated by the Buddhist clergy. In the West gold-writing reached its highest development among the Byzantines (cf. V. Gardthausen, *Buchwesen im Altertum,* p. 214).

The Tibetans do not rub ink on a stone, as the Chinese do, but carry it dissolved in brass ink-pots, together with a pen-case, which are suspended from the girdle. The pens are bamboo styles placed in pen cases of brass, copper, silver, or iron inlaid with silver. In the collections of Field Museum, Chicago, writing materials from Tibet, including ink, pens, specimens of paper, manuscripts and prints, and all implements used in printing, are on exhibition.

In Tibet as well as Mongolia, the pupils in the schools use as slates slabs of black painted wood, dusted over with white chalk, on the surface of which the writing is done with a style.

The Lolo, an aboriginal tribe inhabiting parts of Se-ch'wan and Yün-nan and distant kinsmen of the Chinese and Tibetans, manufacture ink from a soft schist of blood color, which is dissolved in water, and also from the ashes of a large mushroom that grows on the trunk of an oak. They use a style of a tender wood for writing, and at the present time avail themselves of Chinese paper. In ancient times they employed tree-bark for this purpose. In the collections of Field Museum are two Lolo documents written on oblong slips of wood.

In the *Mo shi* of Lu Yu, referred to above (p. 19), there is a brief notice of ink in Chinese Turkestan. A Buddhist monk named Su T'ai-kien is quoted as saying that in Turkestan there are neither ink-slabs nor writing-brushes (a wooden style was in use there), but only excellent ink which is not surpassed by that of China. It was prepared from old pine-trees growing in the Ki-tsu ("Chicken-foot") Mountains. T'ai-kien would keep leaves of the palmyra-palm inscribed with several hundred Sanskrit letters, the ink being twice as glossy as that of China. When at the time of the autumn rains the windows covered with such paper were wetted, the writing even though rubbed could not be wiped out.

Speaking of the documents inscribed on leather and discovered by him in Turkestan, Sir Aurel Stein (*Ancient Khotan,* Vol. I, p. 347) observes, "Owing to its exposed position on the outside surface, the writing of the address has often become faint or been partly rubbed off. But the ink on the obverse has in most cases retained remarkably well its original black color, and makes the writing clearly legible even in those cases where the leather itself has become discolored or stained. I regret not to have found an opportunity for arranging for a chemical examination of this ancient ink. But, judging from its appearance, it seems probable that it was Chinese (or Indian) ink, such as that of which a small stick was actually found by me among the rubbish layer inside the Endere Fort. The ink used on the tablets, both Kharoshthi and Chinese, varies considerably in quality and thickness, but I did not observe any indication pointing to a difference in the composition of the ink."

The Mongols appear to have borrowed their ink from China at a comparatively early date, as is proved by their word *bäkhä,* which is based on Old Chinese *mak, bak,* or *bäk.* The Manchu adopted the same word, presumably from the Mongols. In this connection it is interesting to note also that Mongol *sir* ("varnish") is borrowed from Old Chinese *tsit* or *tsir,* and Mongol *bir* ("writing-brush") from Old Chinese *bir, bit.*

In 1848 the great Finnish linguist, A. Castrén, wrote from Kiachta, "In the art of printing, the Mongol Lamas are comparatively less skilled than in writing; but it is curious enough that this very art is practised in this barbarous country. The Lamas, in accordance with their regulations, are obliged to know how to cut printing-blocks, to prepare printer's ink, and to print from the blocks."

CHAPTER V

THE HISTORY OF INK IN INDIA

IN considering the history of writing materials in India we are at the outset confronted with a psychological situation radically distinct from that in China and even almost the opposite to it. In China the written word and everything connected therewith were regarded with fervent reverence and treated as a fetish. Among the Brahmans of ancient India, it was not the written, but the spoken word which was looked upon as a fetish. The hymns of the Veda were memorized and transmitted for ages from generation to generation merely by memory; even at a time when an alphabet was in existence, the Brahmans first steadfastly refused to commit their sacred texts to writing, and but slowly and reluctantly yielded to this far-reaching innovation which threatened to break down the prerogatives of their caste. China never labored under a caste system; China has always been democratic, and placed the means of learning and education in the hands of whoever endeavored to learn and to read.

In India learning was the privilege of an exclusive sacerdotal class which kept in splendid isolation. In the Mahabharata it is said that those who sell, forge, and write the Veda are condemned to hell. In a society where such an aversion to writing prevailed it is not likely that much interest was evinced in the production and perfection of writing materials. It is striking also that despite her close contact with China, which set in from the first century A.D., India did not adopt paper and printing. Paper was introduced only in the Mohammedan period by the Arabs (there is no Sanskrit word to designate paper), and the first printing press in India was set up by the Portuguese at Goa in the sixteenth century. Whatever progress was made in India in the direction of writing must have been due to the caste of nobles and warriors, the Kshatriya, and to the merchants.

In the sixth century B.C. there were schools for methodically teaching the art of writing. Wooden writing-boards (*phalaka*) were in use, but these were incised with a stylus. The whole terminology relating to script and scribes hints at the fact that the letters were scratched in hard objects. There is no vestige of the use of ink in the early period. In the fourth century B.C. we learn from the Greek writers that prepared cotton-stuffs and birch-bark were employed in India, like papyrus, for writing letters. From this fact G. Bühler (*Indische Palæographie*, p. 91) is inclined to infer that ink was presumably used, and he confirms his supposition by palæographic evidence.

From the second century B.C. we have the oldest extant specimen of ink-writing on a stone vessel recovered from the tope (stupa) of Andher. In post-Christian times we have manuscripts written on birch-bark, the oldest being the small leaves folded and fastened with yarn (so-called "twists") discovered by Masson in the stupas of Afghanistan, followed by the famous Bower Manuscript which goes back to the fourth century A.D. Hoernle, in his edition and translation of the Bower Manuscript, says nothing concerning the ink. Aside from birch-bark, the leaves of several species of palm were enlisted as writing material in early times. Hüan Tsang, the famed Chinese Buddhist pilgrim, who visited India in 629-645, states that the leaves of the *tala* palm, which are long and broad and bright in color, are everywhere used for writing on in all countries of India.

In the Horiuji Monastery of Japan is preserved a Buddhistic palm-leaf manuscript inscribed with ink, and numerous such manuscripts of the ninth and later centuries from Nepal, Bengal, Rajputana, Gujarat, and the northern Dekhan demonstrate that in northern, eastern, central, and western India ink was used in writing upon palm-leaves. In Orissa and Dravidian India, however, the letters are incised in the leaf with a metal style, and are subsequently blackened with soot or charcoal. The use of palm-leaves for manuscripts is still common in southern India; the oldest manuscript extant there comes down from A.D. 1428. Palm-leaves were also used in southern India for letters, as well as official and private documents, and are still so used; there, and in Bengal likewise, they serve for writing in school. In the schools of Bengal banana-leaves also are said to be used and inscribed with lampblack ink.

In the Vasavadatta, a Sanskrit romance written by Subandhu in the seventh century A.D. and translated into English by L. H. Gray, occurs the passage, "The pain that has been felt by this maiden for thy sake might be written or told in some wise or in some way in many thousands of ages if the sky became palm-leaves, the sea an ink-well (*melamanda*), the scribe Brahma, and the narrator the Lord of Serpents."

John Fryer, who travelled in India and Persia for nine years (1672-81), informs us that the Persians "use Indian ink, being a middling sort betwixt our common ink and that made use of in printing: instead of a pen they make use of a reed, as in India."

We have a well-authenticated testimony for the existence of ink in India in the first century of our era in the

Periplus of the Erythrean Sea, a Greek work from the hand of an unknown author, probably written between A.D. 80-89, roughly about A.D. 85. In chapter 39 of this book *Indikon melan* ("Indian black") is given as one of the articles exported from the Indian port Barbarikon. The earlier commentators have explained this term as indigo, and B. Fabricius (in his edition of the *Periplus*, p. 152) is even inclined to interpret it as textiles made in India and dyed black. These opinions are not to the point, for the indigo of India is called in Greek simply *indikon*, in Latin *indicum* (cf., further, *Sino-Iranica*, pp. 370-371), while *melan* is the common Greek designation for ink: *Indikon melan*, consequently, means "India ink." Moreover, Pliny (*Hist. nat.*, XXXV, 25), in his chapter on ink (*atramentum*), points out "indicum, a substance imported from India, the composition of which is at present unknown to me," and says expressly that good ink prepared from dried wine-lees will bear comparison with that of indicum. Indigo is discussed by Pliny in a separate chapter.

There is hence no doubt that indicum signifies "ink of India," which was exported from India into the Roman Empire, and as confirmed by the Periplus, shipped together with other Indian goods from Barbarikon. Old Beckmann (*Geschichte der Erfindungen*, Vol. IV, 1799, pp. 489-496; cf. also Blümner, *Technologie*, Vol. IV, p. 517) has devoted a profound and ingenious investigation to the whole question, and has arrived at the same result. He thinks it very probable that the "Indian black" of the ancients was nothing but what is now termed India ink which approaches the finest ivory-black and lees-black so closely that by this means some still imitate it and actually delude ignorant buyers. "Ink in India," he concludes, "is in general use, and has presumably been so from earliest times; for in India almost all products of art are extremely ancient, but I do not mean to say that ink is a new Indian invention; it may have been improved, above all, by the Chinese."

There is no evidence to the effect that the Indian ink of the ancients was Chinese ink. In the first century A.D., as we have seen, it was still in its initial stages and very far from the perfection of the later days. All that we are permitted to assert safely is that the Indian ink of the ancients was an ink manufactured in India.

Blümner argues that the manufacture of Chinese ink is exceedingly old, and that in the same manner as Chinese silk was traded to the West, also ink might have arrived in Europe by way of India. Its native country being unknown, it was designated as Indian. There is, however, not a trace of documentary evidence for such a trade in ink, either in Chinese or in Western sources; and Blümner also adds cautiously that Chinese ink has not yet been traced in any paintings or pigments of classical antiquity; all investigations of black pigments have only yielded substances consisting of pure carbon.

The oldest Sanskrit designation for ink is *masi* or *mashi*. The word is indigenous, and according to Bühler, originally means "something ground, powder." It then came to denote several kinds of powdered charcoal which was mixed with gum-arabic, water, and sugar, and thus served as an ink. Another name for ink, *mela*, has been derived by some scholars from Greek *mélas* ("black"), but Bühler rejects this view and connects the word with Prakrit *maila* ("dirty, black"). According to L. D. Barnett (*Antiquities of India*, p. 231), ink was made in early times of charcoal mixed with water, sugar, gum-arabic, etc., and was applied with pens of wood or reed. A solution of chalk was also used as writing fluid, and was conveyed to the tablet by a wooden style.

In modern times the ink used for writing on paper is compounded of lampblack with an infusion of roasted rice, with the addition of a little sugar and sometimes the juice of a plant called *kesurte* (*Verbesina scandens*). It requires several days' continued trituration in a mortar before the lampblack can be thoroughly mixed with the rice infusion, and want of sufficient trituration causes the lampblack to settle down in a paste, leaving the infusion on top unfit for writing. Occasionally, acacia gum is added to give a gloss to the ink; but this practice is not common, sugar being held sufficient for the purpose. Of late, an infusion of the emblic myrobalan, prepared in an iron pot, has occasionally been added to the compound; but the tannate and gallate of iron formed in the course of preparing this infusion are injurious to the texture of paper, and Persian manuscripts sometimes written with such ink suffer much from the chemical action of the metallic salts.

The ink for palm-leaf consists of the juice of *Verbesina scandens* and a decoction of *alta* (cotton impregnated with lac dye). It is highly esteemed, as it sinks into the substance of the leaf and cannot be washed off. Both these inks are very lasting, and being perfectly free from mineral substances and strong acids, do not in any way injure the paper or leaf. They never fade and retain their gloss for centuries (after A. E. Gough, *Papers rel. to the Collection and Preservation of the Records of Ancient Sanskrit Literature in India,* p. 18, Calcutta, 1878).

Colored inks with which especially the Jaina produced beautiful manuscripts are frequently mentioned in Brahmanic literature, e.g., in the Puranas when donations of manuscripts are mentioned. Chalk and minium served as substitutes for ink in ancient times.

According to G. Watt's *Dictionary of the Economic Products of India,* at present various substances are used by the natives of India in making ink, the usual process being to mix some astringent principle such as galls or myrobalans with one of the iron salts or oxides. In Madras charcoal of the rice plant is employed in combination with lac and gum-arabic, and the Mohammedans generally prepare their ink from lampblack, gum-arabic, and the juice of the aloe. The following are the plants specially mentioned as adjuncts in the formation of inks:

(1) *Alnus nepalensis,* D. Don. Bark forms an ingredient in native red inks.

(2) *Cordia myxa* L. The unripe fruit is said to be used as a marking ink, though its color is less enduring than that from *Semecarpus.*

(3) *Phyllanthus emblica* L (Sanskrit *amaleka*). Fruits are largely employed in making black ink.

(4) *Semecarpus anacardium* L. (cf. Sino-Iranica, p. 482). The marking-nut tree bears a fruit with fleshy receptacle which contains a bitter and astringent substance universally used in India as a marking-ink, the juice being mixed with lime water as a mordant. Without the addition of lime it is often employed as ordinary writing-ink. As it is apt to cause severe inflammation, it has to be used with caution.

(5) *Terminalia belerica* and *T. chebula,* the unripe fruit of either species, or indeed of any *Terminalia,* is combined with iron in making ink.

The Siamese largely make use of Chinese ink, with which they write on long strips of gray paper made from tree-bast. A professional class of writers, called *alak,* avails itself ordinarily of a gum-resin dissolved in water, writing in yellow script on black paper. The Buddhist scriptures of the Siamese composed in Pali are written in Indian fashion on palm-leaves, the characters being incised by means of a style.

In ancient Camboja, according to the account of Chou Ta-kwan, who visited the country in the thirteenth century, official and private documents were written on pieces of deer-skin dyed black. They availed themselves for writing of a white clay, probably chalk, resembling the kaolin of China, moulding it into sticks, which were handled like pencils. Paper and ink were introduced from China and were used at an early date.

CHAPTER VI

THE HISTORY OF INK IN EGYPT, PALESTINE, GREECE, AND ITALY

IN a very interesting article entitled "The Physical Processes of Writing in the Early Orient and Their Relation to the Origin of the Alphabet" (*Journal of Semitic Languages,* Vol. XXXII, 1916), Professor J. H. Breasted observes with reference to an Egyptian representation of a noble of the thirteenth century B.C. with writing outfit, "In the use of this outfit the scribe made his own ink, mixing soot or lampblack with an aqueous solution of vegetable gum, which kept the insoluble black in suspension. This was done in one of the circular recesses shown on the little palette, and the pen was replenished from there. In the outer recess the scribe produced red ink in the same way, only using a red iron oxide instead of black. It was for this reason that we so often see the scribe with two pens behind his ear, one for the red and the other for the black ink. The red was used for the introductory words of a paragraph, and it was from this custom, as is well known, that the manuscripts of Europe received the so-called rubric, which has passed over into modern typographical usage."

Professor Breasted's researches endeavor to prove also that papyrus, pen, and ink were introduced from Egypt into Western Asia, beginning after 1100 B.C. In this respect Egypt's position in the West is identical with that of China in the East.

Ink is mentioned in the Old Testament but once: Jeremiah dictated his prophecies, and Baruch, his secretary, recorded them on a roll with ink (Jeremiah, XXXVI, 18: "He pronounced all these words unto me with his mouth, and I wrote them with ink in the book"), the word for the latter being *deyo*. Ezekiel (IX, 2, 3, 11) speaks of the ink-well of the scribe ("a man clothed with linen, with a writer's inkhorn by his side"). An allusion to writing without reference to ink occurs in Numbers (V, 23): "And the priest shall write these curses in a book."

In the New Testament ink is mentioned in three passages, as follows: "Forasmuch as ye are manifestly declared to be the epistle of Christ ministered by us, written not with ink, but with the Spirit of the Living God" (II Corinthians, III, 3). "Having many things to write unto you, I would not write with paper and ink: but I trust to come unto you and speak face to face" (II John, 12). "I had many things to write, but I will not with ink and pen write unto thee" (III John, 13). It is supposed, and with good reason, too, that the Jews during their sojourn in Egypt acquired their writing-materials from the Egyptians, and that the ink used by them was identical with that of ancient Egypt and Greece.

Among the Greeks ink was called *mélan* ("black") or *énkauston*. It varied according to the writing-material, and was distinct for parchment and papyrus. With the latter the Greeks adopted from the Egyptians both black and red ink. Two black pigments were known,—*trygi-non mélan* made from dried wine-lees, and *elefántinon mélan* ("elephant's ink") made from burnt ivory. The latter method, according to Pliny, was invented by the painter Apelles, while Polygnotus and Micon, the most celebrated painters of Athens, made their black from grape-husk. In both cases it was soot pulverized with gum and dissolved in water. The proportions, according to the Materia Medica of Dioscorides (V, 182), were three parts of soot to one of gum.

This ink was more unctuous than that of modern times, and was, perhaps, more durable, resembling our printer's ink. Like Chinese ink it was solid and kept dry. Demosthenes reproaches Æschines for having been so poor in his youth that he allowed himself to sweep the school-building, to scrub the benches with a sponge, and rub the ink. According to Diocletianus' edict of Megalopolis, ink was sold in a dry state by the pound, the price being comparatively high, as the pound cost twelve denars.

Pliny (*Hist. nat.* XXXV, 25) writes that *atramentum* (literally, "black coloring substance") must be reckoned among the artificial pigments, but that it is also derived in two ways from the earth. Sometimes it is found exuding from the earth like the brine of salt-pits, while at other times an earth itself of a sulphurous color is sought for the purpose. Painters have been known to go so far as to dig up half-charred bones from the graves for the same purpose. This would make an inferior ivory-black. The earth mentioned afore is considered by Ajasson to be a deuto-sulphate of copper, a solution of which in gallic acid is still used for dyeing black. Beckmann (*Geschichte der Erfindungen*, Vol. IV, 1795, p. 491) regards these earths as two vitriolic products, a mud (*salsugo*) and a yellow vitriolic earth otherwise styled *misy*. Others think of oxide of iron or mangan; others, of brown-coal. It is evident that the Plinean account exhibits a most striking analogy with the earliest Chinese attempt to derive an ink from a black earth and other minerals.

It appears from Pliny's further data that the mineral ink was no longer used in his time, but was superseded by several artificial preparations from the soot yielded by the combustion of resin or pitch. This process had advanced to such an extent that factories were built on the principle of not allowing an escape for the smoke. The most esteemed black was prepared from the wood of the torch-pine. Vitruvius, in his work on Architecture (VII, 10) describes the factories alluded to by Pliny, in the translation of M. H. Morgan, thus:—

"A place is built like a Laconicum ('Laconian hall,' a room in a bathing establishment), and nicely finished in marble, smoothly polished. In front of it, a small furnace is constructed with vents into the Laconicum, and with a stokehole that can be very carefully closed to prevent the flames from escaping and being wasted. Resin is placed in the furnace. The force of the fire in burning it compels it to give out soot into the Laconicum through the vents, and the soot sticks to the walls and the curved vaulting. It is gathered from them, and some of it is mixed and worked with gum for use as writing ink, while the rest is mixed with size, and used on walls by fresco painters.

"But if these facilities are not at hand, we must meet the exigency as follows, so that the work may not be hindered by tedious delay. Burn shavings and splinters of pitch pine, and when they turn to charcoal, put them out, and pound them in a mortar with size. This will make a pretty black for fresco painting.

"Again, if the lees of wine are dried and roasted in an oven, and then ground up with size and applied to a wall, the result will be a color even more delightful than ordinary black; and the better the wine of which it is made, the better imitation it will give, not only of the color of ordinary black, but even of that of India ink."

According to Pliny, the lampblack ink was adulterated by mixing it with the ordinary soot from furnaces and baths, and this substance was also employed for writing.

Others, again, calcined dried wine-lees, saying that if the vine was originally of good quality, it will bear comparison with that of indicum (the "Indian ink" already discussed). The dyers prepared an ink from the black inflorescence that adheres to the brazen dye-pans. It was made also from logs of torch-pine burnt to charcoal and pounded in a mortar. The preparation of every kind of ink was completed by exposure to the sun; the black for writing receiving an admixture of gum; and that for coating walls, an admixture of glue. Black pigment that has been dissolved in vinegar, he concludes, is not easily effaced by washing. Beckmann annotates that our ink too is much improved by being exposed to the sun-rays in shallow vessels, and that our cotton-printers are familiar with the fact that vinegar solidifies the black. He himself made good ink by taking clear brewed beer-vinegar.

Alluding to the sepia, Pliny remarks that it has a wonderful property of secreting a black fluid, that, however, no color is prepared from it. He obviously means that no pigment for the use of painters was made from it, for Persius mentions sepia ink for purposes of writing. It was used as an ink especially in Africa. Cicero calls the animal *atramentum* ("ink"), in the same manner as the Chinese speak of the "ink-fish." The Greeks of the earlier period never mention the sepia ink; Aristotle knew the cuttle-fish well, but not its ink. In all probability its use was then unknown.

Both lampblack and sepia ink were principally used for papyrus, and could easily be removed completely by cleansing. Sepia ink can be almost entirely wiped out, and chemical reagents remain without effect. Haubenreisser, however, has sometimes employed a varnishing process with success (V. Gardthausen, *Das Buchwesen im Altertum*, p. 204). Hence in the epigrams of the Roman poets the sponge plays a conspicuous part; it was one of the regular implements of the scriba librarius: Martialis sends to his patron his latest verses accompanied by a sponge, in case they should not find favor with him. Augustus, when interrogated by his friends as to what had become of his tragedy "Ajax," responded that his Ajax had rushed not into his sword, but into the sponge ("waste-basket," as we would say in these days).

This feature shows plainly that the ink of the ancients must have been different in composition from that of the Chinese, which cannot be washed off or destroyed. An inkstand containing some ink, thick but still fluid, was found at Pompeii. Its viscous character was sometimes a ground of complaint, yet it was well adapted for writing on papyrus. For the smooth and permanent parchment, an ink prepared from oak-galls (Greek *kekis*, Latin *galla*) was preferred. In the course of centuries this ink assumes a fine yellowish brown rust tinge which is esteemed as a symptom of great age.

This ink marks another fundamental divergence between the East and the West, for it is not known in the East. The Chinese became acquainted with oak-galls as late as the T'ang period when they were introduced from Persia (cf. *Sino-Iranica*, pp. 367-369), and used the ink only occasionally under the Sung. In all probability gall-ink was invented in the anterior Orient, for the species of oak (chiefly *Quercus lusitanica* var. *infectoria*) on which the gall-wasp deposits its ova that form the excrescences known as galls grows in Asia Minor, Armenia, Syria, and Persia. Pliny is not yet acquainted with this ink, and it seems to have come into existence only during the first centuries of our era.

The use of galls for ink is mentioned by Philo of Byzantium in the second century, in a description of sympathetic ink, and by Martianus Capella in the fifth century. It has, moreover, been established by Sir H. Davy's experiments in the Herculanean manuscripts (*Phil. Trans.*, Vol. II, 1821, p. 205). Chardin (*Voyages en Perse*, 1721, Vol. II, p. 108) writes that the ink of the Persians is very black and made from galls, pounded carbon, and lampblack. It is greasy and thick like our printer's ink. They use inks of all colors, red and blue, and also write with gold.

The inkstands of the ancients were of various shapes, cylindrical or hexagonal, and of various materials, as terra-cotta, bronze, or bronze inlaid with silver and gold, and sometimes highly decorated. Some are provided with rings for attachment to the girdle. There are single and double inkstands, the latter being intended to contain both black and red ink.

Arabic science is largely based on that of the Greeks, and the Arabs' formula for ink is derived from Dioscorides. His work on materia medica was translated by Ibn al Baitar (1197-1248) in his *Treatise of Simples*, translated into French by L. Leclerc. Ink is treated in Vol. III, p. 297. It is called *midad* in Arabic, and in accordance with Dioscorides, is prepared from lampblack collected from pines (*dadi*), one ounce of gum being taken and mixed with three ounces of lampblack. The ink for painters' use is also made of resin-black.

177

驯养研究中的各种方法

THE SCIENTIFIC MONTHLY

EDITED BY J. McKEEN CATTELL

VOLUME XXV
JULY TO DECEMBER

NEW YORK
THE SCIENCE PRESS
1927

METHODS IN THE STUDY OF DOMESTICATIONS

By Dr. BERTHOLD LAUFER

FIELD MUSEUM OF NATURAL HISTORY

It is a fond delusion of many to believe that ethnology, because it deals essentially with living peoples of the present time, can contribute little or nothing toward reconstructing the cultures and culture movements of the past. What I wish to demonstrate by a few practical examples from the chips of my workshop for the benefit of the younger generation is that it is possible to reconstruct by means of purely ethnological data and methods mental processes and culture phases of the past which can not be reached by historical or archeological methods. I choose my examples from the domestication of animals, because this is a subject of fundamental importance for the history of civilization and but little scrutinized and elucidated by ethnologists.

Our domestic animals can be studied from many different angles—zoological, anatomical, biological, economic, geographical, historical, archeological—and these various sciences have made many excellent contributions to our knowledge of the subject. It is obvious also that only by a combination of the results gained from so many different fields can a satisfactory solution of the problems be attained. Yet the most interesting question relating to our domestic animals has hardly been touched upon, and this is man's relation to them, or in other words the domestic animals from an ethnological point of view. What motives prompted primitive man to go to the trouble of domesticating animals and by what means and by what mental attitude was the primeval process of domestication brought about? To the layman the answer seems easy enough. We hear it almost daily and read it in our school-books that man keeps sheep for the sake of their wool, cows for their milk, chickens for their eggs, swine for the sake of flesh, etc. But all the material advantages which we now derive from domesticated animals are but the effect and result of prolonged activity in matters of domestication, and cause and effect can not be identical. Wild fowl, *e.g.*, do not propagate to a large extent, nor do they lay eggs in great numbers. The egg-laying habit of our chickens, to such an extent that it was of some economic advantage to man, was only developed after many hundreds or perhaps thousands of years of a gradual evolutionary process of domestication. When primitive man first adopted the wild fowl into his household, he could not foresee or anticipate any substantial benefit, no more than the alchemist of a thousand years ago could foresee that his alchemical lore would develop into the science of chemistry in the nineteenth century. Moreover, the utilitarian viewpoint is contradicted by plain observations of present-day conditions. There are, for instance, many tribes in southeastern Asia, Polynesia and Melanesia who keep chickens and have kept them for millenniums, but do not use their eggs or their flesh; there are other tribes among whom the pig is a sacred and sacrificial animal and is eaten but once a year in the solemn act of a religious ceremony. The Chinese have raised sheep and goats for thousands of years, but have never utilized their wool for making textiles for clothing. All East-Asiatic nations keep cows and buffalo, but never milk these and never consume animal milk. Primitive man is not rationalistic, but emotional, imaginative, impulsive, and it is in vain to look for rational motives in his first contact with the animal world.

Years ago I pointed out that Asia with its European annex is split into two well-defined and sharply contrasted economic camps, the boundary line running along the frontier of China and Tibet. All West-Asiatics, including the Indo-European nations, Semites and all nomad tribes of central Asia, have a highly developed dairy economy, and dairy products, such as milk, butter and cheese, form an important part of their daily sustenance. The entire East-Asiatic world, however, including the Chinese, Koreans, Japanese, Annamese, Siamese, Burmese, Malayans and many peoples of India, do not take animal milk for food and evince a deep-rooted aversion toward it. A Chinese looks upon a milk-drinker with the same dislike as we may feel toward people who subsist on monkeys, snakes and insects. The amazing feature, of course, is not the mere fact that East-Asiatics do abstain from milk, for the natives of the South Seas, Australia, America, etc., do or did the same simply for lack of milk-producing animals, but the point at issue is that the Chinese and their followers adhere to this negative practice despite an abundance of milk-furnishing beasts (they rear cows, buffalo, mares, donkeys, camels, sheep and goats) and despite a long and constant intercourse with neighboring milk-consuming peoples whose habits and mode of life were well known to them; yet they never acquired the habit of milk-drinking. Now what does this fact mean? In the first place it means that our consumption of milk can not be looked upon as a self-evident and spontaneous phenomenon for which it is usually taken, but that it is merely a matter of educated force of habit historically evolved. Objectively it is not natural at all. As "natural" as it appears to us in consequence of time-honored tradition and custom, so just as unnatural, abnormal and barbarous does it impress the Chinese and other nations of eastern Asia. Above all, however, these plain ethnological observations and considerations may lead us to distinguish two ancient culture periods in the early history of Asia which lie beyond all recorded history and archeological monuments. In the first period, which denotes the primary stage in the domestication of cattle, the milking faculty of the cow was unknown to man. The ox was exclusively the sacred animal trained in the service of agriculture solely for drawing the plough, and simultaneously he was the highest sacrifice to the gods of heaven. The invention of the plough and the wheeled cart, as well as the cultivation of cereals, are events closely affiliated with the domestication of cattle. This is the very point which the cultures of eastern Asia have in common with western Asia and Europe. In the second period, western Asia advanced to the stage of dairy economy; this was a slow and complex movement operative on the one hand in the producer, the animal, in which the productive power was gradually trained as the result of a domestication extending over millenniums; on the other hand, in the consumer, man, who just as gradually acquired the habit of taking to milk. This new development remained confined to the West, but it did not affect eastern Asia, and must therefore have taken place at a time when the East was definitely settled in its culture pattern and was no longer ready to absorb extraneous ideas. It bespeaks a lengthy prehistoric cultural development in the East independent of the West. I do not continue this discussion, as I merely wish to cite this case as an illustration of how ethnological methods may carry us into the remotest past and help us to discover and unravel ideas of which no record has been preserved.

The following case is still more instructive, as it will bear out some unexpected results in regard to the domestication of the cock and the pig. The first striking fact in their distribution is the parallel occurrence or non-occurrence of the two. Both pig and chicken are sedentary animals and consequently make their domicile only in sedentary, never in migratory communities. I say advisedly sedentary, not agricultural, because neither pig nor chicken bears any relation to the stage of agriculture and were introduced into the great agricultural civilizations of continental Asia from the more primitive, marginal or peripheral culture-sphere of southeastern Asia. In Tibet with its division into two social groups, an agricultural one in the fertile valleys and a migratory-pastoral one on the high plateaus, pig and chicken are found only among the farmers, but never among the nomads who raise cattle, horses and sheep. Throughout eastern Asia cock and hog are met with as parallel factors for reasons which will become clear after awhile. The sole exception is presented by ancient Japan which had only the cock, not the pig. In Formosa both cock and pig were found by the Chinese at the time of the discovery of the island in A. D. 605. In Polynesia the two occur everywhere, but the easternmost isle, Easter Island, reflects the same situation as Japan; it had chickens, but no pigs, and chickens were the only domestic animals encountered there. New Zealand had neither the pig nor the chicken. Such factors of geographical distribution are the result of historical events, and each case of this kind merits a special investigation.

Coming back to my previous proposition, I now wish to offer an explanation of the motives and circumstances which prompted the original phase of the domestication of these animals, and I shall explain briefly how I arrived at my idea.

A fundamental of culture in eastern Asia is divination. In this respect we have four distinct culture provinces which despite their difference of method share one important point inasmuch as divination is based on the bones of certain animals. Among several ancient and now extinct tribes inhabiting Korea and Manchuria there was a system of divination from the hoofs of cattle and horses and the bones of oxen. In central Asia divination was practiced from the shoulder blade of a sheep which was scorched over a fire, and from the cracks thus arising in the bone the future was predicted. In ancient China the carapace of a tortoise was utilized in a similar manner. The tortoise was regarded as a sacred animal imbued with a knowledge of the future. In 1899 a deposit of several thousand fragments of bones, chiefly tortoise shell, was discovered at Chang-te fu, Honan. These bones are engraved with inscriptions of a very archaic style, representing the earliest form of Chinese script we now possess, and were used for purposes of divination. The oracles and in some cases the answers were carved into the bones. We meet, *e.g.*, inscriptions such as these: "We consulted the oracle to ascertain whether the harvest will be abundant," or "The oracle was consulted, as we wish to know whether God will order a sufficient rainfall so that we may obtain an adequate food-supply," or "If we go a-hunting to-morrow, shall we capture any game?" Divination has always dominated the whole life of the Chinese from the cradle to the coffin, and no business was transacted, no marriage concluded, no burial undertaken, without consulting a fortune-teller. These ancient augurs formed a special profession, in their social position comparable to the lawyer of our society. In the same manner as the modern financier and captain of industry consults his lawyer on all important questions, so the Chinese did not make a step in the most trivial matters without asking a diviner's advice.

South of the Yangtse and widely diffused all over southeastern Asia we find from ancient times a complex system of divination based on the femora or thigh bones of a cock. Formerly it had a much wider geographical distribution than at present, and was found among the Yüe, a non-Chinese tribe which occupied what is at present the southern portion of Che-kiang and the province of Fu-kien, and among the widely extended Miao group in southern China. We have numerous Chinese records of this form of divination ranging from 110 B. C. down to recent times. At present this system of chicken-bone divination is still flourishing among the Lolo, the Karen tribes of Upper Burma, the Ahom and other tribes of the Tai or Shan family, and the Palaung and other tribes of the Mon-Khmer stock. Among all these peoples the cock is a divine or sacred bird, a supernatural creature, a messenger of the gods endowed with the gift of prophecy, with a knowledge of the future and of good and evil. He plays a fundamental rôle in mythology, in tribal migration stories, in prayers and rituals. Among all these people chickens are not primarily kept for utilitarian purposes, and the eggs are hardly, if ever, consumed. The most primitive conditions still prevail there in regard to fowl-rearing, and it is a curious and corroborating coincidence that the same area, Upper Burma, had previously been claimed by naturalists and orientalists alike as the primeval center for the domestication of the cock. Thus it seems to me there are good reasons for concluding that it was the practice of cock divination which was instrumental or at least the principal agency in bringing about the domestication of the bird. The Palaung still capture the wild jungle fowl and cross it with the domestic breed. Eggs of jungle fowl are often brought from the jungle and set under the village hen. The pullets when large enough to look after themselves are sometimes carried to the jungle and there set at liberty. These people are perfectly aware of the relationship of the wild species to the domesticated variety, and the interbreeding of the two continues under their care. They keep cocks essentially as time-keepers and as sacred birds in divination. According to their traditions it was the captive jungle-cock who was first employed for this purpose. Chinese accounts give us the same information with reference to the Miao tribes.

There is a peculiar reason why the Palaung constantly take recourse to the jungle in order to replenish their domestic stock. The thigh bones of fowl have fine perforations, the foramina of the blood-vessels, and these play a prominent part in the procedure of divination. Fine needle-like bamboo splinters are inserted into the foramina and project at various angles from the sides of the bones; according to their position (whether slanting, straight, or upright) good or ill omens are decided. There are books illustrated with over one hundred diagrams of chicken bones with the splinters inserted and accompanied by an oracle in old Shan. In the wild jungle fowl and in domestic fowl bred from jungle stock there is a much greater variation in the position and number of the foramina than in the pure domestic stock. For this reason the bones of the jungle cock are deemed preferable for divination to those of the purely domestic breed. This whole process goes to show that the Palaung are actuated by no other motive than the desire to have the service of the bird's divinatory qualities and that any utilitarian viewpoint is alien to their thought. Unconsciously and involuntarily they have contributed to the improvement and eugenics of their domestic stock by rejuvenating it with the blood of the jungle fowl, but the sole motive was and is to obtain a bird of a superior degree as the messenger of divine will.

It is a curious coincidence that among the majority of the tribes under consideration, the pig is likewise enlisted for purposes of divination. Among the Karen, the pig is an important sacrificial animal, and its gall-bladder is examined for favorable omens. In their elaborate divination rituals cock and pig function simultaneously. A peculiar condition prevails among the Khasi who draw auguries by examining the lungs, liver, spleen and gall-bladder of a pig; if the liver is spotless and healthy, *e.g.*, the augury is good; if the reverse, it is bad. On the other hand, the Khasi also revere the cock, regarding him as a mediator between the deity and man, yet in opposition to all other tribes of the area they do not use the femora of the cock, but its entrails. Here evidently the basic ideas of the pig complex were grafted upon the ideas of the cock complex, and this example illustrates well that the two complexes are closely interrelated and cultural parallels, and that correlations and adjustments have taken place between the two. The same condition as among the Khasi prevails in Borneo, and traces and survivals of this whole complex go over the entire Malayan, Polynesian and Melanesian area.

Another interesting point is that those primitive tribes to whom cock and pig are essentially sacred animals of religious significance have never developed cock-fights and pig-fights—quite naturally, for such cruel sports are reserved for the so-called higher civilizations. Again, the coincidence is striking that wherever in southeastern Asia we meet cock-fights there are also pig-fights; the two, *e.g.*, occur together in ancient Camboja and ancient Java. Much has been written on cock-fights, but no one seems to have seen what it is all about. This institution has also grown out of a form of divination, save that it is a sophisticated and materialized offshoot of it. In the beginning it was not individuals who set their roosters to fight, but villages and communities entered into a contest against each other to decide the question of superiority by relying on divine judgment that would manifest itself in the prowess of their divine roosters.

I hope that this brief outline is sufficient to show that through intensive ethnological study the history of our domesticated animals can be disentangled and that we may hope in course of time to be able to reconstruct culture sequences and periods in well-defined areas for a number of domestications. Like numerous other culture traits, domestications have been evolved progressively on a line from the irrational to the rational, or to express it more guardedly in terms of relativity, from what we are inclined to regard as irrational to what we are wont to regard as rational. For the Palaung who abstains from eggs, but trusts in the efficacy of the chicken-bone oracle is just as rational and honest and sincere in his way as we are in maintaining chicken-farms and eating cold-storage eggs.

178

卡特《中国印刷术的发明及其西传》书评

JOURNAL

OF THE

AMERICAN ORIENTAL SOCIETY

EDITED BY

MAX L. MARGOLIS
Dropsie College

W. NORMAN BROWN
University of Pennsylvania

VOLUME 47

PUBLISHED BY THE AMERICAN ORIENTAL SOCIETY
ADDRESS, CARE OF
YALE UNIVERSITY PRESS
NEW HAVEN, CONNECTICUT, U. S. A.
1927

REVIEWS OF BOOKS

The Invention of Printing in China and Its Spread Westward. By THOMAS FRANCIS CARTER. New York: COLUMBIA UNIVERSITY PRESS, 1925.

Although the subjects and problems treated in this volume are quite familiar to the majority of orientalists, Professor Carter is the first who gives us in a learned book an intelligent summary of the past research in this interesting subject, particularly based on the important manuscript-discoveries in Turkistan. This general survey of the whole field enables us to discern clearly the many gaps in our knowledge that remain to be filled; and, while the history of rag-paper and printing may be well outlined in its essential features, there are still many problems awaiting solution or wanting further elucidation. The plan of the book is admirably conceived and consistently carried through. The entire work testifies to assiduous study at home and abroad, both in Chinese and European sources; it is attractively written, and the volume is well printed and well gotten up, being illustrated by 37 half-tones, a graphic chart in colors demonstrating the development of rag-paper and printing in China and the West, and a map showing the migration of rag-paper from China to Europe.

While I am grateful to the author for having written this useful book, I feel obliged to dissent from his opinions and conclusions in certain points. First, as to method, I do not share his optimism in regard to Chinese encyclopaedias as being reliable (p. 189); they are, in my opinion, not more trustworthy than our own; they are assuredly helpful for ready reference as a first aid, but a real study must be based on the original texts whenever available. A twenty years' occupation with the *T'u shu tsi ch'eng* has convinced me that numerous quotations in it are incomplete, corrupt, or even senseless and that very important texts are entirely omitted; still less do I have great confidence in the *Ts'e yüan* published by the Commercial Press of Shanghai, over which so much fuss is now being made. As to another point of method, Professor Carter has a fondness for evolutionizing and correlating things as being derived one from another; thus, the charm was the transition from the seal to the block print (p. 11), which is merely an unproved speculation. Considerations of similar phenomena in other cultures tend to make one skeptical about evolutionary reconstructions of this character just in this particular field. Seals were also known in the ancient Near East; the Babylonian inscribed seal-cylinders, which rolled over soft clay and left in it an imprint of the text, represent a method analogous to our book-printing. In India wooden blocks for making impressions on textiles were known, but they were never applied to books; likewise the Polynesians who are ignorant of seals and letters utilize blocks for printing designs on their tapa, and the Dayak of Borneo use them as well for impressing tattoo marks on their bodies. In ancient Mexico paper was manufactured from maguey fibres, but no advance was made toward printing. Carter is still inclined to presume that European typography resulted from block-printing, but he overlooks the fact that wooden types were never made in Europe and that the alleged development from wood-engraving to typography has been successfully contested (bibliography in G. Jacob, *Einfluss des Morgenlands auf das Abendland,* 1924, p. 42).

It is to the merit of Professor Carter that he has elucidated the text of Lu Shen to which Julien's statement of the initiation of block-printing in A. D. 593 goes back (p. 202), and the conclusion is quite plausible that this passage, traceable to an older text which contains nothing about printing, is due to a misunderstanding. It seems to me rather hasty, however, to assert that "there is apparently nothing about printing in the Annals of the Sui Dynasty." To enable one to make such a positive assertion would require reading of a considerable portion at least of the Sui Annals. Julien, by the way, is not the only one who ascribed printing to the Sui or who can be held responsible, as Carter thinks, for the repetition of this statement in most histories of China in western languages. A. Wylie wrote, "Printing was known in the time of the Sui, and practised to a limited extent during the T'ang; but the early efforts at the art do not seem to have been sufficiently successful to supersede the manuscripts." Even Palladius (in his Chinese-Russian Dictionary, I, 264) remarks, "It is supposed that printing began from the Sui dynasty; it is perfectly credible from the Sung dynasty." This point requires further investigation.

In discussing the history of movable type in China, Professor Carter translates a text said to have been written in 1314 under the Mongol dynasty by Wang Cheng; this text, however, is preserved only in an appendix to a work on agriculture by this author edited in the K'ien-lung period (1736-95). Carter reproduces from this book the illustration of a revolving wheel alleged to have been contrived by Wang Cheng as a type-setting device in 1314, but here he remarks cautiously, "Whether this illustration goes back to the original edition of 1314 or whether it is a reconstruction by K'ien-lung's editors, is uncertain." But this suspicion is ripe for the whole text: the wooden movable types ascribed to Wang Cheng are strikingly similar to a font of wooden types made under K'ien-lung in 1773 for printing the catalogue of his library (not mentioned by Carter), and there is a well-illustrated Chinese book extant which describes the various stages in the manufacture of this type. There are striking coincidences between the descriptions of this book and those of Wang Cheng, and a critical comparison of the two texts would probably clear up the problem in part.

In the biography of Pi Sheng (p. 160) it is justly denied that, as Julien has it, he was a smith (note on p. 251); nevertheless, on p. 181, the author speaks of "Pi Sheng the smith."

The date 1403 as denoting the first use of movable type in Korea is probably correct, but there is a statement in the Annual Report on Reforms and Progress in Chosen 1914-15 by the Government-General of Chosen (Keijo, 1916, p. 17) which would merit investigation: "It is said that a Chinese Book of Etiquette was printed with movable type by Koreans in the reign of Kō-jong (1214-66), the twenty-third king of Koryu." The interesting information is also given there that "old types, whether made of metal, earth, or wood, in the possession of the former imperial household of Korea, numbering about 500,000 pieces, were transferred to the care of the Government-General, and arranged in better order by classifying them according to the Chinese dictionary of K'ang-hi."

For the fact that wall-paper is a Chinese invention, the reader is referred to Grande Encyclopédie, while in our own American literature we have an excellent book on this subject by Kate Sanborn, *Old Time Wall Papers, an Account of the Pictorial Papers on our Forefathers' Walls* (Greenwich, Conn., 1905), with many excellent colored plates. Chinese wall-papers were first introduced into Europe by Dutch traders at the end of the seventeenth century under the name "pagoda-papers." As early as 1735 they were brought to America. Specimens of Chinese wall-papers are still to be found in colonial houses of Massachusetts, some even imported in 1750 and in good state of preservation. Many of the older American papers exhibit their relationship to the Chinese in that the decoration is not repeated, but runs continuously about the entire room or contains a scenic representation. It is not correct that as stated in the Introduction (p. xii), the scientific study of the invention of paper in the West was begun by F. Hirth. Hirth's article "Die Erfindung des Papiers in China" (*T'oung Pao,* 1890, pp. 1-14) can hardly be called scientific; it is a compilation based on previous studies by J. Edkins and A. Wylie in which most of Hirth's data are anticipated. Another article by Hirth, "Western Appliances in the Chinese Printing Industry" (1886), would have supplied Carter with some useful data.

No reference whatever is made in the book to the name of A. Wylie, and the introduction to his Notes on Chinese Literature, which contains a valuable and critical history of printing in China, has not at all been utilized,—an almost unintelligible omission. A careful perusal of Wylie's study would have made many a slip unnecessary. There is no foundation for ascribing the invention of the writing-brush to the general Mung T'ien in the third century B. C. (p. 2). This is a tradition merely found in the late and apocryphal *Po wu chi;* the contemporaneous records (Se-ma Tsien's *Shi ki*) have nothing to this effect. It is surprising also that a brochure entitled *The Rise of the Native Press in China* by Y. P. Wang (Columbia University, 1924, 50 p.) has not been consulted. Mr. Wang gives very interesting information on the old Peking Gazette, the oldest newspaper in the world, which dates back to the days of the T'ang dynasty. The question as to when and how this newspaper was first printed ought to have been ventilated in a book devoted to printing in China, but the subject is not even touched upon (cf. Mayers, "The Peking Gazette," *China Review,* III, p. 16).

The activity of the Ming and Manchu in printing numerous Tibetan, Mongol and Manchu books is passed over in silence, nor is the Islamic press mentioned with its numerous editions in Arabic, Chinese, and Arabic-Chinese. Ibn Baṭūṭa's account of China is strongly overvalued (pp. 114, 233), and is very far from "containing a true picture of China." Whether G. Ferrand is right in his assertion or not that he may never have visited China, many of his data concerning China are unintelligible, absurd, or fictitious. Rashid-eddin is called an Arabic writer on p. 197 and correctly a Persian historian on p. 219.

"Whether picture or text, practically all the earliest block prints on paper that have been preserved are religious. On the other hand, . . . none of the textile prints, whether in Asia or Europe, has a religious motive" (p. 149). Yet, in Tibet, prayer formulas and incantations with or without religious pictures are printed on cotton and hemp cloth, many examples of which are in the Field Museum, Chicago. Printing on textiles has survived longest among the secret societies of China. In the Heaven and Earth League (T'ien ti hui), a political organization of anti-Manchu tendency, certificates of membership issued after initiation were generally of white cloth on which the characters were printed in black and laid out in the form of an octagon, with the seals stamped in vermilion in the centre; sometimes they were of yellow silk with characters printed in black (W. Stanton, *The Triad Society,* Hongkong, 1900, pp. 71, 76, 78, 85).

In chapter 19 playing-cards are considered as a factor in the westward movement of printing. The author's information on the history of games, however, is rather vague. It is not correct that polo spread from Persia to India and China about the same time; it reached India only under the rule of the Moghuls, while it was in full swing in China under the T'ang dynasty. A sinologue should not be content to refer his readers to the article Polo in the Encyclopaedia Britannica if his nearest colleagues like Parker, Giles, Chavannes (not to speak of myself) have made contributions to the history of the game from Chinese sources. According to Carter, the earliest reference to dice, "which form the background of Chinese playing-cards," is in the year 501. Dice are mentioned as early as A. D. 406 in the Chinese version of the Brahmajālasūtra (§ 33), translated by Kumārajīva, under the name *po-lo-sai* (anciently *pa-la-sak*), which is a transcription of Sanskrit *prāsaka* or *pāçaka* ("die, dice"); and Giles, in his Chinese Dictionary (No. 9658), even remarks that dice date from the third century A. D., and were first made of baked clay. With reference to another term, *shu p'u,* which occurs in the text quoted by Carter on p. 243, Giles observes, "Said to be of Indian origin, first mentioned by Ma Jung, second century A. D." There is no doubt that dice in China are of Indian origin: they are not referred to in any ancient Chinese system of divination. In India, they are of immemorial antiquity, being used both for divination and gambling (cf. Lüders, *Würfelspiel im alten Indien,* 1907). A standard book on Indian dice is mentioned in the literature of the Sui dynasty. As playing-cards are of Chinese origin, it is at the outset not probable that they had dice as their background or, as Carter also puts it, that a transition from dice to cards took place. The two games, in my opinion, represent two distinct developments. The above term *prāsaka* (*po-lo-sai*) denoted in particular the game of backgammon (Persian *nard*), which was introduced into China in the first part of the sixth century (not during the T'ang or a little before, as said on p. 139). There is no doubt that the Arabs transmitted playing-cards to Europe, for Spanish-Portuguese *naipe,* Old Italian *naibi,* are of Arabic origin, according to G. Jacob (*ZDMG,* 1899, 349 and *Geschichte des Schattentheaters,* 1925, p. 206) from Arabic *la'ib* ("play"). This rather plausible derivation has unfortunately not been entered in Meyer-Lübke's *Romanisches etym. Wörterbuch.* The date A. D. 969 which Carter quotes as an early reference to playing-cards in China does not mean much; the game was fully developed in the course of the ninth century, as could have easily been ascertained from Schlegel's doctor's thesis of 1869, *Chinesische Bräuche und Spiele in Europa,* p. 20, and in this point Schlegel is right. Carter emphasizes the fact that playing-cards are not mentioned in ancient Arabic records. This may be correct, as gambling games are forbidden by Islam; the Chinese also indulge in many gambling games, no record of which is preserved. The fact remains, however, that the Arabs do play cards (cf. Lane, *Manners and Customs of the Modern Egyptians,* 5th ed., II, p. 46). For myself I do not believe that playing-cards were instrumental in transmitting the Chinese method of block-printing to Europe or that they exerted any tangible influence on the art of printing.

These various points of criticism bearing on details do not detract much from the real value of the book. As a whole it is excellent and serves the interests of both the layman and the scholar in furnishing a guide into a difficult subject which offers attractions to every cultured mind.

B. LAUFER.

Field Museum, Chicago.

179

玛瑙——与之相关的考古学与民间传说

AGATE

Physical Properties and Origin

BY

OLIVER C. FARRINGTON

CURATOR OF GEOLOGY

Archaeology and Folk-lore

BY

BERTHOLD LAUFER

CURATOR OF ANTHROPOLOGY

GEOLOGY

LEAFLET 8

FIELD MUSEUM OF NATURAL HISTORY

CHICAGO

1927

Agate—Archaeology and Folk-lore

The Sumerians, the earliest inhabitants of Mesopotamia, were the first nation in history, as far as we know at present, that recognized the ornamental value of semiprecious stones and that understood and practised the art of stone-cutting for the purpose of making cylinder seals, signet-rings, beads, and other articles of jewelry. In the excavations undertaken by Field Museum at Kish in cooperation with Oxford University under the auspices of Captain Marshall Field, great quantities of beads of various substances and forms have been brought to light. These beads were worn by both sexes, and the materials commonly used for their manufacture were agate, carnelian, and lapis lazuli, which occur in almost every necklace. It appears from their relative number that carnelian and agate beads were more popular than those of lapis lazuli. Many are of oblong, cylindrical shape, up to two and two and a half inches long with perforations firmly and evenly drilled. Many examples of such beads may be viewed in the exhibits of Kish antiquities in Stanley Field Hall (Cases 6 and 20). In Plate XIII one of the finest necklaces from Kish is reproduced. It consists of agate and lapis-lazuli beads alternating, and also contains beads of gold foil made of the same shape as those of agate. Perrot and Chipiez figure a cylinder of veined agate on which are portrayed winged quadrupeds seizing and devouring gazelles. It was found by De Sarzec at Tello, and is now in the Louvre of Paris. The source of the agates and carnelians used by the Sumerians has not yet been traced.

Aside from beads, the Sumerians used agate also for making ceremonial axe-heads. One of these, with a three-line inscription, is in the American Museum of Natural History, New York, and is dated by J. D. Prince between 3000 and 2300 B.C., probably nearer the former than the latter date. It is illustrated and described in Journal of the American Oriental Society, XXVI, 1905 (pp. 93-97), also in Bulletin of the American Museum of Natural History, XXI (pp. 37-47). Another Babylonian axe-head of agate, inscribed with characters of an early form, is in the Metropolitan Museum of Art, New York.

Agate is first mentioned in literature by Theophrastus (372-287 B.C.) in his small treatise *On Stones*. He briefly refers to it as a beautiful stone which is sold at a high price, and he derives its name from the river Achates in Sicily, where such stones are said to have been found for the first time. This etymology is repeated by Pliny, and has been generally accepted by the ancients. A derivation of the word from the Semitic has been attempted recently, but is not convincing.

Pliny, in his Natural History, discusses agate to some extent, but gives no description of it. He writes that "Achates is a stone which was formerly held in high esteem, but is not so now; it was first found in Sicily, near a river of that name, but has since been discovered in numerous other localities." We may assume that because it was found in numerous localities, it had lost its former appreciation. Besides Sicily, Pliny gives Crete, India, Phrygia, Egypt, Cyprus, the Oeta Mountains, Mount Parnassus, Lesbos, Messenia, Rhodus, and Persia as places where agate occurred. A number of varieties are named by him; such names as iaspachates ("jasper-like agate"), smaragdachates ("emerald agate"), haemachates ("blood agate"), and leucachates ("white agate") apparently refer to color varieties, while dendrachates ("tree agate") alludes to the designs in the stone and may correspond to our moss agate. Corolloachates ("coralline agate") was spotted all over, like sapphirus, with drops of gold, and was commonly found in Crete, where it was also known as "sacred agate." It was regarded as capable of healing wounds inflicted by spiders and scorpions, a property which Pliny says might really belong to the stones of Sicily, as scorpions in that island lose their venom. The agates found in Phrygia have no green bands, and those of Thebes in Egypt lack red and white veins. The Egyptian agate was reputed as an antidote to the poison of the scorpion, and the stones of Cyprus were credited with the same property. By some people the highest value was set upon those stones which present a transparency like that of glass.

Pliny, further, relates a number of superstitious notions entertained by the Magi of Persia with reference to agates. Stones covered with spots like a lion's skin were believed to be an efficient protection against scorpions. In Persia, agates were used, by a process of fumigation, for stopping storms and hurricanes, as well as the course of rivers; if they were thrown into a boiling cauldron and turned the water cold, this property was regarded as a proof of their efficacy. A similar notion, it will be seen below, turns up in China. To be really efficacious, Pliny adds, the stones must be fastened with hair from a lion's mane; hyena's hair is rejected in this case, as it is apt to arouse discord in families. An agate of one color renders the athletes invincible, the Magi argue, on the ground that if thrown into a jar filled with oil together with pigments and boiled for two hours, it will impart a uniform color of vermilion to all the pigments.

In this case, it was not the agate which in the opinion of the Magi received new colors, but it was the coloring matters which through the agency of agate changed all their colors into one—vermilion. And since vermilion was the color proclaiming victory, and agate had the effect of producing this color in other pigments of a different nature, the Magi reasoned that an agate carried by an athlete would lead him to victory. The question here is not of a technical process, but is merely one of a purely imaginary, magical superstition. The doctrines of the Magi are frequently quoted by Pliny, but as a rule with disapproval.

Another passage in Pliny's Natural History has been interpreted by some writers as referring to the artificial coloring of agates. In fact, however, the question here is neither of agates nor of artificial coloring. Pliny in this case speaks not of *achates*, but of *cochlides*, a word derived from *cochlea* ("snail"), which may refer to shells or, according to others, to petrified shells, or to stones of snail-like shape. Pliny informs us that cochlides are now very common, being rather artificial than natural productions, which were found in Arabia in large masses. These, it is said, are boiled in honey for seven days and nights uninterruptedly. By this process all earthy and faulty particles are removed; and thus cleaned, the mass is adorned by the ingenuity of artists with variegated veins and spots, and cut into shapes to suit the taste of purchasers. These articles were formerly made of so large a size that they were used in the East as frontals and pendants for the trappings of the horses of kings.

Pliny, accordingly, speaks merely of purifying a certain substance of unknown character in honey, but says nothing about new colors being brought out in it by means of a chemical process. On the contrary, he states expressly that veins and spots were added by the hands of artists. Nöggerath, a German scholar, who was familiar with the artificial coloring of agates as practised in Idar and Oberstein, has simply interpreted this process into the above passage of Pliny, and this speculative hypothesis has been adopted by many others without reason. There is no evidence whatever to the effect that the method of coloring agates artificially was known to the ancients, and the fact remains that no such agate of classical antiquity has ever been found.

The Physiologus, a very popular Greek natural history, which originated at Alexandria in the second century A.D. and was subsequently translated into all European languages, contains a story about the agate and the pearl, which does not occur elsewhere. It is said that the divers avail themselves of an agate in searching for pearls. They fasten a piece of agate to a rope which is let down into the sea. The agate turns into the direction of where a pearl is hidden, and remains there steadfast, so that they find the pearl by diving alongside the rope.

Pliny mentions a valuable agate in the possession of Pyrrhus, the king who was so long at war with the Romans. On this agate were to be seen the Nine Muses and Apollo holding a lyre, not as a work of art, but as the spontaneous produce of nature, the veins in the stone being so arranged that each of the Muses had her own peculiar attribute. We must confess that either it must have required a high flight of imagination to recognize these pictures in the veins of this agate, or that nature had been considerably aided by art.

It is not stated by Pliny or other ancient writers that agate was cut into gems, but a number of cut gems of agate have come down to us, and are preserved in museums or private collections. They go back as far as the Aegaean or Mycenaean age, agate gems with mythological subjects having been discovered at Vaphio. A few cameos of agate and carnelian are on view in Case 2 (upper left section) of the Gem Room

(H. N. Higinbotham Hall). Aside from cut gems, agate was wrought into beads, scarabs, rings, and figures.

Ointment bottles, cups and bowls were also occasionally made of agate, but few of these have survived. The best known example is a precious agate bowl preserved in Vienna and measuring 28½ inches in diameter. It was brought to Europe by the crusaders after the conquest of Constantinople. Another famous agate vessel in existence, presumably made at the time of Nero, is a two-handled cup holding over a pint and covered with Bacchanalian subjects. It was presented by Charles the Bald in the ninth century to the abbey of St. Denis, and was used to hold the wine at the coronation of the kings of France. In the Treasury of Vienna there is an agate bowl with a diameter of 30 inches, which is traditionally believed to have been made about A.D. 1204.

The Persians, Armenians, and Arabs, like all Oriental nations, do not clearly discriminate between agate, carnelian, chalcedony, and related stones. The most esteemed kind is called *yamani* ("originating from Yemen"), as it is chiefly found in Yemen, but, according to Arabic authors, also came from India and Maghreb (northwestern Africa). The stone was chiefly utilized for finger and signet rings in which the wearer's name was engraved. A verse from the Koran or also a magical figure was sometimes carved in such an agate which then served as a talisman. It was believed that those wearing a *yamani* ring were guarded against the danger of being killed by a collapsing wall or house.

From ancient times India has been celebrated for the beauty of its agates. Pliny narrates that the agates of India possessed great and marvelous properties, as they present the appearance of rivers, woods, beasts of burden, and forms even like ivy and the trappings of horses,—alluding to undulated and moss agates. The druggists of his time, according to Pliny, used these as stones for grinding drugs, and the very sight of them was regarded as beneficial for the eyes. Held in the mouth, they were believed to allay thirst.

In the sixteenth century Limodra in Guzerat was the principal seat of the agate industry, the mines being situated four miles from the town. This locality was visited early in the sixteenth century by Duarte Barbosa, a Portuguese traveller, who reports, "Here is found an agate (*alaquequa*) rock, which is a white, milky, or red stone, which is made much redder in the fire. They extract it in large pieces, and there are cunning craftsmen here who shape it, bore it and make it up in divers fashions; that is to say, long, eight-sided, round, and olive-leaf shapes, also rings, knobs for hilts of short swords and daggers, and other ways. The dealers come hither from Cambaya to buy them, and they sell them on the coast of the Red Sea, whence they pass to our lands by way of Cairo and Alexandria."

Barbosa found also that a great amount of work was done at Cambay in coral, agate, and other stones. In the beginning of the seventeenth century the headquarters of the agate industry appear to have been transferred from Limodra to Cambay, in the Bombay Presidency. Henceforth only the preliminary operations of sorting the stones and exposing them to fire to develop their color were performed at Limodra, and this is the case even now. They are then taken to Cambay to be cut, polished, and worked up.

The Portuguese word *alaquequa* or *alaqueca,* also *laqueca,* is derived from Arabic *al' aqīq,* and refers to the red carnelians exported from India. The Portuguese settled in India called *olhos de gato* ("cat's eyes") what is known as Indian eye-stone or eye-agate,—small pieces of agate cut *en cabochon* with a flattish, circular, or oval back to show the "eye" or "eyes." Nicolo Conti, a Venetian, who travelled in India during the first part of the fifteenth century, writes that some regions of India have no money, but instead use for exchange stones which we call cat's-eyes.

As is evident from Barbosa's account, the art of coloring agates artificially was partially understood in India. At the present time the stones collected near the village of Rotanpur near Cambay are classified into two sorts,—those that should be baked and those that should not be baked. The object of baking the stones is to bring out their colors. After exposure to the sun or by being baked in a cow-dung fire, light browns become white, and dark browns deepen into chestnut. Of yellows, straw colors become rosy, and orange is intensified into red; other shades of yellow become pink. Pebbles with cloudy shades turn into brightly veined stones in red and white. The deeper and the more uniform the color, the greater the value. Again, the larger and thicker the stone, the more is it valued. White carnelians, when large, thick, even-colored, and free from flaws, are precious; yellow and variegated stones are worth little.

Barbosa also mentions at Limodra, or as he calls the town Limadura, "much chalcedony, which they call *babagore;* they make beads with it and other things which they wear about them." This is the white agate of Cambay, called in Anglo-Indian *babagooree,* from Hindustani *babaghuri.* It is so called from the patron saint or martyr of the district in which the mines are located, under whose special protection the miners place themselves before descending into the shafts. According to tradition, he was a prince of the great Ghori dynasty, who was killed in a battle in that region; but this prince is not known from historical records. By command of Akbar, the Moghul emperor, grain weights of babaghuri were made to be used in weighing. All the weights used at court for weighing jewels were made of transparent white agate.

Agates are much used in India for ornamental purposes, being made into brooches, rings, seals, cups, and other trinkets. A considerable trade is still carried on in the raw material which is obtained from the amygdaloidal flows of the Deccan trap, chiefly from the State of Rajpipla, where the main source is a conglomerate near the village of Ratanpur. Here the right to collect the stones is leased for a period of five years at an annual rental. Aside from Cambay, which is the most important place for cutting agate, this industry is also carried on at Jabbalpur (or Jubbulpore) and a few other places within range of the Deccan trap. Much of the agate sold in Europe is exported from Cambay, and large quantities are also shipped to China.

The French traveller and gem-merchant, Jean Baptiste Tavernier (1605-89), mentions the beautiful agates cut at Cambay into cups, knife-handles, beads, and other objects.

Moss agates were formerly known also as tree-stones (French *agates arborisées*). John Fryer, who travelled in India and Persia from 1672 to 1681, describes the precious stones found in India in his time, among these tree-stones with the lively representation or form of a tree thereon.

It was a wide-spread belief among the Mohammedans of India that agate had the power of stopping the flow of blood, presumably because of its blood-red color. The white carnelian was regarded as a "milk-stone," and was beneficial to women in increasing their supply of milk.

It is curious that agate is not referred to in ancient Sanskrit literature, either in medical texts or in mineralogical treatises. On the other hand, great quantities of agate objects have been discovered on very ancient archaeological sites of southern India, not only in the shape of beads, but also in the form of cores, flakes, scrapers, and strike-a-lights; numerous color varieties like white, gray, red and white, brown and gray, banded gray, deep red, dull red, orange-red, etc., are represented among these antiquities. It may hence be inferred that the ancient aboriginal inhabitants of India were well acquainted with the stone and utilized it for every-day implements in times anterior to the Aryan conquest and that the Aryan invaders learned its use and adopted it from the aborigines.

Agate is appreciated by the Tibetans, and is used to some extent, though not so largely as turquois, coral, and amber, their favorite jewels. It is partially imported from India, partially from China, and some is found in the country itself. Large pieces of red agate attached to cloth are worn by the Panaka women in the Kukunor region in their hair which is plaited in numerous small braids falling over their shoulders. Agate is frequently used by the Tibetans in finger-rings (examples in Case 61, West Gallery).

Ancient agate beads, rings, and seals were discovered by Sir Aurel Stein at Khotan and other localities of Chinese Turkestan; also an intaglio of agate with the figure of a lion.

The ancient Chinese herbalists and Taoist doctors, who were chiefly interested in the healing properties of organic and inorganic substances, classified agate as a species of its own. Under this term, which is *ma-nao* in their language, they included also carnelian. Their definition of *ma-nao* is formulated to the effect that it is neither a common stone nor jade, but that it holds a rank inferior, but next to their highly prized jade. Agate, accordingly, was appreciated, though not the equal of jade and not like the latter a sacred substance. It was recognized as a hard stone, being capable of resisting cutting instruments. Red, white, and black varieties were distinguished. Those which after carving and polishing offered pictures of men, animals, birds, or objects were most highly esteemed. In southern China a kind of agate of a pure red and without veins was found; it was made into cups and vases. A dark green variety was obtained in the northwestern parts of the country. Moss agate is designated "cypress-branch agate," also "nettle-hemp agate"; undulated agate, "cloud agate." Other terms like "brocade-red agate, silk-thread agate, rice-water agate" refer merely to color varieties. "Lampwick agate" is a variety with white veins. "Dark-like-gall agate" is what we call bloodstone. "Bamboo-leaf agate" came from Yi-chou in Shan-tung Province, and was used for inlaying in screens and tables; as implied by the name, it displays designs like bamboo leaves. The same locality produced another kind termed "jade agate."

Chinese authors speak of a kind of agate that is brilliant white of color if looked at straight, but that appears like coagulated blood if looked at from the side. It was called "double foetus agate" (*kia t'ai ma-nao*). "Purple-cloud agate" was found at Ho-chou in An-hui Province.

The ancient Chinese conceived the origin of several stones and salts as marvelous transformations from other substances; thus, white rock-crystal was believed to be thousand years old water changed into ice. By a similar process of naive reasoning agate was interpreted as a transformation of the blood of the manes or departed spirits, also of malignant devils.

Another theory was based on the name for agate, *ma-nao,* which means literally "horse's-brain." In writing the two characters, each is usually preceded by the classifier "jewel" or "precious stone." The significance "horse's brain" is regarded by most Chinese authors as the origin of the word, and may have been elicited by a certain outward resemblance of the veins and striation in agate with the brain of a horse. Hence a popular notion arose that agate beads were spit out of the mouths of horses.

For the purpose of testing agate the following recipe is given: "Rub it with a piece of wood; if it does not become heated, it is genuine; if it will be heated, it is not genuine." This test is based on the notion that the nature of agate is cold and that its coldness is unchangeable. The Chinese formula is practically identical with what Pliny ascribes to Persia: there the efficacy of an agate was determined by throwing it into a cauldron of boiling water and turning the water cold. It may be that the Chinese derived this idea from Persia.

In the first centuries of our era the Chinese became acquainted with the fact that agate and many other valuable stones were abundant in the Near East. As numerous articles were then traded from the Hellenistic Orient to India and China, while Chinese silk found its way to the West, it is very probable that agate was included among the export products of western Asia.

It is even possible that agate first became known to the ancient Chinese as an importation from abroad, for it is not mentioned in the literature of pre-Christian times, and the earlier authors inform us that it came from western countries, from the countries in the south-west, or from the western and southern barbarians. In one source it is even stated that it was a product of the country of the Yüe-chi, known to us as the Indo-Scythians. A tribute of agate was sent from Samarkand to the Chinese Court in the beginning of the eighth century. It is specifically asserted that agate was imported into China by the Arabs, but this product in all probability was carnelian, as it is described as "standard red in color without flaw." It was the raw material which was imported and which was wrought into objects by the Chinese. Subsequently, however, agate was discovered by them in many localities of their country, especially near Ninghia, Kwa-chou, and Sha-chou in Kan-su Province and in the outlying deserts, also in some mountains of northern Shan-si, Chi-li, and Shan-tung.

Toward the end of the Ming dynasty, in the first part of the seventeenth century, as we learn from a Chinese cyclopaedia completed in 1632, agate was imported into China from Europe; this became known as "foreign agate." That of red color was most highly appreciated; in its interior it displayed branches of cypresses and veins of various colors, as fine as silk threads; a variety with white veins was regarded as superior. Soon afterwards the same stone was discovered in Yün-nan, and was termed "native agate." The Ai-lao Mountains in the prefecture of Yung-ch'ang of that province enjoy a special reputation for their agates. Agate was also imported into China from Japan in three varieties—red, black, and white. As stated above, rough agates are exported from India to China in considerable quantities, particularly to Canton.

Small flat disks of red agate, usually covered under the surface with fine white lines of clayish origin, have been found in graves of the Han period (206 B.C.-A.D. 220). There are several early records of wine-vessels of agate having been discovered in tombs. Horse's-bits of agate are also mentioned. Large agate beads of circular and cylindrical shapes, rings and bangles, as well as small disks of translucent moss agate are traceable in graves of the T'ang (A.D. 618-906) and later periods. An ancient necklace or rosary found in a grave of Shen-si Province and shown in Case 38, East Gallery, consists of beads carved from agate, lapis lazuli, jade, and jujube-stones.

In A.D. 662 a tree three feet high made of agate in the shape of a lamp was sent by the country To-khara as a gift to the Chinese emperor. The branches of this agate tree were presumably fashioned in such a manner that they could hold an oil-lamp or candle. In more recent times jade trees were made by the Chinese as wedding gifts. In many of these leaves and flowers are carved from jade, but agate and carnelian are much used for the petals of the blossoms, as may be seen in a good example of the Blackstone Chinese Collection (Case 1). A paper-weight of white and red agate in which eight lizard-shaped dragons are carved is on view in the same case.

Formerly agate was also wrought in China into hair-pins, fish-hooks, and chessmen; and large slabs were used for desk-screens and table tops. The art-loving emperors of the Sung dynasty (A.D. 960-1278) had a high appreciation of agate. In A.D. 1113 some large agate blocks were found and transported into the imperial atelier, where they were wrought into precious objects like vases and ornamental plaques for girdles which were preserved in the imperial treasury for more than a century. Finally the colors are said to have faded away, and the stones assumed the color of white bone, whereupon the objects were discarded and disposed of to the people. In A.D. 1272 the Mongol emperors established in their capital Ta-tu and other places an "Agate Bureau" in charge of a director who supervised five hundred workmen.

In the fourteenth century there were made finger-rings inlaid with a piece of agate in which were engraved the twelve horary characters corresponding to the twelve signs of the zodiac. A contemporary author describes the work of engraving as fine as hair and conveying the impression as though it were not an artifact of man; it was therefore styled "devil's work stone" or "stone of the devil's country."

Powdered agate is said to have been used together with copper oxide and other ingredients in the production of a red glaze on porcelains.

The great force of the Chinese lapidary is the carving of snuff-bottles in which he strives at bringing out the colors of the stone to best advantage, or cuts the designs in layers so that the different colors stand out in relief as in antique cameo-work (Plate XIV, Fig. 3). Agate snuff-bottles are on view in the case illustrating the use of tobacco in China (south end of West Gallery). Some of these are reproduced in Leaflet 18 of the Department of Anthropology (Plates VIII and IX). The sentiment attached to the gift of a snuff-bottle of moss agate is that it should be a disperser of melancholy.

During the K'ien-lung period (1736-95) and somewhat later fine agate carvings were also made to be worn as pendants in the girdle. Three such ornaments are illustrated in Plate XIV, Figs. 1-2 and 6. The pendant in Fig. 1 represents a carp with lotus leaves; that is, the carp is conceived as swimming in a lotus pond. That in Fig. 2 shows a bird with a fruit, leaves, and blossom. That in Fig. 6 is carved into three jujubes (*Zizyphus vulgaris*) with two small peanuts (only one is visible in the illustration). The snuff-bottle in Fig. 3 is of milk-white agate with relief carvings in black, brown, and yellowish layers. These represent two monkeys, a spotted deer (*Cervus manda-*

rinus), and a magpie flying into the open from a pine-tree. Fig. 4 is a plain agate bottle of various colors, brown in the upper portion and green in the lower one. Fig. 5 is a ring of moss agate, 1¾ inches in diameter.

Agate was traded by the Chinese to their neighbors, the Tibetans, Mongols, Manchu, and Japanese, all of whom have adopted their word *ma-nao* (in Japanese *meno*). The Japanese, like the Chinese, manufacture agate and carnelian into beads for rosaries, paper-weights, ink-stones for rubbing the cakes of ink on, fruits, buttons, seals, tea and wine cups, and in particular into the small ornaments known as *netsuke*.

It is known in Japan that agate becomes more opaque on being exposed to sunlight or subjected to an intense heat in a closed jar, but the methods of coloring agate artificially, as employed in Europe, were unknown both in China and Japan.

The agate found in the province of Kaga was regarded as very precious. A red variety of it was called "vine-grape stone," and served for plaques to be inlaid in girdles in the place of jade. The provinces Mutsu, Echiu, Suruga, and Kai have the highest reputation for their agates and the skill of their lapidaries. Agate was formerly also imported into Japan from China.

Agate, being found in numerous localities of America, attracted the attention of the aboriginal inhabitants at an early date. In North America and Mexico agate was wrought into arrow-heads and spear-heads. A beautiful agate spear-head, for instance, was found in one of the Hopewell mounds of Ohio. The Museum has numerous agate beads recovered from prehistoric graves of Colombia, South America.

B. LAUFER.

"FORTIFICATION" AGATE. URUGUAY.
ARTIFICIALLY COLORED.

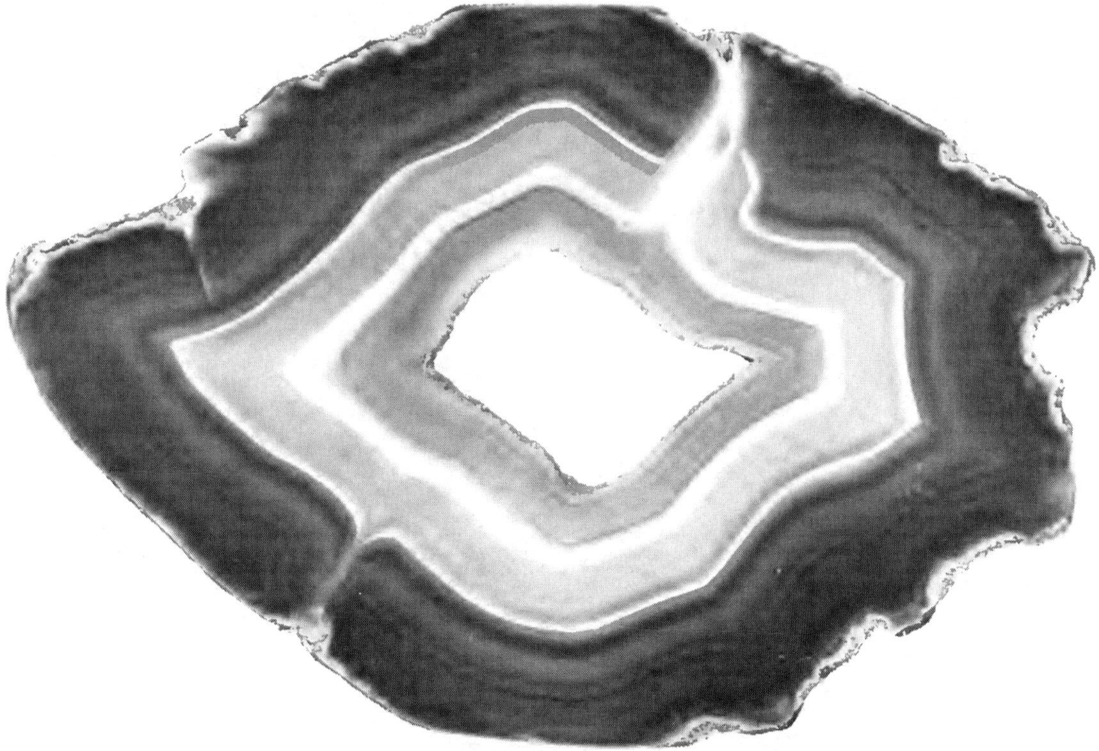

"FORTIFICATION" AGATE SHOWING SUPPOSED ENTRANCE CANALS.

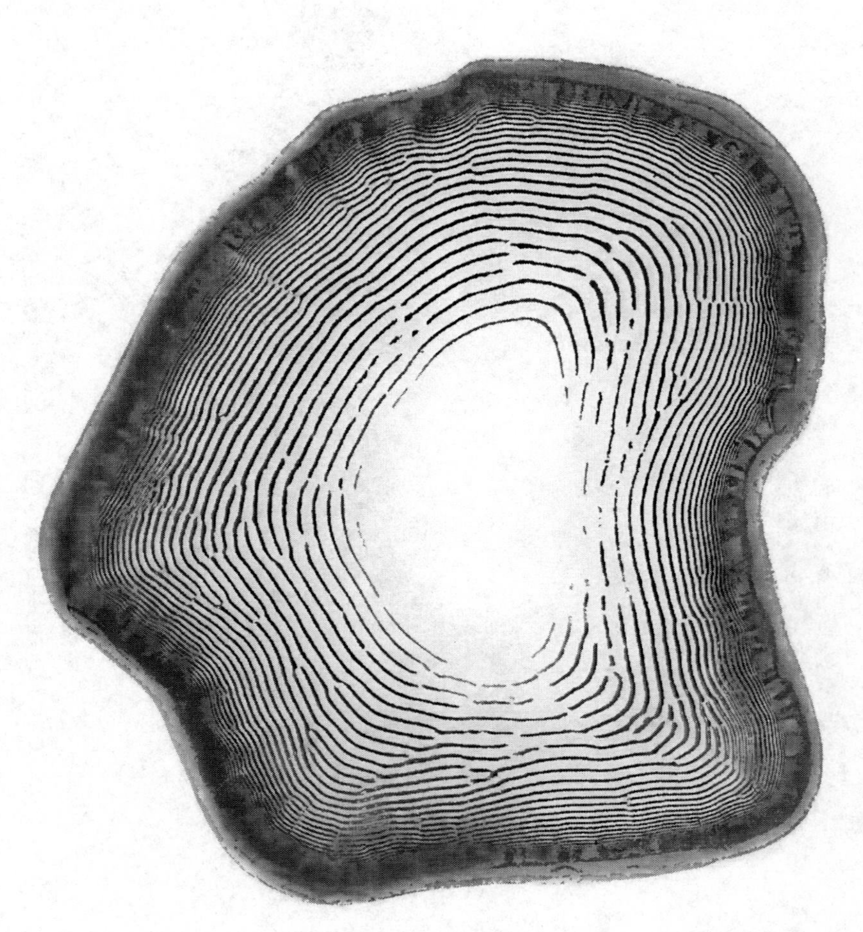

AGATE-LIKE BANDING PRODUCED BY DIFFUSION IN GELATINE.
''LIESEGANG'S RINGS''

"FORTIFICATION" AGATE WITH LAYERS IN TWO DIRECTIONS.

"RUIN" AGATE.

MOSS AGATE. CHINA.

IMITATION OF MOSS AGATE.
PRODUCED BY ADDING IRON VITROL TO WATER GLASS.

SECTION OF AGATE.

SHOWING NATURAL COLOR AT THE UPPER END AND ON THE REMAINDER VARIOUS ARTIFICIAL COLORS.

SHAPING AGATES. IDAR, GERMANY.

PITS FROM WHICH AGATE IS DUG. CATALAN GRANDE, URUGUAY.

Photograph by H. W. Nichols. Capt. Marshall Field Expedition of 1926.

NEAR VIEW OF AGATE PIT. CATALAN GRANDE, URUGUAY.

Photograph by H. W. Nichols. Capt. Marshall Field Expedition of 1926.

MAKING AGATE BEADS.
IDAR, GERMANY.

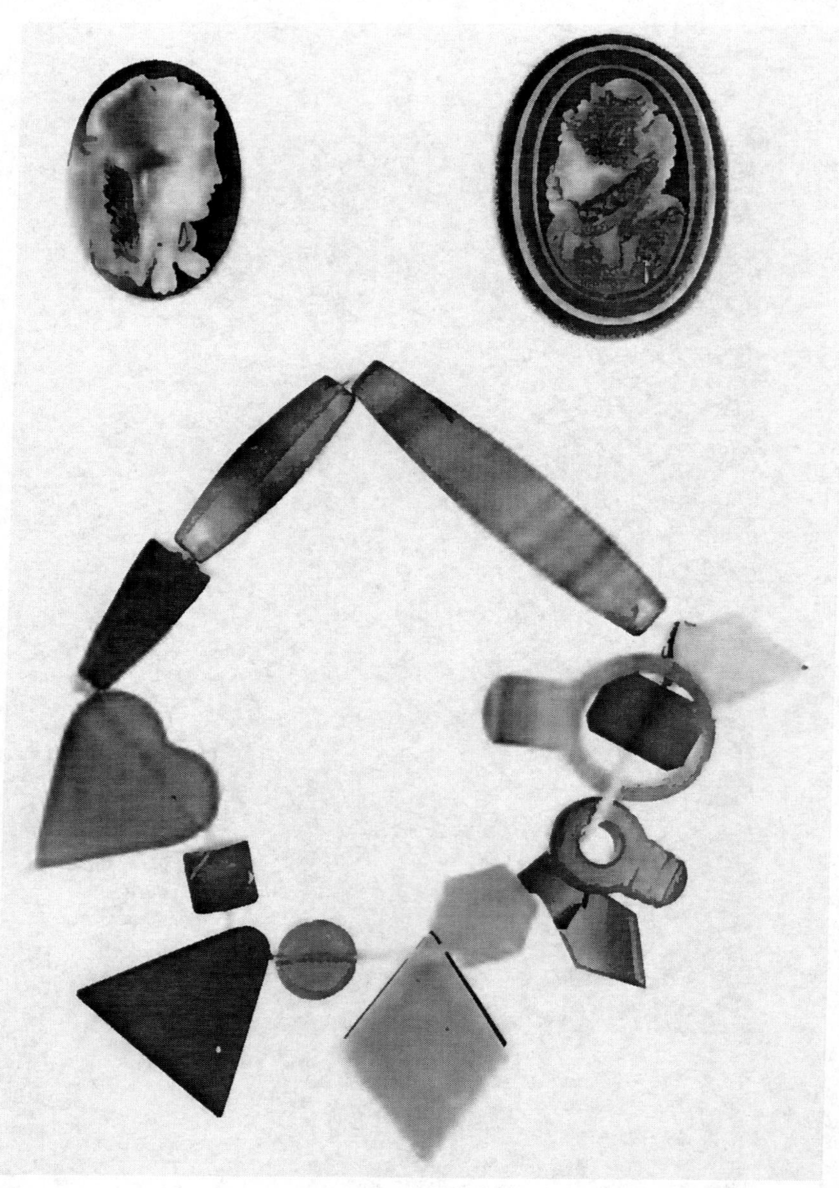

CAMEOS MADE FROM AGATE (Above).
AGATE CHARMS AS SOLD TO THE NATIVES OF SENEGAL (Below).

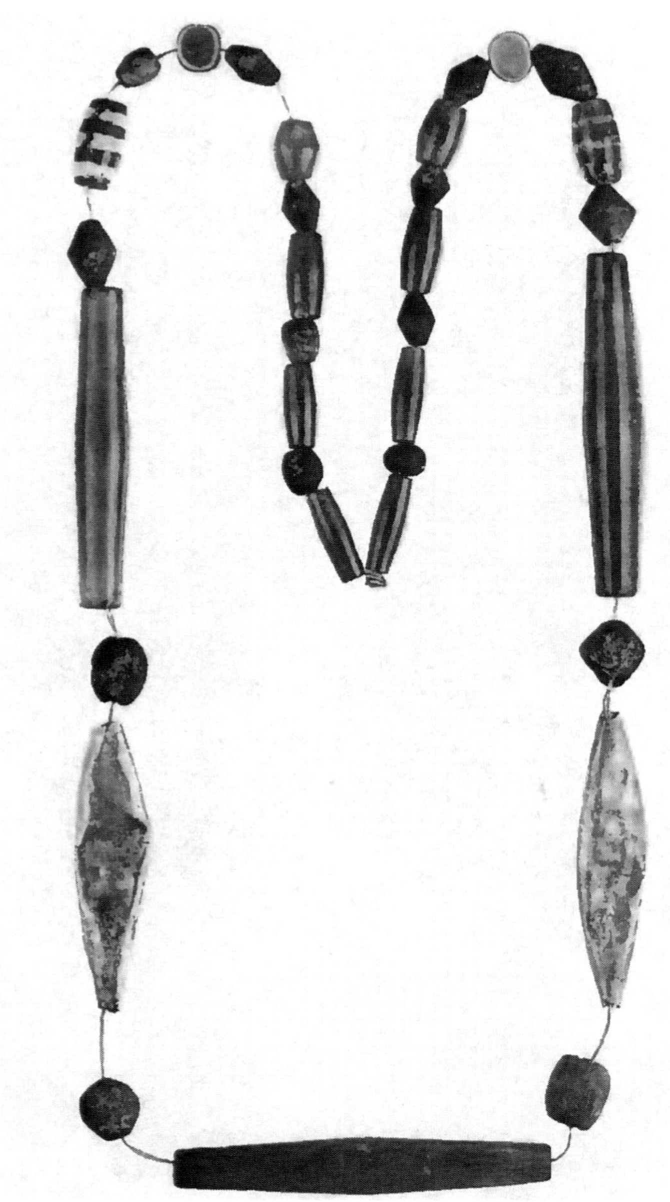

NECKLACE OF AGATE, LAPIS LAZULI AND GOLD.
About 3000 B. C. From Excavations at Kish by Field Museum-Oxford University Joint Expedition (Capt. Marshall Field Fund).

AGATE ORNAMENTS AND SNUFF-BOTTLES, CHINA.

1-2, 6, GIRDLE PENDANTS; 5, RING OF MOSS AGATE.
Capt. Marshall Field Expedition to China, 1923.

3-4, SNUFF-BOTTLES.
From the Collection of Mrs. George T. Smith, Chicago.

180

鸣虫与斗虫

Insect-Musicians and Cricket Champions of China

BY

BERTHOLD LAUFER
CURATOR OF ANTHROPOLOGY

12 Plates in Photogravure

ANTHROPOLOGY
LEAFLET 22

FIELD MUSEUM OF NATURAL HISTORY
CHICAGO
1927

The Anthropological Leaflets of Field Museum are designed to give brief, non-technical accounts of some of the more interesting beliefs, habits and customs of the races whose life is illustrated in the Museum's exhibits.

LIST OF ANTHROPOLOGICAL LEAFLETS ISSUED TO DATE

1. The Chinese Gateway . . . . $ .10
2. The Philippine Forge Group 10
3. The Japanese Collections 25
4. New Guinea Masks 25
5. The Thunder Ceremony of the Pawnee 25
6. The Sacrifice to the Morning Star by the Skidi Pawnee 10
7. Purification of the Sacred Bundles, a Ceremony of the Pawnee 10
8. Annual Ceremony of the Pawnee Medicine Men . .10
9. The Use of Sago in New Guinea 10
10. Use of Human Skulls and Bones in Tibet 10
11. The Japanese New Year's Festival, Games and Pastimes 25
12. Japanese Costume 25
13. Gods and Heroes of Japan 25
14. Japanese Temples and Houses 25
15. Use of Tobacco among North American Indians . .25
16. Use of Tobacco in Mexico and South America . . .25
17. Use of Tobacco in New Guinea 10
18. Tobacco and Its Use in Asia 25
19. Introduction of Tobacco into Europe 25
20. The Japanese Sword and Its Decoration 25
21. Ivory in China . . . . . . , ...75
22. Insect-Musicians and Cricket Champions of China . .50
23. Ostrich Egg-shell Cups of Mesopotamia and the Ostrich in Ancient and Modern Times 50
24. The Indian Tribes of the Chicago Region with Special Reference to the Illinois and the Potawatomi 25
25. Civilization of the Mayas 75
26. Early History of Man 25

D. C. DAVIES, DIRECTOR

FIELD MUSEUM OF NATURAL HISTORY
CHICAGO, U. S. A.

BOYS PLAYING WITH CRICKETS (p. 10).
Scene from Chinese Painting of the Twelfth Century in Field Museum.

FIELD MUSEUM OF NATURAL HISTORY
DEPARTMENT OF ANTHROPOLOGY
CHICAGO, 1927

LEAFLET NUMBER 22

Insect-Musicians and Cricket Champions of China

Of the many insects that are capable of producing sound in various ways, the best known and the most expert musicians are the crickets, who during the latter part of summer and in the autumn fill the air with a continuous concert. They are well known on account of their abundance, their wide distribution, their characteristic chirping song and the habit many of them have for seeking shelter in human habitations.

Crickets belong, in the entomological system, to the order Orthoptera (from the Greek *orthos*, "straight," and *pteron*, "a wing"; referring to the longitudinal folding of the hind wings). In this order the two pairs of wings differ in structure. The fore wings are parchment-like, forming covers for the more delicate hind wings. The wing-covers have received the special name *tegmina;* they are furnished with a fine network of veins, and overlap at the tip at least. There are many species in which the wings are rudimentary, even in the adult state. The order Orthoptera includes six families,—the roaches, mantids, walking-sticks, locusts or short-horned grasshoppers, the long-horned grasshoppers including the katydids, and the crickets (Gryllidae). Of crickets there are three distinct groups,—known as mole-crickets, true crickets, and tree-crickets. The first-named are so called because they burrow in the ground like moles;

they are pre-eminently burrowers. The form of the body is suited to this mode of life. The front tibiae, especially, are fitted for digging; they are greatly broadened, and shaped somewhat like hands or the feet of a mole. The mole-crickets feed upon the tender roots of various plants. The true crickets are common everywhere, living in fields, and some species even in our houses. They usually live on plants, but are not strictly vegetarians; sometimes they are predaceous and feed mercilessly upon other insects. The eggs are laid in the autumn, usually in the ground, and are hatched in the following summer. The greater number of the old insects die on the approach of winter; a few, however, survive the cold season. The tree-crickets principally inhabit trees, but they occur also on shrubs, or even on high herbs and tall grass.

Like their near relatives, crickets have biting mouth parts, and, like the grasshoppers and katydids, rather long hind legs which render them fit for jumping. Although many of them have wings when full grown, they move about mainly by jumping or hopping. When the young cricket emerges from the egg, it strongly resembles the adult, but it lacks wings and wing-covers, which gradually appear as the insect grows older and larger. The final development of wings and wing-covers furnishes the means whereby the male cricket can produce his familiar chirping sound. It is only the adult male that sings; the young and the females cannot chirp.

On examining the base of the fore wings or wing-covers of the male cricket, it will be noticed that the veins at the base are fewer, thicker, and more irregular than those on the hind or lower wings. On the under side of some of these thick veins will also be seen fine, transverse ridges like those on a file. The wing-covers of the female have uniform, parallel veins, without a trace of ridges. The male cricket produces

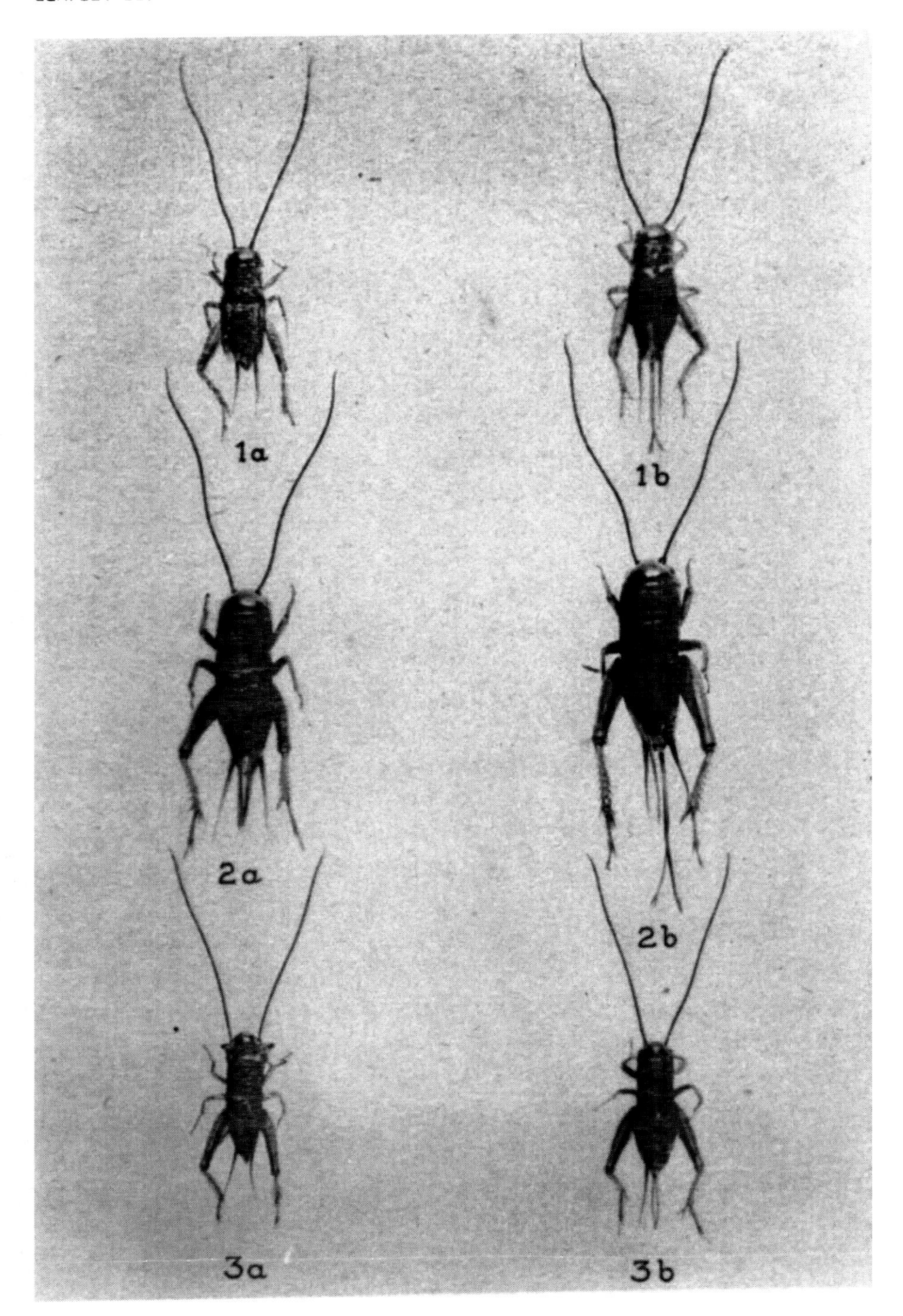

CRICKETS OF CHINA (a, male; b, female).

1. Besprinkled Cricket, *Gryllus conspersus* Schaum. 2. Mitred Cricket, *Gryllus mitratus* Burmeister. Chinese *si-so* or *ts'u-chi*, Peking Colloquial *ch'ü-ch'ü*. 3. Broad-faced Cricket, *Loxoblemmus taicoun* Saussure. Chinese *pang-t'ou* ("Watchman's Rattle").

his chirping sound by raising his wing-covers above his body and then rubbing their bases together, so that the file-like veins of the under surface of the one wing-cover scrape the upper surface of the lower.

Only the wings of the male cricket have sound-producing attachments, and the males have them only when their wings are fully developed at the age of maturity. The young cricket has no wings.

Since crickets produce a characteristic sound, it is natural to suppose that both males and females are able to hear it. On the lower part of the fore legs of both sexes is found a little drum-like surface, which serves as the tympanum of an ear. The sound-producing organ and the ear of the katydids, which rank next to the crickets in their singing ability, are somewhat similar in structure and location.

The sound made by crickets is, of course, not a true song, but a mechanical production, as are all of the sounds produced by insects. The object of the chirping or stridulating is somewhat conjectural. It may be a love-song, mating-call, or an expression of some other emotion. The fact that the crickets are able to sing only when they are full grown and capable of mating would seem to suggest that their chirping is a love-song.

This commonly held view, however, is contested by Frank E. Lutz in a recent article on "Insect Sounds" published in *Natural History* (1926, No. 2). Dr. Lutz starts from the opinion that not everything in nature has a practical or utilitarian purpose and that many striking characters and characteristics of animals and plants are of no use to their possessors or to any other creature; they seem to him to be much like the figures in a kaleidoscope, definite and doubtless due to some internal mechanism, but not serving any special purpose. This highly trained entomologist then proceeds to observe, "The most familiar example of insect sounds made by friction is the chirping of crickets. Now, only the males do this. Chirping is distinctly a secondary sexual character, the stock explanation of which is that it is a mating-call developed by sexual selection. The adult life of a male cricket lasts a month or so, and he chirps most of the time, but he spends little of that time in mating. Why does he chirp when there is no female around? Possibly hoping that one will come; I do not know. When he has mated, his sexual life is done, but he keeps on chirping to his dying day. I do not know why; possibly to pass the time. I do not know this, however, and my knowledge is based on the breeding of literally thousands of crickets, while I was using them in a study of heredity: a female cricket pays but little attention to a chirping male. She may wave her antennae in his direction, but so will she when he is not chirping, and so will she at a stick or a stone." And the general conclusion Lutz arrives at is, "The significance of insect sounds is still an open subject and, while it is altogether probable that some of these sounds do have a biological significance, I firmly believe that many of them have none, being merely incidental to actions that are not intended to make a noise and to structures that have arisen for some totally different purpose or for no purpose at all."

The Chinese, perhaps, have made a not uninteresting contribution to this problem. Of the many species of crickets used by them, the females are kept only of one,—the black tree-cricket (*Homoeogryllus japonicus*: Plate III, Fig. 2), called by them *kin chung* ("Golden Bell," with reference to its sounds), as they assert that this is the only kind of cricket that requires the presence of the female to sing. The females of all other species are not kept by the Chinese. As soon as the insects are old enough that their sex can be determined, the females are fed to birds or sold to

CRICKETS OF CHINA (a, male; b, female).

1. Yellowish Tree Cricket, *Oecanthus rufescens* Serville. Chinese *kwo-lou*, Peking Colloquial *kwo-kwo*. 2. Black Tree Cricket, *Homoeogryllus japonicus* Haan. Chinese *kin chung* ("Golden Bell"). 3. Infuscated Shield-backed Katydid, *Gampsocleis gratiosa infuscata* Uvarov. Peking-Chinese *yu-hu-lu*.

bird-fanciers. Accordingly, the males of all species kept in captivity by the Chinese, with a single exception, sing without the presence of the female. But whether captive insects are instructive examples for the study of the origin and motives of their chirps is another question. Our canaries and other birds in confinement likewise sing without females. Whatever the biological origin of insect sounds may be (and it is not necessary to assume that the sounds of all species must have sprung from the same causes), it seems reasonable to infer that the endless repetition of such sounds has the tendency to develop into a purely mechanical practice in which the insect indulges as a pastime for its own diversion. It is conceivable that insect music has little or nothing to do with the sex impulse, but that it is rather prompted by the instinct to play which is immanent in all animals.

The relation of the Chinese to crickets and other insects presents one of their most striking characteristics and one of the most curious chapters of culture-historical development. In the primitive stages of life man took a keen interest in the animal world, and first of all, he closely observed and studied large mammals, and next to these, birds and fishes. A curious exception to this almost universal rule is presented by the ancient Chinese. In accordance with their training and the peculiar direction in which their imaginative and observational powers were led, they were more interested in the class of insects than in all other groups of animals combined; while mammals, least of all, attracted their attention. Their love of insects led them to observations and discoveries which still elicit our admiration. The curious life-history of the cicada was known to them in early times, and only a nation which had an innate sympathy with the smallest creatures of nature was able to penetrate into the mysterious habits of the silkworm and present the world with the discovery of silk. The cicada as an emblem of resurrection, the praying-mantis as a symbol of bravery, and many other insects play a prominent role in early religious and poetical conceptions as well as in art, as shown by their effigies in jade.

In regard to mammals, birds, and fishes, Chinese terminology does not rise above the ordinary, but their nomenclature of insects is richer and more colorful than that of most languages. Not only do they have a distinct word or even several terms for every species found in their country, but also numerous poetic and local names for the many varieties of each species for which words are lacking in English and other tongues.

Corresponding to their fondness for crickets, the Chinese have developed a special literature on the subject. The first of these works is the *Tsu chi king* ("Book of Crickets") written by Kia Se-tao, a minister of state, who lived in the first part of the thirteenth century, under the Sung dynasty. His book, continued and provided with additional matter by Chou Li-tsing of the Ming period, is still in existence, and has remained the most important and authoritative treatise on the subject, which has been freely drawn upon by all subsequent writers. The author, a passionate cricket fancier himself, gives minute descriptions and subtle classifications of all species and varieties of crickets known to him and dwells at length on their treatment and care. Under the title *Tsu chi chi* ("Records of Crickets") a similar booklet was produced by Liu Tung under the Ming dynasty. During the Manchu period, Fang Hü wrote a *Tsu chi p'u* ("Treatise on Crickets"), and Ch'en Hao-tse, in his *Hua king* ("Mirror of Flowers") written in 1688, offers several interesting sections on crickets.

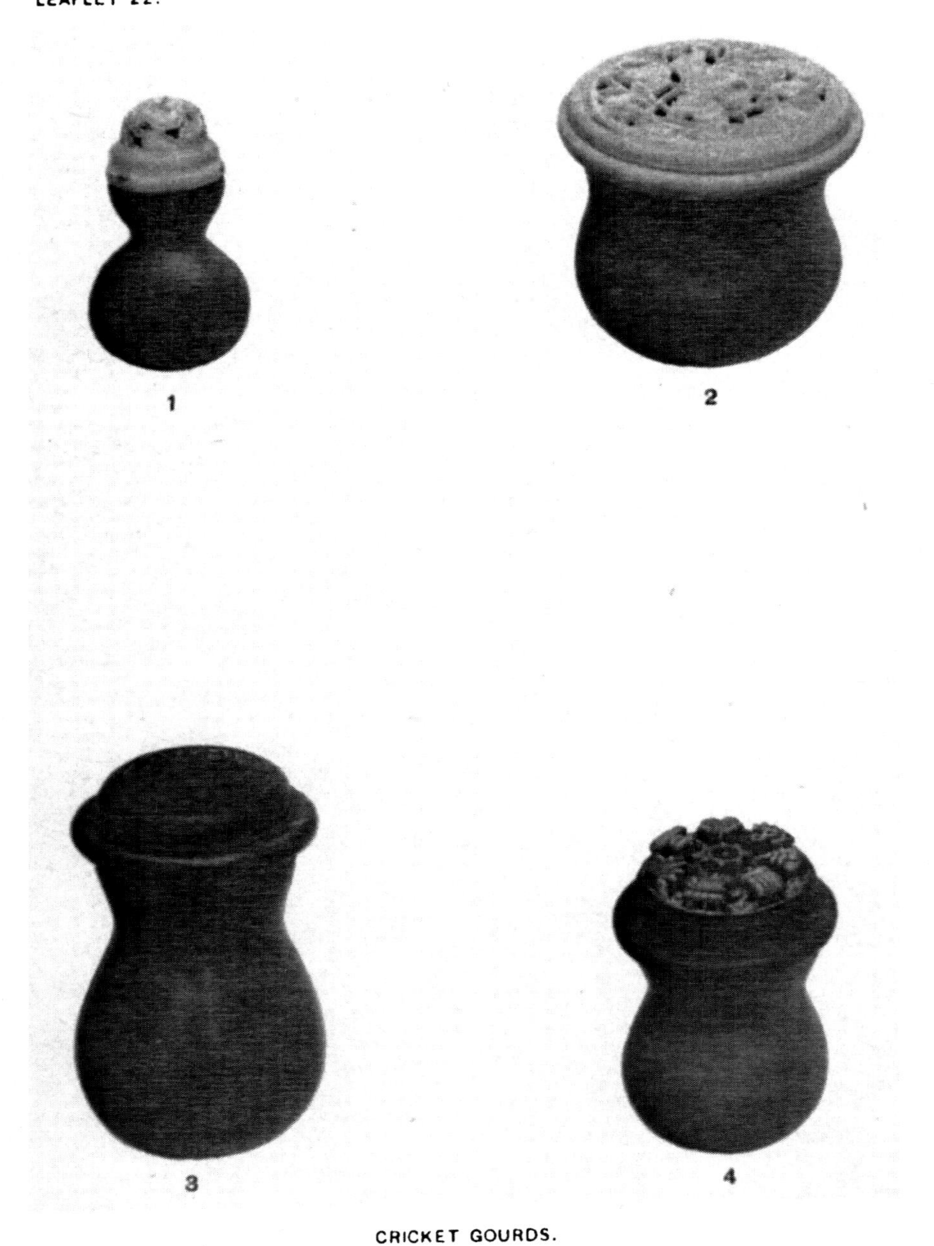

CRICKET GOURDS.

The Winter Habitations of the Insects.

1-2. Covers of ivory carved in open work. 3. Cover of carved coconut-shell. 4. Cover of sandalwood decorated with eight auspicious Buddhistic emblems.

In their relations to crickets the Chinese have passed through three distinct periods: during the first period running from the times of early antiquity down to the T'ang dynasty, they merely appreciated the cricket's powerful tunes; under the T'ang (A.D. 618-906) they began to keep crickets as interned prisoners in cages to be able to enjoy their concert at any time; finally, under the Sung (A.D. 960-1278) they developed the sport of cricket-fights and a regular cult of the cricket.

The praise of the cricket is sung in the odes of the *Shi king,* the earliest collection of Chinese popular songs. People then enjoyed listening to its chirping sounds, while it moved about in their houses or under their beds. It was regarded as a creature of good omen, and wealth was predicted for the families which had many crickets on their hearths. When their voices were heard in the autumn, it was a signal for the weavers to commence their work.

The sounds produced by the mitred cricket (*Gryllus mitratus:* Plate II, Fig. 2) recall to the Chinese the click of a weaver's shuttle. One of its names therefore is *tsu-chi,* which means literally "one who stimulates spinning." "Chicken of the weaver's shuttle" is a term of endearment for it.

One of the songs in the *Shi king* consists of three stanzas each of which begins, "The cricket is in the hall." The time intended is the ninth month when the year entered on its last quarter. In another song of the same collection it is said, "The *se chung* [a kind of cricket] moves its legs; in the sixth month, the spinner [another species of cricket] sounds its wings; in the seventh month it is in the wilderness; in the eighth month it is under the eaves; in the ninth month it is around the doors; in the tenth month the cricket enters under our beds."

At this point the Chinese are not distinguished from other nations. Our word "cricket" is imitative of the sound of the insect (literally, "little creaker," derived from French *criquer,* "to creak"). In old England it was considered a sign of good fortune to have a cricket chirping by the hearth, and to kill one of these harmless little creatures was looked upon as a breach of hospitality. Their cheerful tunes suggested peace and comfort, the coziness of the homely fireside. They were harbingers of good luck and joy. Gower, in his *Pericles,* offers the verse:—

> And crickets sing at the oven's mouth,
> E'er the blither for their drouth.

Ben Jonson *(Bartholomew Fair)* alludes to the insect's tunes thus: "Walk as if thou hadst borrowed legs of a spinner and voice of a cricket." Shakespeare has several references to this lover of the fireside whose note is so suggestive of cozy comfort. Milton (*Il Penseroso,* 81) has the line:—

> Far from all resort of mirth
> Save the cricket on the hearth.

On the other hand, the tunes of the hidden melodist were regarded by many persons with superstition and awe, and were believed to be an omen of sorrow and evil; its voice even predicted the death of a member of the family (see J. Brand, Observations on the Popular Antiquities of Great Britain, 1888, Vol. III, p. 189).

No one, however, has depicted the cricket's chirping with more poetic insight and charm than Charles Dickens in his immortal story *The Cricket on the Hearth,* in describing the competition between the cricket and the boiling kettle.

"And here, if you like, the Cricket did chime in! with a Chirrup, Chirrup, Chirrup of such magnitude, by way of chorus; with a voice, so astoundingly disproportionate to its size, as compared with the Kettle; (size! you couldn't see it!) that if it had then and there burst itself like an overcharged gun, if it had fallen a victim on the spot, and chirruped its little body into fifty pieces, it would have seemed a natural and inevitable consequence, for which it had expressly laboured.

"The Kettle had had the last of its solo performance. It persevered with undiminished ardour; but the Cricket took first fiddle and kept it. Good Heaven, how it chirped! Its shrill, sharp, piercing voice resounded through the house, and seemed to twinkle in the outer darkness like a Star. There was an indescribable little trill and tremble in it, at its loudest, which suggested its being carried off its legs, and made to leap again, by its own intense enthusiasm. Yet they went very well together, the Cricket and the Kettle. The burden of the song was still the same; and louder, louder, louder still, they sang it in their emulation.

"The cricket now began to chirp again, vehemently.

" 'Heyday!' said John, in his slow way. 'It's merrier than ever, to-night, I think.'

" 'And it's sure to bring us good fortune, John! It always has done so. To have a Cricket on the Hearth, is the luckiest thing in all the world!'

"John looked at her as if he had very nearly got the thought into his head, that she was his Cricket in chief, and he quite agreed with her. But it was probably one of his narrow escapes, for he said nothing.

" 'The first time I heard its cheerful little note, John, was on that night when you brought me home—when you brought me to my new home here; its little mistress. Nearly a year ago. You recollect, John?'

" 'Oh yes,' John remembered. 'I should think so!'

" 'Its chirp was such a welcome to me! It seemed so full of promise and encouragement. It seemed to say, you would be kind and gentle with me, and would not expect (I had a fear of that, John, then) to find an old head on the shoulders of your foolish little wife.'

" . . . . 'It spoke the truth, John, when it seemed to say so; for you have ever been, I am sure, the best, the most considerate, the most affectionate of husbands to me. This has been a happy home, John; and I love the Cricket for its sake!' . . . .

" 'I love it for the many times I have heard it, and the many thoughts its harmless music has given me'."

The Chinese book *T'ien pao i shi* ("Affairs of the Period T'ien-pao," A.D. 742-756) contains the following notice:—

"Whenever the autumnal season arrives, the ladies of the palace catch crickets in small golden cages. These with the cricket enclosed in them they place near their pillows, and during the night hearken to the voices of the insects. This custom was imitated by all people."

As it happened in China so frequently, a certain custom first originated in the palace, became fashionable, and then gradually spread among all classes of the populace. The women enshrined in the imperial seraglio evidently found solace and diversion in the company of crickets during their lonesome nights. Instead of golden cages, the people availed themselves of small bamboo or wooden cages which they carried in their bosom or suspended from their girdles.

The Museum owns a valuable painting in the form of a long roll depicting the games and pastimes of a hundred boys and attributed to Su Han-ch'en, a renowned artist of the twelfth century: one of the scenes shows six boys surrounded by cricket jars, one of them holding a tickler and letting a cricket out of a trap-box into a jar (see Plate I).

In Plates II and III the principal species of crickets kept by the Chinese in Peking are illustrated from actual specimens obtained, which will be found on exhibition in the case illustrating the cricket cult (West Gallery, second floor). The scientific identifications

1 2 3 4

CRICKET GOURDS.

1. Cover of ivory. 2. Cover of white jade. 3. With moulded designs of dragons. 4. Coated with carved red lacquer in two layers. Cover of Ivory with carving of three lions playing ball.

were kindly made by Dr. James A. G. Rehn of the Academy of Natural Sciences of Philadelphia. Most of the genera belong to the family Gryllidae, only two to the family Tettigoniidae: *Gampsocleis inflata* Uvarov and *G. gratiosa*, subspecies *infuscata* Uvarov, the latter illustrated in Plate III, Fig. 3. The besprinkled cricket (*Gryllus conspersus* Schaum, Chinese *si-so*), figured in Plate II, Fig. 1, is common all over China, and is also known from the Luchu Islands, Hawaii, and the East Indies. The mitred cricket (*Gryllus mitratus* Burmeister) in Plate II, Fig. 2, is known from most countries of Eastern Asia, particularly China, Korea, Japan, Tonking, and the Malay Archipelago. The broad-faced cricket (*Loxoblemmus taicoun* Saussure) in Plate II, Fig. 3, has also been described from Japan and Java.

The yellowish tree-cricket (*Oecanthus rufescens* Serville: Plate III, Fig. 1) is a favorite with the people of both Peking and Shanghai; it occurs also in the East Indies, but is quite distinct from *O. longicauda* Matsumura of Japan. The black tree-cricket (*Homoeogryllus japonicus* Haan: Plate III, Fig. 2), the "Golden Bell" *(kin chung)* of the Chinese because its sound is compared with that of a bell, is very popular in Peking; it is also known from Japan, Java, and northern India. It is evident that the large, glossy black insect in Plate III, Fig. 3, is quite different from the crickets and, as mentioned, is placed by us in a separate family. The Chinese also distinguish it from the cricket and bestow on it the peculiar name *yu-hu-lu* which is imitative of its sound; this word belongs to the colloquial language, there is no literary name for this insect.

As to color, green, black, yellow, and purple crickets are distinguished by the Chinese, the green and black ones taking the first rank.

The notes of the Golden Bell are described as being like the tinkling of a small bell, and its stridulation is characterized with the words *teng ling ling*. The Japanese designate this species "bell-insect" *(suzumushi)*. Lafcadio Hearn, who in his essay "Insect-Musicians" describes the various kinds of crickets favored by the Japanese, says that the bell of which the sound is thus referred to is a very small bell, or a bunch of little bells, such as a Shinto priestess uses in the sacred dances. He writes, further, that this species is a great favorite with insect-fanciers in Japan, and is bred in great numbers for the market. In the wild state it is found in many parts of Japan. The Japanese compare it with a watermelon seed, as it is very small, has a black back, and a white or yellowish belly. This insect, according to the Chinese, stridulates only at night and stops at dawn; the concert produced by a chorus causes a deafening din which is characterized by Hearn as a sound like rapids, and by a Chinese author as the sound of drums and trumpets.

Chinese authors know correctly that the "voices" of crickets, as they say, are produced by the motion of their wings. The stridulatory sounds are described by them as *tsa-tsa* or *tsat-tsat,* also as *tsi-tsi.* The term *kwo-kwo* for the yellowish tree-cricket (Plate III, Fig. 1) also is onomatopoetic. Terms of endearment for a cricket are "horse of the hearth, chick of the hearth, chick of the god of the hearth."

There are various methods of catching crickets. They are usually captured at evening. In the north of China a lighted candle is placed near the entrance of their hole, and a trap box is held in readiness. Attracted by the light, the insects hop out of their retreats, and are finally caught in the traps made of bamboo or ivory rods. Some of these ivory traps are veritable works of art: they are surmounted by carvings of dragons, and the trap doors shut very accurately (Plate XII, Figs. 1-2). The doors are shown open in the illustration.

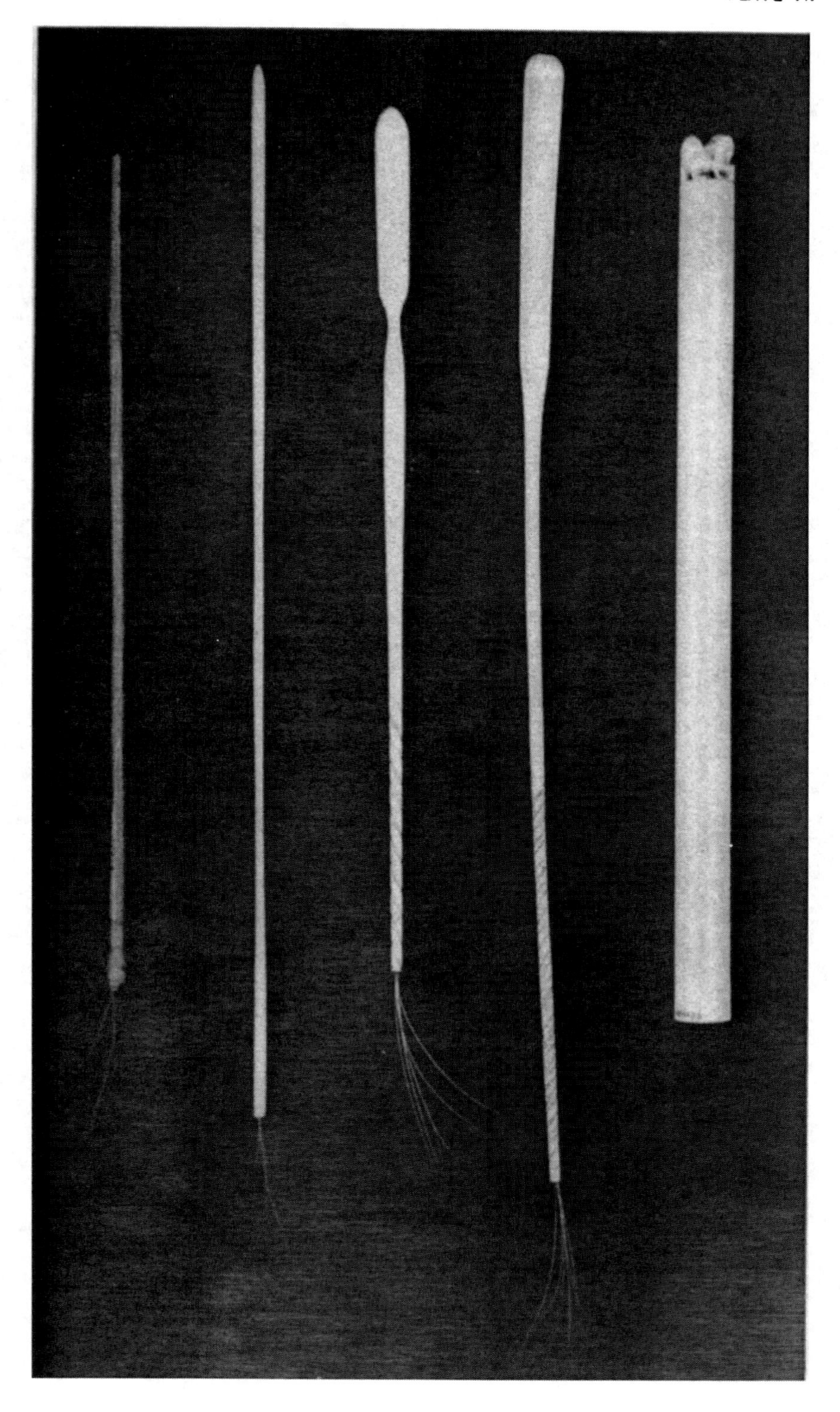

CRICKET TICKLERS.

For inciting crickets to sing or fight. In Peking they are made from rat or hare whiskers inserted in a reed, bone, or ivory handle. On the right an ivory tube, with cover surmounted by the figure of a lion, for keeping ticklers.

The trap shown in Fig. 4 of the same Plate is an oblong, rectangular wooden box, as used in central China; the trap door at the end of the box is a plain wooden slip fitting into a groove, which may be lifted and lowered in a few seconds.

In the south, men avail themselves of what is called a fire-basket *(fo lam)* which is made of iron rods and in which a charcoal fire is kept burning. This fire drives the insects out of their dens. Sometimes the cricket-hunters reach their object by pouring water into the holes where the insects hide. Sometimes they endeavor to entice them from the nest by placing at its entrance the fruit of *Nephelium longana* (*lung yen*, "dragon's eyes").

In Shanghai and Hangchow grasshoppers are also held captives and enclosed in wooden cages, usually of the shape of a chair, stool, or table (Plate XI).

Cicadas were formerly also kept in small cages which were suspended at the eaves of houses or from the branches of trees, but this custom is no longer practised. The cicada is at present not offered for sale in the markets like the cricket. It may occasionally be caught by boys and caged by them for their amusement temporarily, but otherwise interest in this insect has waned. The same holds good for Japan, where cicadas are never caged. Japanese poets, as Lafcadio Hearn observes, are much more inclined to praise the voices of night-crickets than those of cicadas; there are countless poems about the latter, but very few which commend their singing.

Many people rear hundreds of crickets in their homes, and have several rooms stacked with the jars which shelter the insects. The rich employ experts to look after theirs. As soon as you enter a house like this, you are greeted by a deafening noise which a Chinese is able to stand for any length of time.

During the summer the insects are kept in circular pottery jars made of a common burnt clay and covered with a flat lid, which is sometimes perforated. Many potters made a special business of these cricket houses, and impressed on them a seal with their names; for instance, Chao Tse-yü, who lived in the first part of the nineteenth century and whose productions still enjoy a special reputation. There are old pots said to go back as far as the Ming dynasty (1368-1643), and these are highly prized. The crickets keep cool in these jars, which are often shaped in the form of a gourd, as the heat does not penetrate the thick clay walls. Tiny porcelain dishes decorated in blue and white or small bits of clay contain food and water for the insects, and they are also provided with beds or sleeping boxes of clay (Plates VIII and IX). Jars of somewhat larger size serve for holding the cricket-fights.

During the winter months the crickets change their home, and are transferred to specially prepared gourds which are provided with loose covers wrought in open work so as to admit fresh air into the gourd. This is said to be a special variety of the common gourd *(Lagenaria vulgaris)*, the cultivation of which was known to a single family of Peking. A Chinese model of the plant,—the flowers of jade, the gourds of turquois,—is placed on exhibition; likewise gourds in their natural shapes and others in the process of being worked. The gourds used as cricket habitations are all artificially shaped; they are raised in earthen moulds, the flowers are forced into the moulds, and as they grow will assume the shape of and the designs fashioned in the moulds. There is accordingly an infinite variety of forms: there are slender and graceful, round and double, cylindrical and jar-like ones. Those formerly made for the Palace, of which the Museum possesses a number, are decorated with figures and scenes in high relief fashioned in the clay mould. The technique employed in these ancient pieces is now lost; at least they are no longer made, though there are poor modern imitations in which the surfaces are carved, not moulded.

The covers of the gourd, flat or tall, are made of jade, elephant or walrus ivory, coconut shell, and sandalwood, all elaborately decorated, partly in high relief, partly in open work, or in the two methods combined, with floral designs, dragons, lions and other animals. Gourd vines with flowers and fruits belong to the most favorite designs carved in the flat ivory covers; gourd and cricket appear to be inseparable companions. A kind of cement which is a mixture of lime and sandy loam is smeared over the bottom of the gourd to provide a comfortable resting-place for the tenant. The owner of the cricket may carry the gourd in his bosom wherever he goes, and in passing men in the street you may hear the shrill sound of the insect from its warm and safe place of refuge. The gourds keep the insects warm, and on a cold night they receive a cotton padding to sleep upon.

Plain gourds are illustrated in Plates IV and V, Figs. 1-2; decorated ones, in Plates V, Figs. 3-4, and X.

In the summer the insects are generally fed on fresh cucumber, lettuce, and other greens. During their confinement in autumn and winter masticated chestnuts and yellow beans are given them. In the south they are also fed on chopped fish and various kinds of insects, and even receive honey as a tonic. It is quite a common sight to see the idlers congregated in the tea-houses and laying their crickets out on the tables. Their masters wash the gourds with hot tea and chew chestnuts and beans to feed them. Then they listen to their songs and boast of their grinding powers. The Chinese cricket books give many elaborate rules for proper feeding which vary with the different species and with every month. The Golden Bell, for instance, should be fed on wormwood (or southern-wood, *ts'ing hao, Artemisia apiacea*), while flowers of the "silk melon" *(Luffa cylindrica)* and melon pulp are prescribed for the Spinning Damsel.

The fighting crickets receive particular attention and nourishment, a dish consisting of a bit of rice mixed with fresh cucumbers, boiled chestnuts, lotus seeds, and mosquitoes. When the time for the fight draws near, they get a tonic in the form of a bouillon made from the root of a certain flower. Some fanciers allow themselves to be stung by mosquitoes, and when these are full of blood, they are given their favorite pupils. In order to stir their ferocity prior to a bout, they are sometimes also compelled to fast. As soon as they recognize from their slow movements that they are sick, they are fed on small red insects gathered in water.

A tickler is used for stirring the crickets to incite them to sing (Plate VI). In Peking fine hair from hare or rat whiskers inserted in a reed or bone handle is utilized for this purpose; in Shanghai, a fine blade of crab or finger grass *(Panicum syntherisma)*. The ticklers are kept in bamboo or wooden tubes, and the rich indulge in the luxury of having an elegant ivory tube surmounted by the carving of a lion (Plate VI). A special brush serves for cleaning the gourds and jars (Plate XII, Fig. 6); and a pair of wooden nippers or tongs is used for handling the food and water dishes (Plate XII, Fig. 5). The insect is held under a wire screen, while its gourd is being cleaned or washed (Plate XII, Fig. 7). A hair net enclosed in a hoop is placed over the jar to watch the doings of the insects (Plate VIII, in upper right corner).

The tympanum of good singers is coated with a bit of wax to increase or strengthen the volume of sound. A small needle about three inches long with blunt end, about the size of a darning needle, is heated

CAGES FOR GRASSHOPPERS IN SHAPE OF WOODEN CHAIR AND TABLE.

Enclosed by glass. Hangchow, China.

over a candle and lightly dipped in the wax. The insect is held between the thumb and forefinger of the closed hand, and the wax is applied to the wing-covers. Specimens of the wax are shown in the case of cricket paraphernalia.

Crickets are imbued with the natural instinct to fight. The Chinese offer the following explanation for this fact: the crickets live in holes, and each hole is inhabited by a single individual; this manner of living gives rise to frictions and frequent combats, for the insects always prefer their old places of refuge, and when they encounter in them another inmate, they will not cede their rights voluntarily, but will at once start to fight over the housing problem. The two rivals will jump at each other's heads with furious bites, and the combat will usually end in the death of one of the fighters. It frequently happens that the victor devours the body of his adversary, just as primitive man did away with the body of his enemy whom he had slain in mortal strife. When driven by hunger, crickets will feed upon other insects and even devour their own relations. When several are confined in a cage, they do not hesitate to eat one another. War and death is a law of nature.

In the course of many generations, the Chinese through long experience and practice, have accomplished what we may call a natural selection of fighting crickets. The good fighters are believed to be incarnations of great heroes of the past, and are treated in every respect like soldiers. Kia Se-tao, the first author who wrote on the subject, says that "rearing crickets is like rearing soldiers." The strongest and bravest of these who are most appreciated at Peking and Tientsin come from the southern province of Kwang-tung. These fighters are dubbed "generals" or "marshals," and seven varieties of them are distinguished, each with a special name.

Those with black heads and gray hair in their bodies are considered best. Next in appreciation come those with yellow heads and gray hair, then those with white heads and gray hair, then those with golden wings covered with red hair, those of yellow color with blood-red hair who are said to have two tails in form of sheep's horns, finally those yellow in color with pointed head and long abdomen and those supposed to be dressed in embroidered silk, gray in color and covered with red spots like fish-scales. The good fighters, according to Chinese experts, are recognized by their loud chirping, their big heads and necks, long legs, and broad bodies and backs.

The "Generals," as stated, receive a special diet before the contest, and are attended to with utmost care and great competence. Observations made for many centuries have developed a set of standard rules which are conscientiously followed. The trainers, for instance, are aware of the fact that extremes of temperature are injurious to the crickets. When they observe that the insects droop their tiny mustaches, they know that they are too warm, and endeavor to maintain for them an even temperature and exclude all draughts from them. Smoke is supposed to be detrimental to their health, and the rooms in which they are kept must be perfectly free from it. The experts also have a thorough understanding of their diseases, and have prescriptions at hand for their treatment and cure. If the crickets are sick from overeating, they are fed on a kind of red insect. If sickness arises from cold, they get mosquitoes; if from heat, shoots of the green pea are given them. A kind of butterfly known as "bamboo butterfly" is administered for difficulty in breathing. In a word, they are cared for like pet babies.

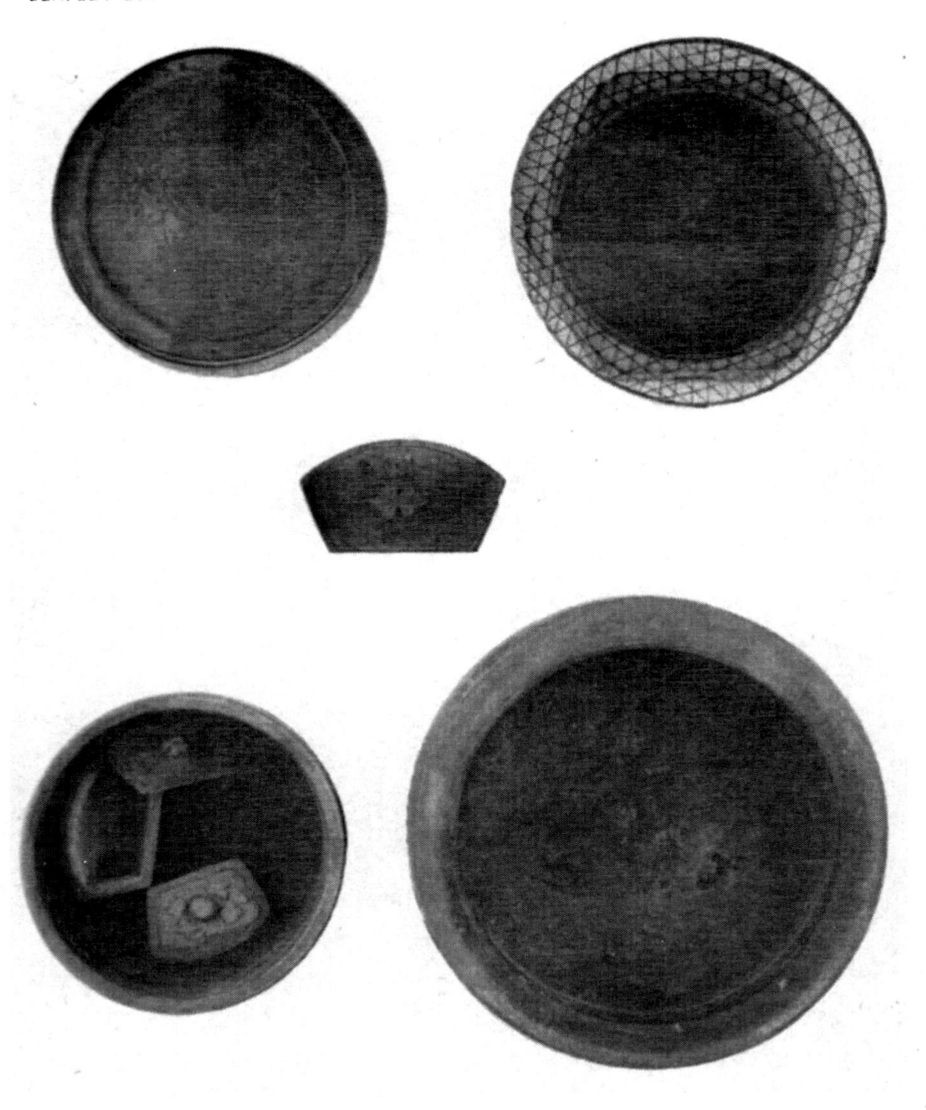

CRICKET POTS.

The Summer Habitations of the Insects.

The one in the lower left contains two clay beds in which the crickets sleep. Another clay bed in the centre.

The tournaments take place in an open space, on a public square, or in a special house termed Autumn Amusements. There are heavy-weight, middle and light-weight champions. The wranglers are always matched on equal terms according to size, weight, and color, and are carefully weighed on a pair of wee scales at the opening of each contest. A silk cover is spread over a table on which are placed the pottery jars containing the warring crickets. The jar is the arena in which the prize fight is staged. A specimen with two crickets in the act of fighting is shown in the exhibition case. As a rule, the two adversaries facing each other will first endeavor to flee, but the thick walls of the bowl or jar are set up as an invincible barrier to this attempt at desertion. Now the referee who is called "Army Commander" or "Director of the Battle" intercedes, announcing the contestants and reciting the history of their past performances, and spurs the two parties on to combat. For this purpose he avails himself of the tickler described above, and first stirs their heads and the ends of their tails, finally their large hind legs. The two opponents thus excited stretch out their antennae which the Chinese not inaptly designate "tweezers," and jump at each other's heads. The antennae or tentacles are their chief weapons. One of the belligerents will soon lose one of its horns, while the other may retort by tearing off one of the enemy's legs. The two combatants become more and more exasperated and fight each other mercilessly. The struggle usually ends in the death of one of them, and it occurs not infrequently that the more agile or stronger one pounces with its whole weight upon the body of its opponent, severing its head completely.

Cricket-fights in China have developed into a veritable passion. Bets are concluded, and large sums are wagered on the prospective champions. The stakes are in some cases very large, and at single matches held in

Canton are said to have sometimes aggregated $100,-000. It happens quite frequently that too ardent amateurs are completely ruined in the game. Gambling is forbidden by law in China as elsewhere, but such laws are usually winked at, and the official theory in this case is that the stakes consist of presents of sweet cakes. Choice champions fetch prices up to $100, the value of a good horse in China, and owners of famous crickets travel long distances to meet their competitors and congregate with them in order to match their champions. Some amateurs delight in raising them by the hundreds in the hope to produce the champion of the champions of the season, who is honored with the attribute of Grand Marshal. These men are by no means low-brows, but highly cultured men and those in responsible government positions are found in this class.

Two localities near Canton, Fa-ti and Cha-pi, not far from Whampoa, enjoy a special reputation for cricket-fighting. At these places extensive mat sheds are erected and divided into several compartments. In each section a contest goes on, the pot which forms the arena being placed on a table. In order to acquaint prospective betters with the merits of the crickets matched against each other, a placard is posted on the sides of the building, setting forth the various stakes previously won by each cricket. Great excitement is manifested at these matches, and considerable sums of money change hands. The sum of money staked on the contest is lodged with a committee who retain ten per cent to cover expenses and hand over the balance to the owner of the winning cricket. The lucky winner is also presented with a roast pig, a piece of silk, and a gilded ornament resembling a bouquet of flowers. This decoration is deposited by him either on the ancestral altar of his house to inform his ancestors of his good luck and to thank them for their protection, or on a shrine in honor of Kwan-ti, a deified hero, who is the personification of all manly virtues and a model of gentlemanly conduct.

The names of the victorious champions are inscribed on an ivory tablet carved in the shape of a gourd (Plate VIII, centre), and these tablets like diplomas are religiously kept in the houses of the fortunate owners. Sometimes the characters of the inscription are laid out in gold. The victory is occasion for great rejoicing and jollification. Music is performed, gongs are clanged, flags displayed, flowers scattered, and the tablet of victory is triumphantly marched in front, the jubilant victor struts in the procession of his overjoyed compatriots, carrying his victorious cricket home. The sunshine of his glory falls on the whole community in which he lives, and his village will gain as much publicity and notoriety as an American town which has produced a golf or baseball champion.

In southern China, a cricket which has won many victories is honored with the title "conquering or victorious cricket" *(shou lip)*; on its death it is placed in a small silver coffin, and is solemnly buried. The owner of the champion believes that the honorable interment will bring him good luck and that excellent fighting crickets will be found in the following year in the neighborhood of the place where his favorite cricket lies buried.

All these ideas emanate from the belief that able cricket champions are incarnations of great warriors and heroes of the past from whom they have inherited a soul imbued with prowess and fighting qualities. Dickens says, "For all the Cricket Tribe are potent Spirits, even though the people who hold converse with them do not know it (which is frequently the case)."

A proverbial saying with reference to a man who failed or has been defeated is, "A defeated cricket,—he gives up his mouth," which means as much as "throwing up the sponge."

The following Chinese stories may give an insight into the cricket rage.

Kia Se-tao, a minister of state and general who lived in the thirteenth century, and who wrote, as mentioned, an authoritative treatise on the subject, is one of the cricket fanciers famous in history. He was completely obsessed with an all-absorbing passion for the cricket cult. The story goes that one day, during a war of the Mongols against the imperial house of Sung, an important city fell into the hands of the foe. When Kia Se-tao received news of the disaster, he was found kneeling in the grass of a lawn and taking part in a cricket match. "In this manner you look out for the interests of the nation!" he was reprimanded. He was not in the least disturbed, however, and kept his attention concentrated on the game.

An anecdote of tragical character is told with reference to an official of Peking, who held the post of director of the rice-granaries of the capital. He once found a cricket of choice quality and exceptional value. In order to secure this treasure, he exchanged his best horse for it and resolved to present this fine specimen to the emperor. He placed it cautiously in a box and took it home. During his absence his prying wife craved to see the insect which had been bought so dearly. She opened the box, and fate ordained that the cricket made its escape. A rooster happened to be around and swallowed the cricket. The poor woman, frightened by the consequences of her act, strangled herself with a rope. At his return the husband learned of the double loss he had suffered and, seized by despair, committed suicide. The Chinese narrator of the story concludes, "Who would have imagined that the graceful singer of the fields might provoke a tragedy like this?"

The "Strange Stories from a Chinese Studio" written by P'u Sung-ling in 1679 (translated into Eng-

BLUE AND WHITE PORCELAIN DISHES FOR FEEDING CRICKETS.

In the centre an ivory tablet in shape of a gourd on which the names of the victorious champions are inscribed.

lish by H. A. Giles) contain the following story of a Fighting Cricket (No. 64) :—

"During the period Süan-te (1426-36) of the Ming dynasty, cricket-fighting was very much in vogue at court (levies of crickets being exacted from the people as a tax. On one occasion, the magistrate of Hua-yin, wishing to befriend the Governor, presented him with a cricket which, on being set to fight, displayed very remarkable powers; so much so that the Governor commanded the magistrate to supply him regularly with these insects. The latter, in his turn, ordered the beadles of his district to provide him with crickets; and then it became a practice for people who had nothing else to do to catch and rear them for this purpose. Thus the price of crickets rose very high; and when the beadle's runners came to exact even a single one, it was enough to ruin several families. In the said village there lived a man named Cheng, a student who had often failed for his bachelor's degree; and, being a stupid sort of fellow, his name was sent in for the post of beadle. He did all he could to get out of it, but without success; and by the end of the year his small patrimony was gone. Just then came a call for crickets. Cheng was in despair, but, encouraged by his wife, went out hunting for the insects. At first he was unsuccessful, but by means of a map supplied by a fortune-teller he at last discovered a magnificent specimen, strong and handsome, with a fine tail, green neck, and golden wings; and, putting it in a basket, he returned home in high glee to receive the congratulations of his family. He would not have taken anything for this cricket, and proceeded to feed it up carefully in a bowl. Its belly was the color of a crab's, its back that of a sweet chestnut; and Cheng tended it most lovingly, waiting for the time when the magistrate should call upon him for a cricket.

"Meanwhile, Cheng's nine year old son, while his father was out, opened the bowl. The cricket escaped instantaneously. The boy grabbed it, seized one of its legs which broke off, and the little creature soon died. Cheng's wife turned deadly pale when her son, with tears in his eyes, told her what had happened. The boy ran away, crying bitterly. Soon after Cheng came home, and when he heard his wife's story, he felt as if he had been turned to ice. He went in search of his son whose body he discovered at the bottom of a well. The parents' anger thus changed into grief, but when they prepared to bury the boy, they found that he was still breathing. Toward the middle of the night he came to, but his reason had fled.

"His father caught sight of the empty bowl in which he had kept the cricket, and at daybreak he suddenly heard the chirping of a cricket outside the house-door. Jumping up hurriedly, there was his lost insect; but, on trying to catch it, away it hopped directly. He chased it up and down, until finally it jumped into a corner of the wall; and then, looking carefully about, he espied it once more, no longer the same in appearance, but small and of a dark red color. Cheng stood looking at it, without trying to catch such a worthless specimen, when all of a sudden the little creature hopped into his sleeve; and, on examining it more closely, he noticed that it really was a handsome insect, with well-formed head and neck, and forthwith took it indoors.

"He was now anxious to try its prowess; and it so happened that a young fellow of the village, who had a fine cricket which used to win every bout it fought, called on Cheng that very day. He laughed heartily at Cheng's champion, and producing his own, placed it side by side, to the great disadvantage of the former. Cheng's countenance fell, and he no longer wished to back his cricket. However, the young fellow urged him,

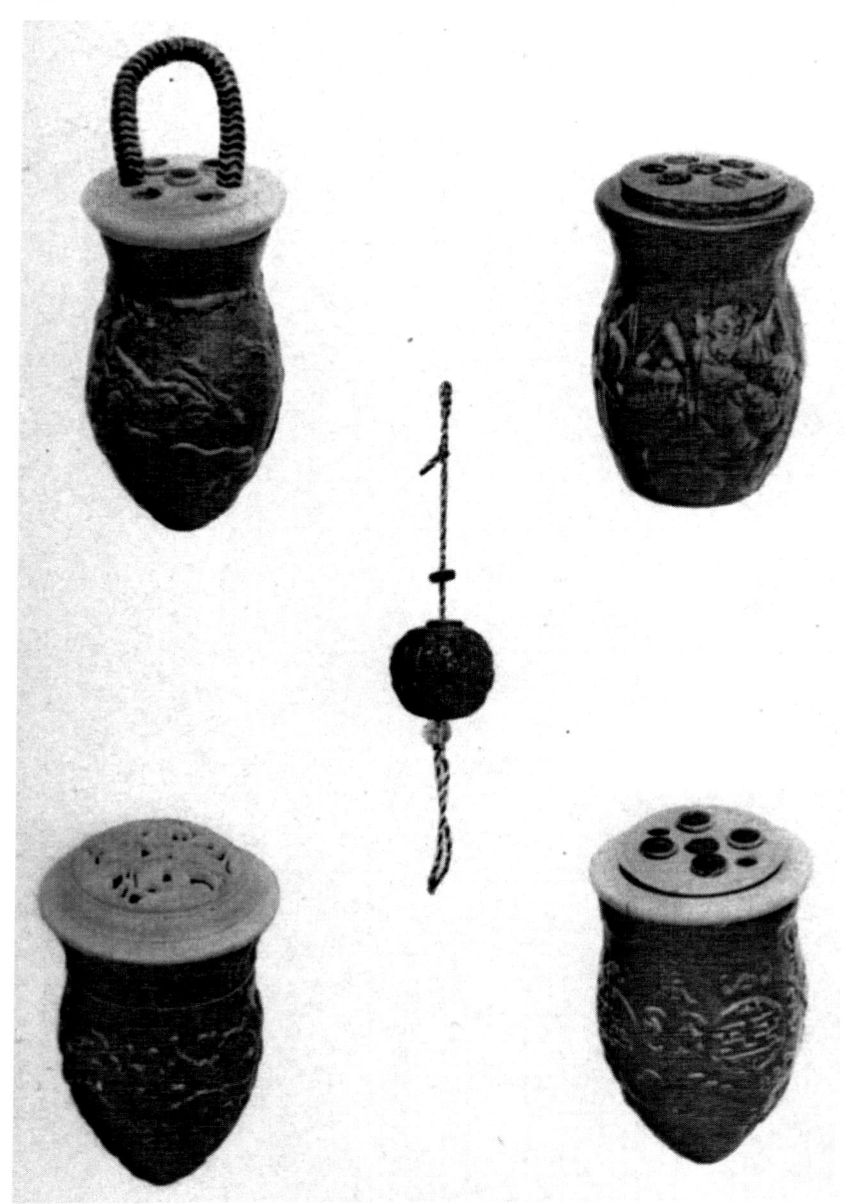

CRICKET GOURDS.

With moulded decorations: scenery, figures, and ornaments. The figure in the centre represents a carved walnut shell an enlargement of which is shown in Plate XI.

and he thought that there was no use in rearing a feeble insect, and that he had better sacrifice it for a laugh; so they put them together in a bowl. The little cricket lay quite still like a piece of wood, at which the young fellow roared again, and louder than ever when it did not even move though tickled with a pig's bristle. By dint of tickling it was roused at last, and then it fell upon its adversary with such fury, that in a moment the young fellow's cricket would have been killed outright had not its master interfered and stopped the fight. The little cricket then stood up and chirped to Cheng as a sign of victory; and Cheng, overjoyed, was just talking over the battle with the young fellow, when a cock caught sight of the insect and ran up to catch it. Cheng was alarmed, but the cock luckily missed its aim, and the cricket hopped away, its enemy pursuing at full speed. In the next moment Cheng saw his cricket seated on the cock's head, holding firmly on to its comb. He then placed it in a cage and sent it to the magistrate, who, seeing what a small one he had provided, was very angry indeed. The magistrate refused to believe the story of the cock, so Cheng set it to fight with other crickets all of whom it vanquished without exception. He then tried it with a cock, and as all turned out as Cheng had said, he gave him a present and sent the cricket on to the Governor. The latter forwarded it to the palace in a golden cage with some comments on its performances.

"It was found that in the splendid collection of his majesty there was not one worthy of being matched with this one. It would dance in time to music and became a great favorite at court. The emperor in return bestowed magnificent gifts of horses and silks upon the Governor. The latter rewarded the magistrate, and the magistrate recompensated Cheng by excusing him from the duties of beadle and by instructing the Literary Chancellor to pass him for the first degree. A few months afterwards Cheng's son recovered his intellect and said that he had been a cricket and had proved himself a very skilful fighter. The Governor also rewarded Cheng handsomely, and in a few years he was a rich man, with flocks, herds, houses and acres, quite one of the wealthiest of mankind."

The interesting point of this story is that the boy's spirit, during his period of temporary mental aberration, had entered into the body of the cricket which had allowed itself to be caught by his father. He animated it to fight with such extraordinary vigor that he might amend the loss caused by his curiosity in letting the other cricket escape.

Cricket-fights are not so cruel as cock and quail fights in which the Chinese also indulge, but the three combined are not so revolting as the bull-fights of Spain and Latin America. The Chinese reveal their sentimental qualities in their fondness of the insect-musicians, in the loving care they bestow on their pets and in lavishing on them the most delicate and exquisite productions their miniature art is able to create. They know how to carve a walnut-shell with the figures of the eighteen Arhat and elaborate ornamental detail (Plates X and XI). A lens is required to appreciate this whole apparatus of intricate design. A walnut like this is suspended at the girdle, and a cricket is enclosed in it just for the purpose of enjoying its musical efforts. Surely people who go to all this trouble must have sentiments and a deep sense of the joy of life and nature.

As far as I know, the Chinese are the only nation that has developed cricket-fights. The Japanese, though fond of chirping insects which they keep as pets in little cages, do not use them for fighting purposes. Kipling writes in his Jungle-book, "The herd-children of India sleep and wake and sleep again, and weave

CARVED WALNUT SHELL (ENLARGED).

Decorated with the figures of the Eighteen Arhat, a pavilion, trees, and the sun emerging from clouds. For keeping singing crickets and carried about in the girdle.
China, K'ien-lung Period (1736-95).

little baskets of dried grass and put grasshoppers in them; or catch two praying-mantises and make them fight." This may be an occasional occurrence in India, but it has not developed into a sport or a national pastime. In regard to Japan the reader may be referred to Lafcadio Hearn's essays "Insect-Musicians," inserted in his "Exotics and Retrospectives," and "Semi" [Cicada] in his "Shadowings."

Field Museum owns a very extensive collection illustrating the Chinese cricket cult and consisting of numerous moulded gourds (many from the Palace and the possession of ancient families of Peking), pottery jars, and all the paraphernalia (altogether about 240 pieces). This collection was brought together by me on the Captain Marshall Field Expedition to China in 1923. A careful selection of this material is placed on exhibit in a case on the West Gallery.

B. LAUFER.

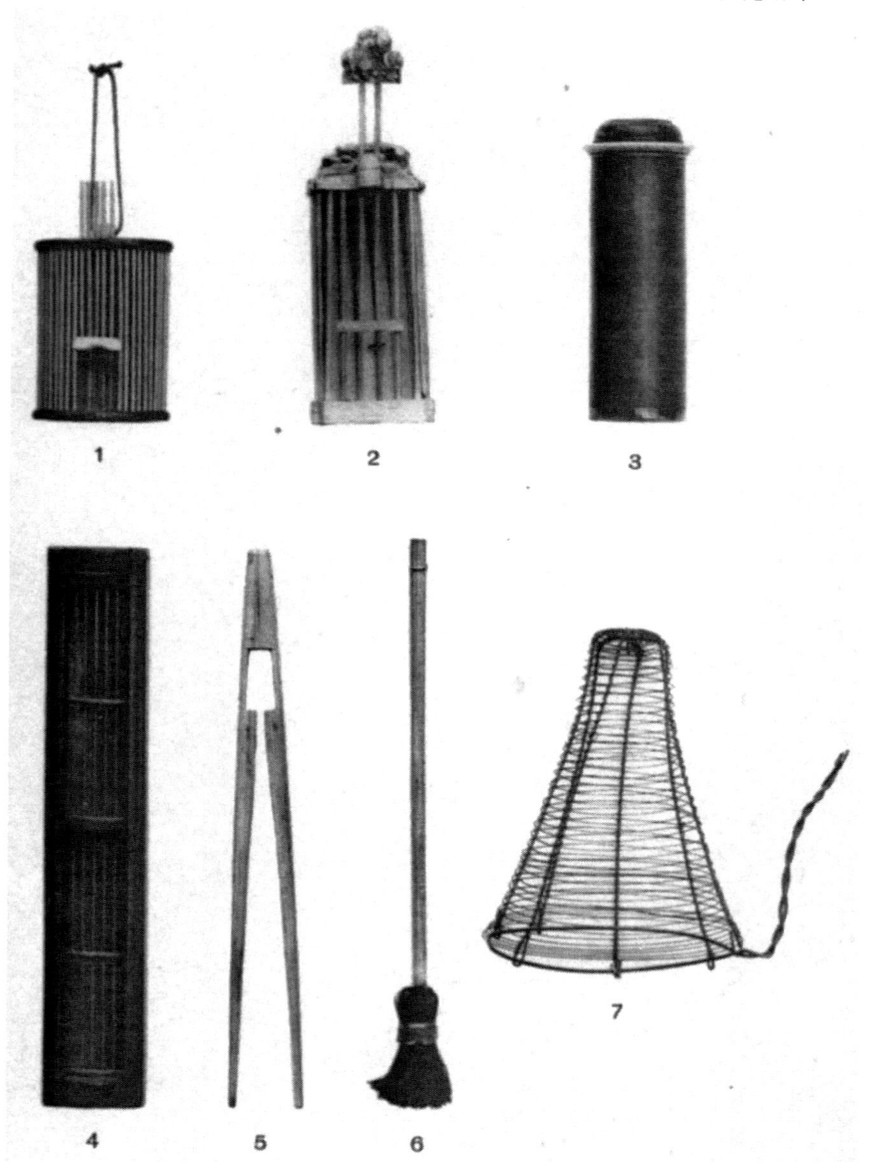

ACCESSORIES.

1, 2, 4. Traps for catching insects, 1 and 4 of bamboo, 2 of ivory. 3. Gourd of cylindrical shape for keeping female crickets to secure eggs. 5. Pair of nippers for taking feeding-dishes out or in. 6. Brush for cleaning cricket-pots and gourds. 7. Wire frame under which crickets are held while their cages are being cleaned.

181

麒麟传入考

The Giraffe in History and Art

BY

BERTHOLD LAUFER

CURATOR OF ANTHROPOLOGY

9 Plates in Photogravure, 23 Text-figures, and 1 Vignette

ANTHROPOLOGY

LEAFLET 27

FIELD MUSEUM OF NATURAL HISTORY

CHICAGO

1928

The Anthropological Leaflets of Field Museum are designed to give brief, non-technical accounts of some of the more interesting beliefs, habits and customs of the races whose life is illustrated in the Museum's exhibits.

LIST OF ANTHROPOLOGICAL LEAFLETS ISSUED TO DATE

1.	The Chinese Gateway (Laufer)	$ .10
2.	The Philippine Forge Group (Cole)	.10
3.	The Japanese Collections (Gunsaulus)	.25
4.	New Guinea Masks (Lewis)	.25
5.	The Thunder Ceremony of the Pawnee (Linton) .	.25
6.	The Sacrifice to the Morning Star by the Skidi Pawnee (Linton)	.10
7.	Purification of the Sacred Bundles, a Ceremony of the Pawnee (Linton)	.10
8.	Annual Ceremony of the Pawnee Medicine Men (Linton)	.10
9.	The Use of Sago in New Guinea (Lewis)	.10
10.	Use of Human Skulls and Bones in Tibet (Laufer)	.10
11.	The Japanese New Year's Festival, Games and Pastimes (Gunsaulus)	.25
12.	Japanese Costume (Gunsaulus)	.25
13.	Gods and Heroes of Japan (Gunsaulus)	.25
14.	Japanese Temples and Houses (Gunsaulus) . . .	.25
15.	Use of Tobacco among North American Indians (Linton)	.25
16.	Use of Tobacco in Mexico and South America (Mason)	.25
17.	Use of Tobacco in New Guinea (Lewis)	.10
18.	Tobacco and Its Use in Asia (Laufer)	.25
19.	Introduction of Tobacco into Europe (Laufer) . .	.25
20.	The Japanese Sword and Its Decoration . . . (Gunsaulus)	.25
21.	Ivory in China (Laufer) . . , ...	.75
22.	Insect-Musicians and Cricket Champions of China (Laufer)	.50
23.	Ostrich Egg-shell Cups of Mesopotamia and the Ostrich in Ancient and Modern Times (Laufer)	.50
24.	The Indian Tribes of the Chicago Region with Special Reference to the Illinois and the Potawatomi (Strong)	.25
25.	Civilization of the Mayas (Thompson)	.75
26.	Early History of Man (Field)	.25
27.	The Giraffe in History and Art (Laufer)	.75

D. C. DAVIES, DIRECTOR

FIELD MUSEUM OF NATURAL HISTORY
CHICAGO, U. S. A.

NORTHERN GIRAFFE.
After Hutchinson, Animals of All Countries.

FIELD MUSEUM OF NATURAL HISTORY
DEPARTMENT OF ANTHROPOLOGY
CHICAGO, 1928

LEAFLET NUMBER 27

The Giraffe in History and Art

CONTENTS

Page
Giraffes. 3
The Giraffe in Ancient Egypt 15
Representations of the Giraffe in Africa outside of Egypt 26
The Giraffe among Arabs and Persians 31
The Giraffe in Chinese Records and Art 41
The Giraffe in India. 55
The Giraffe among the Ancients 58
The Giraffe at Constantinople 66
The Giraffe during the Middle Ages 70
The Giraffe in the Age of the Renaissance 79
The Giraffe in the Nineteenth Century and After 88
Notes 95
Bibliography 98

ACKNOWLEDGMENTS

In issuing this booklet I wish to express my thanks and gratitude to many friends who have aided me with photographs and information, above all, to the firm Carl Hagenbeck of Stellingen for a number of photographs of live giraffes and many useful data, to Professor James H. Breasted for photographs of the Nubian rock-carvings taken by him and published here for the first time, to the Pierpont Morgan Library of New York for the photograph of the Persian painting, to Mr. A. W. Bahr for the loan of the Chinese painting reproduced in Plate IV, and to the Art Institute of Chicago for the photograph of the cotton print in Plate VI. To Professor Lucy H. Driscoll of the University of Chicago I am indebted for references to Italian paintings and important literary sources; and to Professor M. Sprengling, for kind assistance in the translation of Arabic and Persian sources.

The twenty-five drawings illustrating this essay were prepared with great care and skill by the Museum artist, Mr. Carl F. Gronemann, who likewise made the wooden block for the colored giraffe-head on the cover.

GIRAFFES

Giraffes constitute a distinct family of ruminants (*Giraffidae*), natives of Africa (Plates I, VII-IX). Owing to the extraordinary development of the neck and legs, the giraffe is the tallest of all mammals, the height of bulls being from fifteen to sixteen, according to some observers, even from eighteen to nineteen feet, and that of cows from sixteen to seventeen feet. Despite its great elongation, the neck contains only the typical number of seven vertebrae as in nearly all mammals, each vertebra itself being elongated, as every visitor to the Museum may convince himself by viewing the mounted skeleton of a giraffe in Hall 17.

During the present geological epoch the family is strictly confined to Africa, but in former periods of the earth it had a much wider extension, and was distributed over many parts of Europe and Asia, especially Greece, Persia, India, and China, where fossil remains have been discovered from the Miocene onward down to the Pleistocene age. Its maximum development in numbers was reached in the Pliocene of Asia. The living species are distributed all over Africa south of the Sahara.

Two species are generally recognized by zoologists, each with a number of subspecies or geographic races distinguished by variations in the arrangement of the spots, especially on the legs and abdomen. The more widely distributed species is *Giraffa camelopardalis* which ranges throughout most of central and southern Africa. The Reticulated giraffe (*Giraffa reticulata*) is chestnut-colored and covered with a network of white lines (Fig. 1). Its distribution is restricted to northeast Africa in Somaliland, Abyssinia, and northern Kenya. This species will engage our special attention with reference to Persian and Chinese pictorial representations of it.

The existence of the giraffe in the southern part of Africa (*Giraffa capensis*) was first made known by Hop and

Fig. 1.
Reticulated Giraffe.
From a photograph of Carl Hagenbeck.

Brink's expedition to Great Namaqualand in 1761, who found giraffes soon after crossing the Great River and shot several. Tulbagh, the Dutch governor of the Cape Colony, sent the skin of one of these giraffes to the museum of the

University of Leiden; it was the first taken to Europe from South Africa. A rude sketch of the animal made by Hop and Brink was inserted by Buffon in the thirteenth volume of his "Histoire naturelle." In South Africa the name "giraffe" is practically unknown, and the Dutch term "kameel" is always used.

The body of the giraffe is short, and its shape is peculiar in that the back slopes gradually downward to the rump. The greater height of the fore parts is not owing to the greater length of the fore legs which are not much longer than the hind legs (the real difference between the two amounts to hardly seven inches), but to processes of the vertebrae which form a basis for the muscular support of the neck and head and make a hump on the shoulders.

The neck of all giraffes bears a short mane extending from the occiput to the withers. The hair is short and smooth, reddish white, and marked by numerous dark rusty spots, which are rhomboid, oval, and even circular in shape. The hide is about an inch thick and very tough. It is used by the natives of South Africa for making sandals and by the Boers to supply whips for the bullock-carts, known as sjambok. With the practical disappearance of the rhinoceros and the approaching extermination of the hippopotamus in South Africa, there is a constant commercial demand for giraffe-hides, which are worth from four to five pounds sterling apiece. As a consequence, giraffes are killed in large numbers by Boer and native hunters, and may soon be threatened with extinction.

One of the most beautiful features of the giraffe are the eyes, which are dark brown, large and lustrous, full, soft, and melting, and shaded by long lashes. The ears are long and mobile. The nostrils can be tightly closed at will by a curious arrangement of sphincter muscles. This is supposed to be a provision of nature against blowing sand and thorns of acacias on the leaves of which the animal browses. The lips are furnished with a dense coating of thick velvety hair, probably as a further protection against thorns.

Giraffes of both sexes carry two "horns" upon the summit of the head. These are permanent bony protuberances or processes growing from the skull, and are covered with yellowish brown hair, which at the tip becomes black. In the skulls of young animals these false horns are easily detachable, but in the adult they are firmly attached to the bony framework of the head, partly to the frontal and partly to the parietal bones. Adults of the Nubian form often have a prominent third horn, rising from the centre of the forehead, between the eyes, to a height of from three to five inches. The "horns," it should be noted, are persistent, not deciduous as the antlers of deer.

The legs are long and slender; the knees are protected by thick pads or callosities. The feet have cloven hoofs; lateral toes are absent. The end of the tail is provided with a long tassel of hair which the animals are in the habit of pulling out. The tail is an article much in favor with eastern Bantu tribes, and has a value of from ten to fifty shillings, while a particularly fine specimen is worth up to five pounds sterling. Giraffe-tails, as will be seen, are figured on an Egyptian monument, and are presented as tribute to Tutenkhamon.

The dentition of the giraffe is bovine: it has altogether thirty-two teeth, six grinders on each side both above and below, and eight teeth in the lower jaw, but none in the upper one. These lower teeth consist of three incisors, and are canine on each side, the canine having a cleft or bilobate crown.

Its food consists almost entirely of the leaves and tender shoots of mimosa-trees and an acacia (*Acacia giraffae*) commonly known as the kameel-dorn. The leaves are plucked off one by one by its long extensile and flexible tongue, which is thrust far out of the mouth, stretching around the leaves and pulling them tight, and then it cuts them with the lower canine teeth. The tongue is about seventeen inches long and covered with a black pigment. The animals feed chiefly in early morning and late evening, resting during the heat of the day. They are able to go for considerable periods without water, and are found in the driest country long distances away from any possible drinking-places. The Bushmen even assert that they do not drink at all; at any rate, they are singularly independent of water.

The giraffe is a gentle, inoffensive, and defenceless creature, and never uses its horns or teeth in self-defence. Gibbon, the historian, justly speaks of "camelopards, the loftiest and most harmless creatures that wander over the plains of Aethiopia." The heels are the animal's only weapon, and these may deal a very powerful kick. Carl Hagenbeck tells in his memoirs that when he loaded giraffes on a steamer at Alexandria bound for Trieste, one of his brothers received from a giraffe so energetic a blow against his chest that he collapsed and remained unconscious for some time. The lion is said to be the giraffe's sole enemy and to lie in ambush for it in the thickets by rivers and pools. Bryden thinks, however, that lions do not very often succeed in killing giraffes, defenceless though they may be; and when they do, it is generally a solitary animal (individuals of either sex are often seen alone) that has been surprised and pulled down by a party of lions.

The steppe and open bush country are the proper home of the giraffe, but occasionally it seeks the forest. The animal associates in herds from seven to sixteen individuals, though sometimes even larger numbers have been observed in a flock. There is usually a single old male in these herds, the others being young males and females. The oldest males are often found solitary. They are fond of company and frequently live in association with zebra, antelope, wilde-beest, and ostrich. They are difficult of approach, being extremely keen-sighted, and their towering height enables them to command a wide view. While their senses of both sight and smell are highly developed and very acute, they have no voice and are totally mute.

They sleep standing, but some individuals, and in some localities all the individuals, habitually lie down to sleep.

The peculiar gait of the giraffe has attracted the attention of early writers, first of all of Heliodorus (below, p. 62). E. Topsell, in his "Historie of Four-footed Beastes" (1607), observes, "The pace of this beast differeth from all other in the world, for he doth not move his right and left foote one after another, but both together, and so likewise the other, whereby his whole body is removed at every step or straine."

The giraffe, in its untrammeled native freedom, has only two distinct gaits,—the walk and the gallop, not three, as in the case of the camel.

"As may be gathered from observation of menagerie specimens, giraffes when walking do not move their fore and hind legs of opposite sides like ordinary mammals, but the fore and hind leg of the same side, like a camel. They have but two paces, a walk and a gallop, breaking at once from one into the other, as I was once fortunate enough to observe in a continental Zoo" (G. Renshaw).

W. Maxwell, who has taken excellent photographs of galloping giraffes from a pursuing motor-car, writes, "The giraffe, in its native surroundings, is one of the most cherished objects to the nature photographer and the camera sportsman alike. To photograph these animals by stalking up to them in open bush country, which is their usual habitat, requires skilful tactics." In his book "Stalking Big Game with a Camera" he has reproduced the gallop of the giraffe in three stages. "The speed at which the giraffe can travel when driven to its utmost," he says, "varies between twenty-eight and thirty-two miles an hour for distances of a couple of miles or so, and is about as much as a car can perform at a breakneck speed for this kind of country. The speed of the giraffe varies, naturally, according to the age and condition of the animal." The young calves are said to be wonderfully fleet and far more nimble than the adult animals. The giraffe, accordingly, is not easily overtaken by a fleet horse, and is game that taxes the skill of experienced sportsmen. Francis Galton (Narrative of an Explorer in Tropical South Africa in 1851) informs us, "Giraffes are wonderful climbers: kudus are the best; but I think that giraffes come next to them, even before the zebras."

The following graphic account of giraffe stalking, which simultaneously presents a good picture of the animal's life-habits, is given by Sir Samuel W. Baker (The Nile Tributaries of Abyssinia, 1886):—

"For many days past we have seen large herds of giraffes and many antelopes on the opposite side of the river, about two miles distant, on the borders of the Atbara, into which valley the giraffes apparently dared not descend, but remained on the table-land, although the antelopes appeared to prefer the harder soil of the valley slopes. This day a herd of twenty-eight giraffes tantalized me by descending a short distance below the level flats, and I was tempted at all hazards across the river. Accordingly preparations were immediately made for a start... The Arabs were full of mettle, as their minds were fixed upon giraffe venison.

"I had observed by the telescope that the giraffes were standing as usual upon an elevated position, from whence they could keep a good lookout. I knew it would be useless to ascend the slope direct, as their long necks give these animals an advantage similar to that of the man at the mast-head; therefore, although we had the wind in our favor, we should have been observed. I therefore determined to make a great circuit of about five miles, and thus to approach them from above, with the advantage of the broken ground for stalking. It was the perfection of uneven country: by clambering broken cliffs, wading shoulder-deep through muddy gullies, sliding down the steep ravines, and winding through narrow bottoms of high grass and mimosas for about two hours, we at length arrived at the point of the high table-land upon the verge of which I had first noticed the giraffes with a telescope. Almost immediately I distinguished the tall neck of one of these splendid animals about a half a mile distant upon my left, a little below the table-land; it was feeding on the bushes, and I quickly discovered several others near the leader of the herd. I was not far enough advanced in the circuit that I had intended to bring me exactly above them, therefore I turned sharp to my right, intending to make a short half circle, and to arrive on the leeward side of the herd, as I was now to windward: this I fortunately completed, but I had marked a thick bush as my point of cover, and upon my arrival I found that the herd had fed down wind, and that I was within two hundred yards of the great bull sentinel that, having moved from his former position, was now standing directly before me. I lay down quietly behind the bush with my two followers, and anxiously watched the great leader, momentarily expecting that it would get my wind. It was shortly joined by two others, and I perceived the heads of several giraffes lower down the incline, that were now feeding on their way to the higher ground. The seroot fly was teasing them, and I remarked that several birds were fluttering about their heads, sometimes perching upon their noses and catching the fly that attacked their nostrils, while the giraffe appeared relieved by their attentions: these were a peculiar species of bird that attacks the domestic animals, and not only relieves them of vermin, but eats into the flesh, and establishes dangerous sores. A puff of wind now gently faned the back of my neck; it was cool and delightful, but no sooner did I feel the refreshing breeze than I knew it would convey our scent direct to the giraffes. A few seconds afterwards, the three grand obelisks threw their heads still higher in the air, and fixing their great black eyes upon the spot from which the danger came, they remained as motionless as though carved from stone. From their great height they could see over the bush behind which we were lying at some paces distant, and although I do not think they could distinguish us to be men, they could see enough to convince them of hidden enemies.

"The attitude of fixed attention and surprise of the three giraffes was sufficient warning for the rest of the herd, who immediately filed up from the lower ground, and joined their comrades. All now halted, and gazed steadfastly in our direction, forming a superb tableau; their beautiful mottled skins glancing like the summer coat of a thoroughbred horse, the orange-colored statues standing out in high relief from a background of dark-green mimosas.

"This beautiful picture soon changed. I knew that my chance of a close shot was hopeless, as they would presently make a rush, and be off; thus I determined to get the first start. I had previously studied the ground, and I concluded that they would push forward at right angles with my position, as they had thus ascended the hill, and that, on reaching the higher ground, they would turn to the right, in order to reach an immense tract of high grass, as level as a billiard-table, from which no danger could approach them unobserved.

"I accordingly with a gentle movement of my hand directed my people to follow me, and I made a sudden rush forward at full speed. Off went the herd; shambling along at a tremendous pace, whisking their long tails above their hind quarters, and taking exactly the direction I had anticipated, they offered me a shoulder shot at a little within two hundred yards' distance. Unfortunately, I fell into a deep hole concealed by the high grass, and by the time that I resumed the hunt they had increased their distance, but I observed the leader turned sharp to the right, through some low mimosa bush, to make direct for the open tableland. I made a short cut obliquely at my best speed, and only halted when I saw that I should lose ground by altering my position. Stopping short, I was exactly opposite the herd as they filed by me at right angles in full speed, within about a hundred and eighty yards. I had my old Ceylon No. 10 double rifle, and I took a steady shot at a large dark-colored bull: the satisfactory sound of the ball upon his hide was followed almost immediately by his blundering forward for about twenty yards, and falling heavily in the low bush. I heard the crack of the ball of my left-hand barrel upon another fine beast, but no effect followed. Bacheet quickly gave me the single 2-ounce Manton rifle, and I singled out a fine dark-colored bull, who fell upon his knees to the shot, but recovering, hobbled off disabled, apart from the herd, with a foreleg broken just below the shoulder. Reloading immediately, I ran up to the spot, where I found my first giraffe lying dead, with the ball clean through both shoulders: the second was standing about one hundred paces distant; upon my approach he attempted to move, but immediately fell, and was dispatched by my eager Arabs. I followed the herd for about a mile to no purpose, through deep clammy ground and high grass, and I returned to our game.

"These were my first giraffes, and I admired them as they lay before me with a hunter's pride and satisfaction, but mingled with a feeling of pity for such beautiful and utterly helpless creatures. The giraffe, although from sixteen to twenty feet in height, is perfectly defenceless, and can only trust to the swiftness of its pace, and the extraordinary power of vision, for its means of protection. The eye of this animal is the most beautiful exaggeration of that of the gazelle, while the color of the reddish-orange hide, mottled with darker spots, changes the tints of the skin with the differing rays of light, according to the muscular movement of the body. No one who has merely seen the giraffe in a cold climate can form the least idea of its beauty in its native land."

K. Moebius, author of a work on the esthetics of the animal kingdom (Aesthetik der Tierwelt, 1908), maintains that the giraffe is regarded as ugly by the majority of people on account of its disproportionate members, but concedes that it makes a deep esthetic impression when it lifts its long neck straight above its massive chest, calmly looking downward or gazing into the distance with its large, black, long-lashed eyes; its form and color, in his estimation, are well adapted to the character of its habitat, yet it conveys to most people the impression of an ugly animal; in his opinion, it is an evident example of the fact that suitable organization does not render animals beautiful, but that besides it they must have other qualities to be pleasing. Aside from the fact that there is nothing ugly in nature and that "foul and fair" are relative notions much depending on our moods and point of view, the giraffe cannot be judged from menagerie specimens to which the impressions of most of us are confined. The free denizen of the wide, open arid plains of Africa will naturally forfeit its best qualities in the narrow enclosures of our animal prison camps. The giraffe must be observed in the freedom of its native haunts. Sir Samuel W. Baker writes, "No one who has merely seen the giraffe in a cold climate can form the least idea of its beauty in its native land."

"The spectacle of a troop of wild giraffe," Bryden writes, "is certainly one of the most wonderful things in nature. The uncommon shape, the great height, the long, slouching stride, the slender necks, reaching hither and thither among the spreading leafage of the camel-thorn trees, the rich coloring of the animal—all these things combine to render the first meeting with the giraffe in their native haunts one of the most striking and memorable of experiences." He further characterizes them as strangely beautiful, grotesquely graceful creatures and withal so harmless. Marco Polo, who was a keen observer and possessed of sound judgement in most matters, calls them "beautiful creatures to look at," and I think he is right.

In perusing the historical sketches to follow the reader should bear in mind that all early descriptions and illustrations of the giraffe (with the sole exception of the

Nubian and Bushmen petroglyphs in Figs. 5 and 10) are based on observation of more or less tame animals who were taken while young and reared in captivity. The study of the wild giraffe in its natural surroundings is of comparatively recent date and due to the vast progress of zoological science and animal photography. We must remain conscious of this distinction between the past and the present, for it has been observed that giraffes in the wild state are in many respects superior, much deeper and richer in coloring than those in captivity, are better nourished, stronger and considerably heavier than those bred in confinement; and Bryden is even inclined to think that there is a greater difference between wild and captive examples of giraffes than in any other animals.

It is not without interest to pass in review the role which so curious a creature has played in its relation to mankind, to record the impressions which it has left on past generations, and to study the question as to how the artists of all ages acquitted themselves of the task to render it justice in portraiture. The Bushmen and the ancient Egyptians, the Persians as well as the Chinese, the ancient Romans as well as the Italian painters of the Renaissance and other European artists furnish interesting contributions to this question, and it has seemed to me worth while to place their work here on record. Ever since in 1908 I obtained in China the Chinese painting of a giraffe, my interest in this subject has been aroused, and it was a pleasant, though not always easy task embodying a great deal of intense research to trace the vicissitudes of the giraffe through all lands and ages down to modern times. This essay is an attempt at a biography and iconography of the giraffe and endeavors to assemble all important historical data that have become known in whatever countries it made its appearance.

THE GIRAFFE IN ANCIENT EGYPT

The giraffe is one of the animals which appears to have been known to the Egyptians from times of earliest antiquity. A pictographic sign for the animal appears in hieroglyphic writing (see Fig. 9 on right side), and is particularly employed to denote the verb "to dispose, to arrange." The old word for the giraffe is *sr* (the vowels of Egyptian are unknown) which Brugsch connects with a Hebrew root and explains from the constantly swinging motion of the animal's body when at rest. It seems more likely that this word bears some relation to Ethiopic *zarat* (compare Arabic *zarafa*), or may even be derived from the latter. The later Egyptian term for the giraffe is *mmy*.

While there is apparently no written account of the giraffe preserved, presumably because it did not rank among sacred animals, we receive from the monuments of Egypt and Nubia the earliest sculptured and pictorial representations of giraffes which belong to the best known in the history of art. Moreover, the Egyptians show us also how the interesting figure of the giraffe may be utilized for the purposes of decorative art.

In the earliest prehistoric period of Egyptian civilization, animal life was much more plentiful in the unsubdued jungles of Egypt than in later times and at present. The great quantity of ivory employed by the people and the representations upon their pottery show that the elephant was still living in their midst; likewise the giraffe, the hippopotamus, and the strange okapi, which was deified as the god Set, wandered through the jungles, though all these animals were extinct in the historical period (Breasted, History of Egypt, p. 30). The animal represented by Set is identified by Schweinfurth with the African ant-bear (*Orycteropus aethiopicus*).

In this primitive epoch giraffes were used as a decorative motives on various objects. Giraffes are possibly intended in the handles of ivory combs (Fig. 2); there are other such combs surmounted by figures of antelopes. A giraffe is clearly outlined on the surface of a painted vase (Fig. 3), and possibly also appears as a mark on pottery (Capart, Primitive Art in Egypt, p. 140).

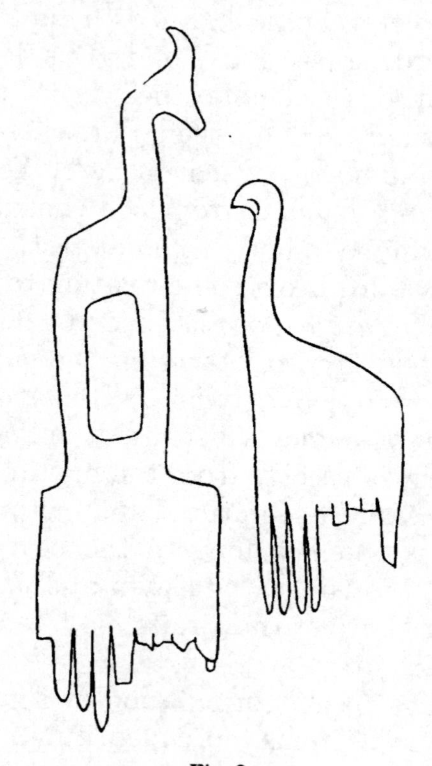

Fig. 2.
Ivory Combs with Figures of Giraffes. Ancient Egypt.
After Capart.

Fig. 4 represents an archaic slate palette carved in relief, from Hieraconpolis, showing the trunk of a palm-tree in the middle and two giraffes standing one on each side of it, apparently browsing. F. Legge, who published a similar slate only the lower part of which is preserved, showing the body and legs of two giraffes (*Proceedings Society of Biblical Archaeology*, 1900, Plate VI), concludes that the scene depicted is taking place in Upper Egypt or rather in the Sudan, the giraffe not being found above the fifteenth de-

PERSIAN PAINTING OF A GIRAFFE (p. 38).
From a Persian Bestiary of the Thirteenth Century in the Pierpont Morgan Library, New York.

gree of latitude. The four dogs around the plaque are defined by Bénédite as Molossian hounds.

On an expedition to Lower Nubia in 1906 Professor Breasted heard a report current among the natives that there is an unknown temple far out in the desert behind Abu Simbel. Various explorers had examined the neighboring

Fig. 3.
Vase with Painting of Giraffe. Ancient Egypt.
After Capart.

desert in the hope of finding it, but were unsuccessful. Accompanied by a native who assured him that he had located this temple, Professor Breasted struck out into the desert. After a two hours' journey his guide pointed to what looked much like a distant building rising out of the sand in the north. "As we drew near," he writes (*Ameri-*

can Journal of Semitic Languages, 1906, p. 35), the supposed building resolved itself into an isolated crag of rock projecting from the sand, and pierced by two openings

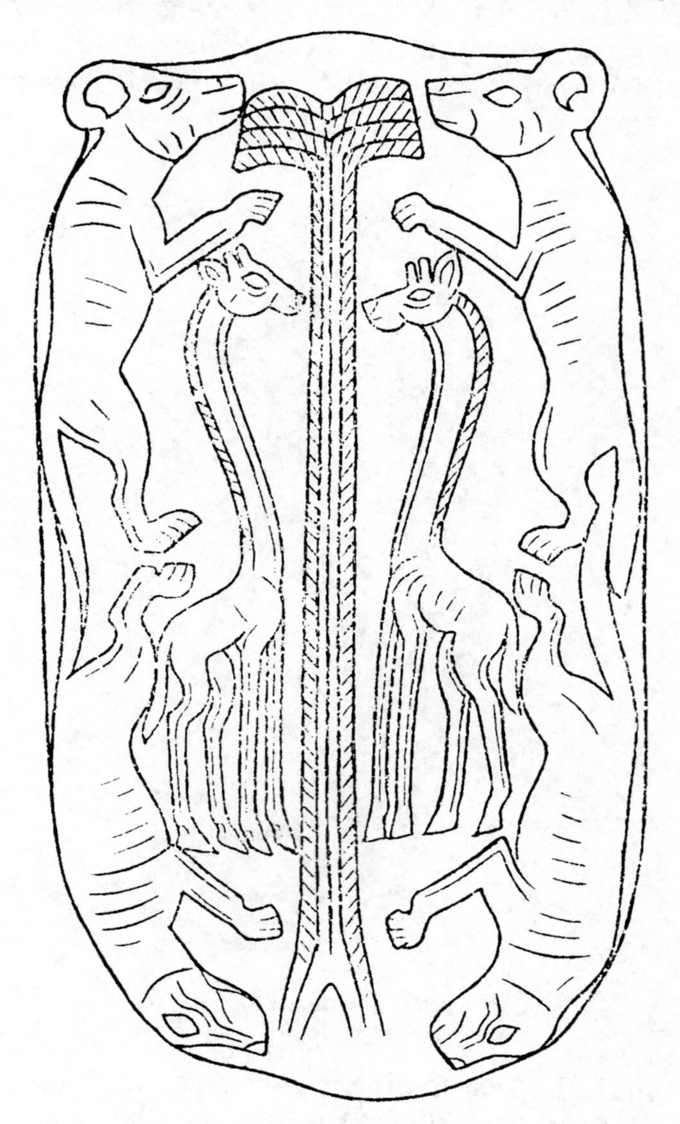

Fig. 4.
Two Giraffes Facing a Palm-tree on a Slate Palette. Ancient Egypt. After Capart.

which passed completely through it, so that the desert hills on the far horizon were clearly visible through them.

One of these openings very much resembles a door, and, to complete the delusion, it bears on one side a number of prehistoric drawings—two boats, two giraffes, two ostriches, and a number of smaller animals—which might be easily mistaken by a native for hieroglyphic writing. There can be no doubt that this curious natural formation and the archaic drawings upon it are the source of the fabled temple in the desert behind Abu Simbel."

Professor Breasted very kindly placed at my disposal two photographs of these rock-carvings taken by him, from

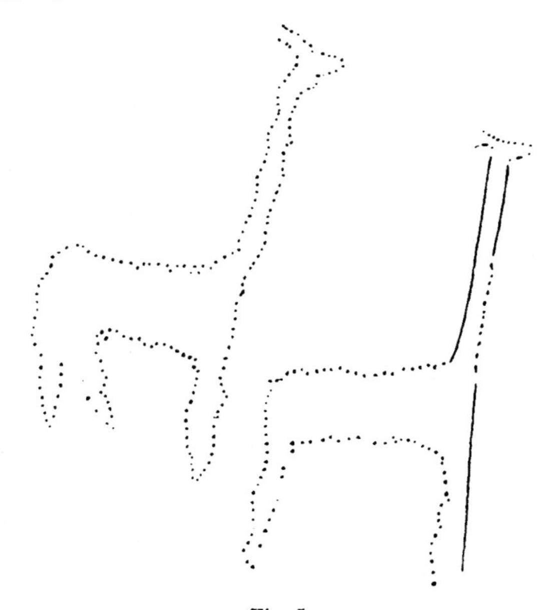

Fig. 5.
Prehistoric Rock-carvings of Giraffes. Lower Nubia.
From photographs by Professor Breasted.

which the giraffes in Fig. 5 have been drawn. These, in all probability, are the oldest representations of giraffes in the world, and by their clever obversation of motion also rank among the best ever made. They are the spontaneous productions of a primitive artist with a keen eye for observation and possessed of great power of expression.

Under the fifth dynasty (2750-2625 B. C.) Sahure continued the development of Egypt as the earliest known naval power in history. He dispatched a fleet on a voyage to Punt, as the Egyptians called the Somali coast at the south end of the Red Sea, and along the south side of the Gulf of Aden. From that region, which, like the whole east, he termed the God's Land, he obtained the fragrant gums and resins so much desired for incense and ointments.

One of the most important events of the reign of Queen Hatshepsut (eighteenth dynasty, about 1501-1480 B.C.) was a naval expedition to the land of Punt with the object to establish commercial relations with peoples of

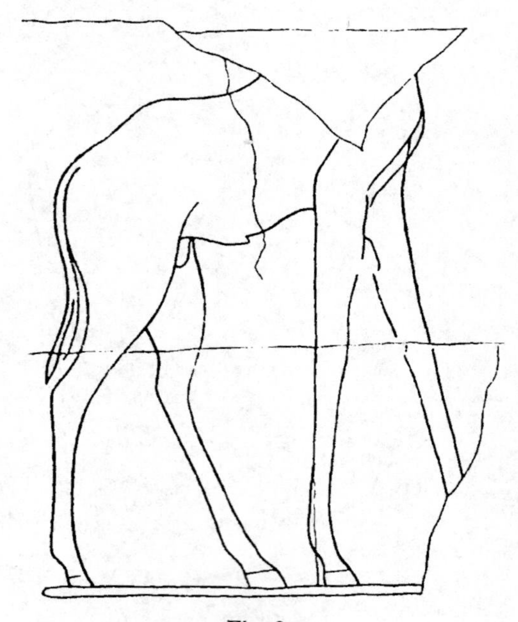

Fig. 6.
Giraffe from a Punt Scene at Der el-Bahri.
From a photograph.

what is now the Somali coast. A sculptured record of this peaceful expedition is preserved on the southern half of the wall stretching behind the middle colonnade of her temple at Der el-Bahri situated on the west side of the river at Thebes. In this procession the giraffe is well represented (Fig. 6), unfortunately mutilated; but even without its head it is a magnificent work of art, body and legs being exceedingly well modeled. According to E. Naville (The Temple of Deir El Bahari, p. 21. Egypt Exploration Fund,

XII, 1894), the giraffe is said to come from the country Khenthennofer, not from the coast. This region is generally distinguished from Punt; the two countries, however, were contiguous, but of somewhat wide and indefinite extent, Punt possessing a coast where vessels could land, while Khenthennofer was located in the mountainous interior. The two countries had a mixed population which included Negroes, and their products were almost identical. Ivory, live panthers, panther-skins, monkeys, gold, ebony,

Fig. 7.
Giraffe from the Presentation of Tribute to Tutenkhamon.
After Nina de Garis Davies.

and antimony were common to both. All these products being typically African, it is evident that Queen Hatshepsut's expedition had been directed to the east coast of Africa. Wealthy Egyptians were fond of keeping live specimens of the fauna of Punt like dogs, monkeys, panthers, leopards, and giraffes.

The illustration in Fig 7, showing a walking giraffe guided by a Nubian, forms part of the Presentation of Tribute to Tutenkhamon, depicted on the walls of the tomb of Huy, viceroy of Nubia under the reign of Tutenkhamon (compare Nina de Garis Davies and A. H. Gardiner, The Tomb of Huy, in The Theban Tombs Series, London, 1926). This tomb is situated high up on the eastern slope of the hill known as Kurnet Murrai which rises from the plain at a little distance north of Medinet Habu. On the west wall of the tomb are depicted scenes of Huy bringing the tribute of Nubia to the Pharaoh. Huy approaches the royal presence from the south, holding in his

Fig. 8.
Giraffes under Palm-trees from the Presentation of Tribute to Tutenkhamon. After Nina de Garis Davies.

left hand a crooked staff betokening his viceregal authority, and with the right waving the ostrich-feather fan which was his perogative as "fan-bearer at the right of the king." Tutenkhamon sits in state under his baldachin. Immediately behind the figure of Huy are shown choice samples of Nubian tribute. Gold in rings and "gold tied up" in bags are there, together with dishes of carnelian or red jasper and of a green mineral. There are tusks of white ivory and jet-black logs of ebony. A model chariot of gold is supported by an attendant Negro, perhaps of ebony, on a gold pedestal. Under the chariot appears to be a golden shrine. Heraldically arranged palm-trees, with monkeys climbing in their branches and giraffes nibbling at their leaves are shown in another scene (Fig. 8), together with kneeling Negroes in an attitude of adoration and with others holding cords attached to the necks of the giraffes. This scene is remarkable for its grace and exquisite realism. There are also Nubians carrying gold, skins, and giraffes' tails (the latter being painted black). Giraffes' tails are highly prized from Kordofan to Uganda (see above, p. 6 and below, p. 87). In an Egyptian story they figure among the presents given to a ship-wrecked sailor by his kindly host, the giant serpent.

The walking giraffe amid the tribute-bearers (Fig. 7) is a very young bull of the Nubian variety. It is light pinkish brown in color, with a few markings on the neck. The immaturity of the animal is denoted by the very slight development of the median horn.

The temples of Nubia contain many references to the Nubian wars of Ramses II (1292-25 B. C). Among the scenes cut on the rock side-walls of the excavated forecourt of the Bet el-Walli temple there is one portraying Ramses enthroned on the right; approaching from the left are two long lines of Negroes, bringing furniture of ebony and ivory, panther-hides, gold in large rings, bows, myrrh, shields, elephants' tusks, billets of ebony, ostrich feathers, ostrich eggs, live animals including monkeys, panthers, a giraffe, ibexes, a dog, oxen with curved horns, and an ostrich (Breasted, Ancient Records of Egypt, Vol. III, p. 203). The giraffe in this rock-carving is of naturalistic style, but is not quite so accurate and true to nature as in other Egyptian monuments. It is reproduced by Professor Breasted in *American Journal of Semitic Languages* (Vol. XXIII, 1906, p. 62).

Fig. 9, illustrating a giraffe with a monkey on its back, is from the tomb of Amunezeh (eighteenth dynasty) at Shekh Abd el-Gurna (compare Max W. Müller, Egyptological Researches, Vol. II, Carnegie Institution of Washington, 1910, p. 52 and colored reproductions in Plate 31). This is also from a series of wall-paintings representing

Fig. 9.
Giraffe with Baboon from the Tomb of Amunezeh.
After W. Max Müller.

tributes of the Nubians. The color of the animal is almost brown dotted with black spots. The hoofs are blue (intended for black). The monkey, probably a baboon, is green-blue with a red face and exaggerated long tail. The uplifted hand of the leader must have held a rope tied to the baboon, and he guides the giraffe by a rope fastened to its right fore leg. To the right of the animal the hieroglyph for the giraffe is added.

Two small green-glazed figurines of the Saitic or Ptolemaic epoch have been published and described by G. Daressy (Deux figurations de giraffe, Annales du Service des Antiquités de l'Egypte, Cairo, Vol. VII, 1906, pp. 61-63, 2 figs.). These represent figures of a headless man with what is explained as a giraffe crouching beside him. It is difficult, however, to recognize giraffes in these animals, as far as the illustrations published in the article are concerned. Crouching giraffes are not known from Egyptian monuments, and no clay figures of giraffes have become known from the Ptolemaic and Graeco-Roman periods.

Ptolemy II Philadelphus (285-247 B.C.) showed a live giraffe to the inhabitants of Alexandria in his triumphal procession through this city. In all periods of history Egypt continued to be the great distributing centre for giraffes, as will be seen in the chapters to follow. It supplied them to the Romans, the emperors of Byzance, the Arab Caliphs, to Spain and Italy in the middle ages, and to Italy, France, and England in more recent times.

REPRESENTATIONS OF THE GIRAFFE IN AFRICA OUTSIDE OF EGYPT

We made the acquaintance of the Bushmen as ostrich-hunters and artists depicting the ostrich (Leaflet 23). They were no less successful in producing rapid and vivid outline sketches of giraffes. At the time of the great artistic development of the Bushmen the whole fauna of South Africa was immensely rich and abounded in animals now extinct, like the oryx which frequented the plains of the Zwart Kei, the giraffe which abounded in the forests of Transval, buffalo, elephant, rhinoceros, hippopotamus, zebra, quagga, gnu, antelopes, and ostrich.

Fig. 10 represents a running giraffe cut in sandstone by the Bushmen in the Orange River Colony. G. W. Stow (Native Races of South Africa) mentions after Barrow a Bushman cave-drawing of a giraffe and writes that he found himself several drawings of it in the Zwart Kei and Tsomo caves, also in the Wittebergen of the Orange Free State. This, according to Stow, indubitably proves that the giraffe was found in the early days over a far wider area of country than at present. Stow also refers to a number of chippings, chiefly representations of animals at Pniel, among these the head and neck of a giraffe which is said to be remarkably fine, both on account of its large size and the correctness of its outline.

G. M. Theal holds that no giraffes have ever been seen by Europeans south of the Orange River, but that as profiles of them are found in Bushman paintings along the Zwart Kei and Tsomo Rivers, it is believed that they must once have existed there. It may be the case, however, that in their artistic efforts the Bus men did not confine themselves to the animals of their habitat, but may also have illustrated animals they encountered during their rovings over the country.

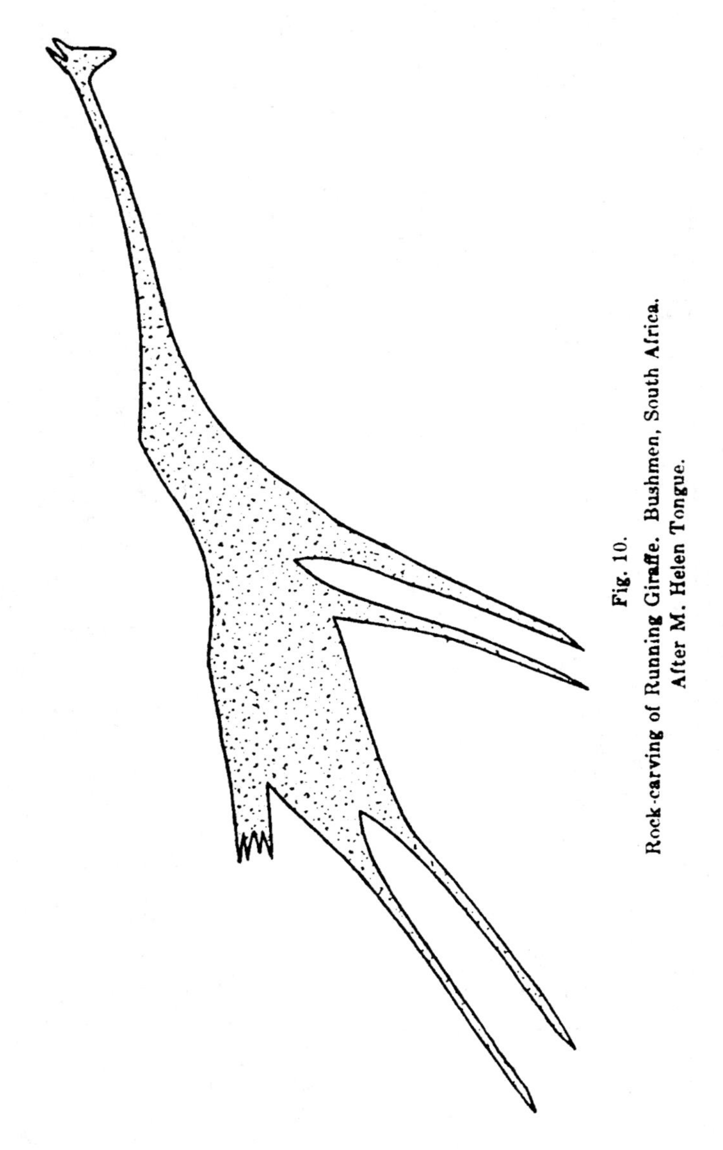

Fig. 10.
Rock-carving of Running Giraffe. Bushmen, South Africa.
After M. Helen Tongue.

Fig. 11.
Three Giraffes and Two Zebras.
Mural Painting in Colors in the Council-house of the Royal Palace at Gawiro, Southeast Africa.

In the folk-lore of the Hottentot the giraffe plays a prominent role.

A wall-painting from a council-room in the royal "palace" at Gaviro, Ubena, in Southeast Africa, shows three giraffes in company with two zebras (Fig. 11). While somewhat stiff and rather inexact in the shape of the body and legs, the movement and action of the animals are well observed, especially in the first, that bends its neck downward and touches one of the zebras, and in the third of which only the front part is represented.

Fig. 12.
Rock-engraving of Giraffe. Tuareg, Sahara.
After E. F. Gautier.

Fig. 12 illustrates a giraffe engraved in a rock in the Tuareg country in the Sahara. This station of rock-carvings among which camels, hunters on camel-back, and many other animals are found, was discovered by E. F. Gautier in 1903 (described by him in *L'Anthropologie*, 1904, p. 497). In his opinion, this picture bears all characteristics of a very great antiquity. The lines are deeply and profoundly cut. It is curious to find a representation of the giraffe in the desert area, where it has never occurred.

According to Gautier, the giraffe is the only animal in the art of Tuareg that does not belong to the fauna of the region, while all other animals do. This problem is not hard to solve, however. Considering the fact that live giraffes were traded by the Arabs to Mediterranean and Asiatic countries and that the commerce in giraffes goes back to the early relations between Egypt and Punt, giraffes could have been brought to Tuareg as well.

THE GIRAFFE AMONG ARABS AND PERSIANS

The giraffe was not known to the Hebrews at the time of Moses, as was formerly believed. This opinion was suggested by the Hebrew word *zamar* or *zemer*, which occurs in Deuteronomy (XIV, 5), and solely in this passage as one of the animals whose flesh was sanctioned by the Mosaic legislation. In the Seventy this Hebrew animal name has been translated into Greek as *kamelopardalis*, and the Vulgate gives *camelopardalus* as the corresponding Latin translation. Edward Topsell, author of "The Hisorie of Four-footed Beastes" (1607), writes that the "flesh of the giraffe is good for meat, and was allowed to the Jews by God himselfe for a cleane beast." J. Ogilby, in his work "Africa" (1607), commits a curious error by writing with reference to the giraffe, "Caesar first shewed him at Rome, though 'tis probable they formerly abounded in Judea, being a food prohibited to the Jews." There is no evidence whatever to the effect that the giraffe ever occurred in Palestine or anywhere in western Asia during historical times, nor is it safe to assume with Joly and Lavocat that Moses might have been acquainted with the animal from pictures on Egyptian monuments. A legislator permits or prohibits an animal known to his nation from real life, but hardly one merely known pictorially. Bochart, in his erudite folio on the animals of the Bible (Hierozoicon), has arrived at the conclusion that the ancient Hebrews were not acquainted with the giraffe, and explains *zamar* as a species of antelope, probably the chamois (*Antilope rupicapra*). "Chamois" was adopted by the English Version as rendering of *zamar*, but this, in all probability, is not correct either, for the chamois does not occur in Palestine. The general consensus of opinion now is that the "camelopardalis" of the Seventy rests on a mistranslation and that the animal intended by the Hebrew word is the wild goat or mountain sheep with curved horns. Professor J. M. Powis

Smith of the University of Chicago informs me, "The best rendering of *zamar* is 'mountain sheep.' The Seventy rendering, I take it, is a mere guess and a wild one at that. The word was probably unknown, and they took a free shot at it."

The Arabs made the acquaintance of the giraffe in Abyssinia at a comparatively late period. Their name for the animal, *zarāfa* or *zurāfa,* is supposed to be derived from Ethiopic *zarat.* In early Arabic poetry the animal is not mentioned, as it never occurred in Arabia.

Masudi, an eminent Arabic traveller and historian, who died in A.D. 956 or 957, writes that the giraffe generally lives in Nubia, but is not found in Abyssinia; there is no agreement as to the origin of the animal; some regard it as a variety of the camel, others assert that it has sprung from the union of the camel and the panther; others, again, hold that it is a distinct species like the horse, the donkey, and the ox, not, however, the product of a crossing like the mule. He emphasizes the giraffe's gentleness and the affection which it displays for the members of its family, and adds that in this species, in the same manner as among elephants, there are wild and tame individuals.

Ibn al-Faqih, an Arabic geographer from Hamadan in Persia, who wrote about A.D. 1022, gives the following account:—

"The giraffe lives in Nubia. It is said that it takes its place between the panther and the camel mare, that the panther mates with the latter who produces the giraffe. There are cases analogous to this one: thus the horse pairs with the ass, the wolf with the hyena, the panther with the lioness from whom the pard issues. The giraffe has the stature of the camel, the head of a stag, hoofs like those of cattle, and a tail like a bird. Its fore legs (literally, 'hands') have two callosities, while these are lacking in its hind legs. Its skin is panther-like and presents a marvellous sight. In Persia the animal is called 'camel-bull- panther' (*ushtur* or *shutur-gāw-palank*), because it has something in common with each of these three. Some scholars assert that the giraffe is generated by stallions of various kinds. This, however, is erroneous, for the horse does not impregnate the camel nor does the camel the cow."

Zakariyā al-Qazwini (1203-83), Arabic author of a cosmography and a work on historical geography, writes in his description of Abyssinia thus:—

"The giraffe is produced by the camel mare, the male hyena, and the wild cow. Its head is shaped like that of a stag, its horns like that of cattle, its legs like those of a nine year old camel, its hoofs like those of cattle, its tail like that of a gazelle; its neck is very long, its hands are long, and its feet are short. A scholar, Timat by name, relates that in the southern equatorial region animals of various kinds congregate during the summer around the cisterns, being driven there by heat and thirst; if an animal of a certain species covers one of another species, strange animals like the giraffe are born: the male hyena mates with the female Abyssinian camel; if the young one is a male and covers the wild cow, it will produce a giraffe."

In another passage Qazwini informs us that the giraffe has knees only in its fore legs, but no knees in its hind legs; in walking it advances its left hind leg first and then its right fore leg, contrary to the habit of all other quadrupeds which advance the right fore leg first and then the left hind leg. Among its natural qualities are affection and sociableness. As Allah knew that it would derive its sustenance from trees, He created its fore legs longer than its hind ones, to enable it to graze on them easily."

This theory of a mongrel origin of a giraffe was merely a popular belief suggested by the peculiar characteristics of the animal, but was not accepted by those who were able to think. An interesting instance to this effect is cited by Damiri (1344-1405) in his Zoological Dictionary (Hayat al-Hayawān, "Life of Animals"), who writes, "al-Jāhiz is not satisfied with this explanation and states that it is the outcome of sheer ignorance and emanates only from people who lack the faculty of discrimination; for God creates whatever He pleases. The giraffe, on the contrary, is a distinct species of animal, independent (sui generis) like the horse or the ass. This is proved by the fact that it is able to produce one like itself, a fact which has been ascertained by observation." Masudi, as mentioned, says also that many regard the giraffe as a particular species, not as the result of any cross-breed.

Dimashki, who wrote a Cosmography about A.D. 1325, commits an odd error by localizing the giraffe in Ceylon (Serendib), but gives a correct description of it. "It is an animal of a remarkable shape," he writes, "it has a neck like a camel, a skin like a leopard and stag, horns like an antelope, teeth like a cow, a head like a camel, and a back like a cock. Its fore legs, as well as its neck, are very long; it measures ten ells and more in height. Its hind legs are very short and without articulation. Only its front legs have knees as among other animals, because the neck is too short in proportion with its fore legs when it grazes on the ground. In walking it sets its right foot ahead and its left foot behind, in distinction from other quadrupeds. It has a gentle disposition, and is sociable toward its companions. It belongs to the ruminants, and its ordure is like that of camels."

Makrizi (1365-1442), in his History of the Mamluk Sultans of Egypt, reports that in the year 1292 a female giraffe in the Castle of the Hill (at Cairo) gave birth to a young one, which was nursed by a cow. This was regarded as an auspicious event which is recorded by three other Arab chroniclers.

The Arabs, like most Oriental nations, paid much attention to dreams, and developed a pseudo-science of divination based on dreams. Thus the appearance of a giraffe in a dream is interpreted by Damiri as follows: "A giraffe seen in a dream indicates a financial calamity. Sometimes it signifies a respectable or a beautiful woman, or the receipt of strange news to come from the direction from which the

CHINESE PAINTING OF A GIRAFFE OF THE YEAR 1485 (p. 47).
In Collections of Field Museum. Blackstone Expedition to China, 1908.

animal is seen. There is, however, no good in the news. When a giraffe appears in a dream to enter a country or town, no gain is to be obtained from it, for it augurs a calamity to your property; there is no guaranty for the safety of a friend, a spouse, or a wife whom you may want to take through your homestead. A giraffe in a dream may sometimes be interpreted to mean a wife who is not faithful to her husband, because in the shape of its back it differs from the riding-beasts."

The flesh of the giraffe is consumed by the Arab hunters of Abyssinia. The long tendons of the legs are highly prized by the Arabs and used like thread for sewing leather, also for guitar strings. The Arab tribes Fazoql and Bertat make shields of giraffe-hide.

The Arabs were the most active dealers in giraffes and traded the animals to the Mediterranean countries as well as to Persia, India, and China. Masudi, in the tenth century, informs us that giraffes were sent as presents from Nubia to the kings of Persia, as in later days they were offered to Arab princes, to the first Caliphs of the house of Abbas and the governors of Egypt.

When Egypt was a province of the Caliphate (A.D. 641-868), Nubia was invaded by the Emir Abdallāh Ibn Sad, and a treaty was concluded in A.D. 652, compelling the Nubians to pay an annual tribute consisting of four hundred slaves, a number of camels, two elephants, and two giraffes. During the reign of the Caliph al-Mahdi (A.D. 775-785) it was ordered again that Nubia be held responsible every year for three hundred and sixty slaves and one giraffe. This tribute was paid for two centuries when it was repudiated in A.D. 854, but this revolt was soon crushed. In 1275, under the rule of the Mamluks, the Sudan was annexed by Egypt, and three giraffes, three elephants, panthers, dromedaries, and oxen were stipulated among the annual tribute.

El-Aziz (A.D. 975-996), a Caliph of the Fatimid empire of Egypt, a bold hunter and a fearless general, was fond of rare animals, and had many strange animals and birds brought to Cairo. Female elephants, which the Nubians had carefully reserved, were at length introduced for breeding under his reign, and a stuffed rhinoceros delighted the crowd. On the occasion of a solemn festival celebrated by the Caliph in A.D. 990, elephants and a giraffe were conducted in front of him, and several giraffes marched before the Caliph on other occasions. Gold vases with figures of giraffes, elephants, and other animals were made for him, also gold statuettes of giraffes and elephants.

Beybars (1260-77), the real founder of the Mamluk empire in Egypt, a native of Kipchak (between the Caspian and the Ural Mountains) and possessor of untold wealth, sent in 1262 giraffes, together with Arab horses, dromedaries, mules, wild asses, apes, parrots, and many other gifts to his ally, the Khan of the Golden Horde.

Ruy Gonzalez de Clavijo, a Spanish knight, who went as ambassador to the court of Timur at Samarkand in the years 1403-06, tells the following interesting story:—

When the ambassadors arrived in the city of Khoi [in the province of Azerbeijan, Persia], they found in it an ambassador, whom the Sultan of Babylon had sent to Timur Beg; who had with him as many as twenty horses and fifteen camels, laden with presents, which the Sultan of Babylon [probably an ambassador from Cairo] sent to Timur Beg. He also had six rare birds, and a beast called *jornufa* (giraffe), which creature is made with the body as large as that of a horse, a very long neck, and the fore legs much longer than the hind ones. Its hoofs are like those of a bullock. From the nail of the hoof to the shoulder it measured sixteen *palmos;* and when it wished to stretch its neck, it raised it so high that it was wonderful; and its neck was slender, like that of a stag. The hind legs were so short, in comparison with the fore legs, that a man who had never seen it before, might well believe that it was seated, although it was standing up; and the buttocks were worn, like those of a buffalo. The belly was white, and the body was of a golden color, surrounded by large white rings. The face was like that of a stag, and on the forehead it had a large projection, the eyes were large and round, and the ears like those of a horse. Near the ears it had two small round horns, covered with hair, which looked like those of a very young stag. The neck was long, and could be raised so high, that it could reach up to eat from the top of a very high wall; and it could reach up to eat the leaves from the top of a very lofty tree, which it did plenteously. To a man who had never seen such an animal before, it was a wonderful sight."

The giraffe which Clavijo observed and described had been sent to Timur in the year 1402 soon after the battle of Angora by the Mamluk Sultan Faraj of Egypt, who dispatched two ambassadors to his court with rich presents, among these a giraffe.

In the History of Timur Begh or Tamerlan written in Persian by Sherefeddin Ali of Yezd in the fifteenth century the presentation of a giraffe is mentioned. When Timur in 1414 celebrated the marriage of his grandchildren, an envoy from the sovereign of Egypt arrived, and had an audience with the emperor, bringing presents of minted silver, precious stones, sumptuous textiles, and among other curiosities a giraffe, which the Persian chronicler writes is one of the rarest animals of the earth, and nine ostriches, of the largest of Africa.

Josafa Barbaro and Ambrogio Contarini, Venetian travellers, saw in 1471 a live giraffe at the court of Persia, and describe it in the old English translation of W. Thomas as follows: "After this was brought forth a Giraffa, which they called Girnaffa [the Italian original in Ramusio has: Zirapha which they also call Zirnapha or Giraffa], a beast as long legged as a great horse, or rather more; but the hinder legs are half a foot shorter than the former, and is cloven footed as an ox, in maner of a violet color mingled all over with black spots, great and small according to their places: the belly white somewhat long haired, thin haired on the tail as an ass, little horns like a goat, and the neck more than a pace long: the tongue a yard long, violet and round as an eele, with the which he grazeth or eateth the leaves from the trees so swiftly that it is scarcely to be perceived. He is headed like a hart, but more finely, with the which standing on the ground he will reach fifteen foot high. His breast is broader than the horse, but the croup narrow like an ass; he seemeth to be a marvellous fair beast, but not like to bear any burden."

The name *surnāpa* or *zurnāpa* for the giraffe is regarded as peculiar to Persian, but it was heard and recorded by P. Belon at Cairo toward the middle of the sixteenth century and a little later by Moryson at Constantinople (cf. pp. 67, 84). This goes to show that the word *surnāpa* was also employed in the colloquial Osmanli and Arabic of the sixteenth century. Yule regards it as a form curiously divergent of *zarāfa*, perhaps nearer the original. A popular Persian etymology analyzes the word into *zurnā* ("hautboy") and *pā* ("foot"), in allusion to the long and thin legs of the giraffe ("having legs shaped like an hautboy"),—assuredly a far-fetched and artificial explanation. Possibly this form may have originated in Ethiopia, presenting a compound of *zur* and Ethiopic *nabun* pointed out by Pliny. Bochart derives this *nabun* from *naba* ("to be elevated").

A very curious picture of a giraffe by a Persian artist is reproduced in Plate II. It is contained in the Manafi-i-Hayawān ("Description of Animals"), an illustrated Persian bestiary of eighty-five folios, completed between the years A.D. 1295 and 1300 and now preserved in the Pierpont Morgan Library of New York. I am under obligation to Miss Belle Da Costa Greene, director of the library, for kindly placing a photograph of the giraffe picture at my disposal. A brief description of this beautiful manuscript has been given by C. Anet (*Burlington Magazine*, 1913, pp. 224, 261) with reproductions of some fine selected specimens of the illustrations, but not of the giraffe which is reproduced here for the first time. The text is a Persian translation of an earlier Arabic manuscript made at the command of Ghazan Khan, a descendant of the Mongol rulers of Persia. In the opinion of C. Anet, the animals of this Persian album are of the highest order, convey an idea of what may be called the primitive period of Persian painting, and show a magnificent originality and a force in style and drawing.

The interesting feature of the Persian painting is that it represents not merely a giraffe in general, but apparently depicts a now well-known particular species, the so-called reticulated giraffe (Fig. 1 on p. 4), which inhabits the Somali country and is chestnut-colored, covered with a network of white lines. The net-work is treated as more or less regular hexagons, but the artist has reproduced the appearance of the characteristic markings of this species quite correctly, as comparison with Fig. 1 will show. Head, neck and body are correctly outlined in general; only the jointless fore legs are stiff. A collar with eight small bells is hung around the animal's neck. Each of its feet appears to be manacled to impede its free motion. It is placed in a surrounding of graceful shrubbery tenanted by three birds. The leaves reach the animal's head, and in this manner the artist has apparently intended to convey a good idea of its extraordinary height.

The picture is accompanied by the following text in Persian which translated is as follows: "This animal is called *shutur-gāw-palank* [see above, p. 32], for the reason that every part or member of it exhibits similarity to a corresponding part of one of these three animals. Its hands (fore legs) and neck are like those of a camel, its skin is like that of a leopard, its teeth and hoofs are like those of an ox. It has long hands (fore legs) and short feet (hind legs). Only its hind legs are provided with knees, not its fore legs. Its head and tail are like those of a deer. Its young ones are said to start eating grass when they put their heads out of their mother's womb. They eat grass until satisfied.

Then the young ones return into their habitation (the womb). When they are severed from the mother, they will run away immediately, for the mother has a rough and flying tongue. When she licks the young one, its flesh and skin will come off, so that it will not approach the mother for three or four days." The statement in regard to the hind legs having knees is a curious inversion of what the Arabs say (above, p. 33).

Colonel Roosevelt (Life-histories of African Game Animals) describes the reticulated giraffe as follows: "The reticulated giraffe is marked on the neck by distinct reticulations, formed by the large rufous squares being set off sharply by narrow lines of white ground-color. This color pattern is so distinctive from the usual blotched coloration of other giraffes that the race has been considered a distinct species by many naturalists. Some specimens of the Uganda giraffe, however, show as narrow reticulations, but the ground-color is seldom so whitish in appearance. The horns of the bull are well developed, the frontal horn being especially large, and is exceeded in height only by the Uganda race. The body is marked by large squares of rufous separated by ochraceous reticulations, and differs decidely from the small size and broken-edged spots of the Masai giraffe. The legs from the knees and hocks downward nearly as far as the fetlocks are reticulated by buffy-whitish ground-color and tawny blotches. One of the distinctive color marks of this race is the carrying forward of the reticulated pattern of the neck over the cheeks and the upper throat to the chin. The mandible shows distinctive characters, being low at the condyles, and having short coronoid processes. The frontal horn is remarkably robust and of great circumference, and is scarcely less in height than in the Uganda race; but the skull itself at this point is much less in height."

CHINESE PAINTING OF A GIRAFFE WITH TWO ARAB GUIDES, FIFTEENTH CENTURY (p. 48).

A. W. Bahr Collection.

THE GIRAFFE IN CHINESE RECORDS AND ART

The giraffe was not known to the ancient Chinese, contrary to what is assumed by certain sinologues. This erroneous conclusion is based on the fact that when live giraffes were first transported into China in the fifteenth century under the Ming dynasty, they were taken by the Chinese for the Kilin (*k'i-lin*), a fabulous creature of ancient mythology, and by way of reminiscence and poetic retrospection received the name *k'i-lin*. This, of course, does not mean that the ancient native conception of the Kilin was based on the giraffe, which in historical times was confined to Africa. In fact, neither the description nor the illustrations of the Kilin bear the slightest resemblance to a giraffe. The Kilin is said to have the body of a deer, the tail of an ox, a single horn, and to be covered with fish-scales. Its horn is covered with flesh, indicating that while able for war, it covets peace. It does not tread on any living thing, not even on living grass. It symbolizes gentleness, goodness, and benevolence. It is said to have appeared just previous to the death of Confucius, and it will appear whenever a benevolent sovereign rules; it was a mythical animal of good omen. The Kilin has a horn with a fleshy basis or fleshy horns, while the giraffe has two bony excrescences on its head which merely resemble horns, but are not. De Groot (see note on p. 96) insists on the good and gentle disposition being ascribed to either creature, but it is obvious that a zoological identification cannot be based on alleged psychological traits; many deer, sheep, and other animals may likewise be characterized in this manner. It is singular that De Groot remained entirely ignorant of the importations of giraffes into China and of what Chinese authors know about the subject.

It is clear that the characteristic features of the giraffe which impress every casual observer—the extraordinary height, the long neck, the proportion of fore and hind legs—

are not found in the Chinese descriptions of the Kilin and that several traits of the latter do not agree with the giraffe. Thus, the voice of the Kilin resembles the sound of a bell, and it walks with regular steps. The giraffe, however, has no voice at all. "It is an interesting fact that giraffes are absolutely mute, and even in their death-agonies never utter a sound" (Hutchinson's Animals of All Countries). Says G. Renshaw, "Giraffes are well hnown to be silent animals. I once heard the Southern giraffe still living in the London Zoo give a kind of coughing sneeze—the only recorded occasion, I believe, of these animals ever having been known to make *any noise at all!* It was, however, probably caused by some irritant in the nasal passage, and cannot be called a *vocal* sound."

The only points of resemblance made by the Chinese between the Kilin and the giraffe are their bodies being shaped like a deer, their tails being like that of an ox, and their gentle disposition. This identification, it should be borne in mind, was established as recently as the fifteenth century when the first live giraffes arrived in China.

The *Sü po wu chi*, a book compiled by Li Shi about the middle of the twelfth century, apparently contains one of the earliest Chinese literary allusions to the giraffe. "The country Po-pa-li [Berbera, on the Somali coast of the Gulf of Aden] harbors a strange animal called camel-ox (*t'o niu*). Its skin is like that of a leopard, its hoof is similar to that of an ox, but the animal is devoid of a hump. Its neck is nine feet long, and its body is over ten feet high."

The designation "camel-ox" corresponds exactly to a Persian designation of the giraffe, *ushtur-gāw* (*ushtur*, "camel"; *gāw*, "ox, cow"), mentioned as early as the tenth century by the Arabic writer Masudi. It may hence be inferred that the information received in regard to the animal had come to China from Persia.

The second reference to the giraffe is made by Chao Ju-kwa in his work *Chu fan chi*, written in A.D. 1225. This author was collector of customs in the port of Ts'üan-chou fu in the province of Fu-kien, where he came in close contact with Arabian merchants and representatives of other foreign nations who then entertained a lucrative commerce with China. From oral information given him by foreign traders and from earlier Chinese sources he compiled his brief book. In his notes on the Berbera or Somali coast of East Africa he mentions as a native of that country "a wild animal called *tsu-la,* which resembles a camel in shape, an ox in size, and is yellow of color. Its fore legs are five feet long, while its hind feet are only three feet in length. Its head is high and looks upward. Its skin is an inch thick." The word *tsu-la* used in the Chinese text is not Chinese, but is of Arabic origin; it is intended to reproduce *zurāfa,* the Arabic term for the giraffe.

African animals were transported to China as early as the thirteenth century under the Yüan or Mongol dynasty. We are informed, for instance, in the Annals of this dynasty that in the year 1287 an envoy from Mabar (Malabar, on the south-west coast of India) presented the emporer with "a strange animal resembling a mule, but larger and covered with hair mottled black and white; it was called *a-t'a-pi.*" Judging from this name, the beast appears to be identical with the topi, the Swahili name for the Topi damaliscus (*Damaliscus jimila*), a kind of antelope peculiar to East Africa, also called bastard hartebeest (see, further, note on p. 96).

In A.D. 1289 the Chinese emperor was presented with two zebras from Mabar, and in the following year another envoy arrived from the same country and offered two piebald oxen, a buffalo, and a tiger-cat. The giraffe, as far as I know, is not mentioned in the Yüan Annals, although there is no reason why it should not have come along with topi and zebra. Malabar, at that time, was in close commercial relations with the ports of southern Arabia, and it was the Arabs who brought these live animals from the Somali coast to southern Arabia and thence transhipped them to India.

There are in the Chinese Annals several records of giraffes being sent alive as gifts to the Chinese emperors during the fifteenth century. In that period a new impetus was given to the exploration of the countries of the Indian Ocean through the exploits of Cheng Ho, eunuch and navigator. In A.D. 1408 and 1412 he conducted, with a fleet of sixty-two ships, naval expeditions to the realms of southeastern Asia, advancing as far as Ceylon, and inducing many states to send envoys back with him to his native country. In 1415 and again in 1421 he returned with the foreign envoys to their countries in order to open trading relations with them. In 1424 he was sent to Sumatra. In 1425, as no envoys had come to Peking, he and his old lieutenant, Wang King-hung, visited seventeen countries, including Hormuz in the Persian Gulf. This was at a time when no European sail had yet been sighted on the Indian Ocean.

In A.D. 1414 (the twelfth year of the period Yung-lo, under the emperor Ch'eng Tsu), Saifud-din, king of Bengal, sent envoys to China with an offering of giraffes and famous horses. The Board of Rites asked permission of the emperor to present an address of congratulation. As the giraffe was termed *k'i-lin,* and the fabulous *k'i-lin* of antiquity was reputed to appear only at the time of a virtuous ruler, the giraffe was obviously regarded as an auspicious omen, and the proposed address of congratulation was chiefly intended as a flattery to the sovereign, who had sense enough to see through the game and denied the request.

In A.D. 1415 the country Ma-lin (Malindi in British East Africa) offered a giraffe to the emperor. On this occasion the President of the Board of Rites, Lü Chen, made a report to the throne, requesting that the officials should offer congratulations to the emperor; the request, however, was denied again.

In the year 1421 the chamberlain Chou travelled for the purpose of purchasing giraffes, lions, and other rare animals, rather to satisfy his own vanity than to make a contribution to knowledge.

In the year 1422 an imperial envoy, the eunuch Li, was sent to Aden with a letter and presents to the king. On his arrival he was honorably received, and on landing was met by the king and conducted by him to his palace. During the sojourn of the embassy, the people who had rarities were permitted to offer them for sale. Cat's-eyes of extraordinary size, rubies, and other precious stones, large branches of coral, amber, and attar of roses were among the articles purchased. Giraffes, lions, zebras, leopards, ostriches, and white pigeons were also offered for sale. An account of this expedition was written by Ma Huan, a Chinese Mohammedan familiar with the Arabic language. He was attached to the suite of Cheng Ho on his cruise in the Indian Ocean, and published on his return (between 1425 and 1432) an interesting geographical work (*Ying yai sheng lan*) in which the twenty countries visited by the expedition are described. With reference to Aden he remarks that the giraffe is found there; it was, of course, not a native of Aden, either at that time or at present, but was transported there by the Arabs from the east coast of Africa. Ma Huan describes the animal "as having fore legs nine feet high and hind legs about six feet; its head is raised, and its neck is sixteen feet long [this, in fact, is the total height of the animal from head to foot]; owing to its fore quarters being high and its hind quarters low it cannot be ridden; it has two short, fleshy horns close to its ears; its tail is like that of a cow, and its body like that of a deer; its hoof is divided into three sections; its mouth is wide and flat, and it feeds on millet, beans, and flour cakes." The last remark shows that the question is of giraffes kept in captivity and receiving cereal food from the hands of men. It appears that a regular trade was carried on by the Arabs in these animals who aroused so much curiosity and that Aden was the centre of this commercial activity.

In the year 1430 Cheng Ho dispatched one of his companions to Calicut in southern India. Having heard that a trading vessel was to sail from that port to Arabia, he commanded this officer to embark and take Chinese goods as presents for the native ruler along. The voyage lasted a year. The Chinese envoy purchased there fine pearls, precious stones, a giraffe (*k'i-lin*), a lion, and an ostrich.

In 1431 giraffes were sent as tribute by embassies from "the countries of the Southern Sea."

Fei Sin, who in 1436 wrote the *Sing ch'a sheng lan*, an account of four voyages made in the Indian Ocean by imperial envoys during the first quarter of the fifteenth century, mentions giraffes under the name *tsu-la-fa* (Arabic *zurāfa*) among the natural products of Arabia, particularly of Zufar on the south coast of the peninsula. He observes that "the ruler of the country and his ministers are very grateful to the Heavenly Dynasty [that is, China], and that their missions are constantly bringing presents of lions and giraffes to offer as tribute."

A noteworthy point is that the giraffes were not sent to China over the land route, as the ostriches, but were conveyed in ships over the maritime route from Aden by way of India. It is a pity that we have no detailed story as to how the animals were transported, for their transportation is a difficult problem even at the present time. Giraffes are very nervous and hence very awkward animals to transport, as they are liable to break their necks by suddenly twisting about in their travelling boxes. It is still more deplorable that the Chinese have not preserved a record of how the animals were cared for in their country, how long they lived, etc.

From an account in the *Wu tsa tsu*, written in 1610, it appears that under the reign of Ch'eng Tsu (1403-25) a painter was directed to make a sketch of a Kilin which had been captured; the artist's picture showed the animal's body shaped like that of a deer, but its neck was very long, conveying the impression that it was three to four feet in

GIRAFFE AT CAIRO, WITH TWO ARAB GUIDES (p. 83).
Drawn by André Thevet, Middle of Sixteenth Century.
After G. Loisel.

length. As at that time giraffes were brought to China, it is possible that they served as models for this picture of a Kilin.

Fig. 13 is a woodcut reproduced, after A. C. Moule, from a Chinese book, entitled "Pictures of Birds and Beasts of Foreign Lands" (*I yü k'in shou t'u*), a copy of which is preserved in the University Library of Cambridge and which may have originated about or after 1420. The animal is designated in the engraving as *k'i-lin;* it is equipped with a headstall, and is guided by a bare-headed foreigner clad only with a skirt. There is a little stump between the animal's ears; the spots are represented by short lines. On

Fig. 13.
Giraffe Guided by a Mohammedan.
Drawing from a Chinese Book of about 1420.
After A. C. Moule.

the whole the artist seems to have endeavored to reproduce the general appearance of a deer; the neck is comparatively too short, the body is not correctly outlined, but the tail is fairly correct.

A Chinese painting representing a giraffe is reproduced in Plate III. It was obtained by me at Si-an fu in 1908. It is a long paper scroll dyed a deep black from which the picture, of circular shape (eleven inches in diameter) is set off in a light brown color. The giraffe is surprisingly well done, the shape of the head with two horns and the outlines of the body are well caught, while no attempt is made at delineating the markings of the skin. The animal is shown freely in nature, surrounded by trees and brushwork,—a unique conception which, as far as I know, does not occur elsewhere.

The picture is inscribed at the top with a stanza of four lines, the characters being neatly written in gold ink. The poem is characterized as an "imperial composition" (*yü ch'i*). It reads,—

> With the tail of an ox and the body of a deer, the animal is seen walking through the wilderness.
> Auspicious clouds are facing the sun, and the prosperity of the government is clearly in evidence.
> The people will meet with great success, and there will be a year of abundant harvest.
> There will be plenty of food, and with songs they will praise the great peace.

Although the animal is not named, it results from the characteristics ("tail of an ox and body of a deer") that the Kilin is implicitly understood. Like the Kilin, the giraffe is considered an auspicious omen, presaging a prosperous government, a good harvest, abundance of victuals for all, and a peaceful reign. The poem, on its left side, is provided with a date which corresponds to our year 1485, and this may also be the date of the picture; or the latter may be somewhat earlier, and the poem was added to it in 1485; at any rate, the picture is a production of the fifteenth century, the age of the importation of giraffes.

The Chinese painting of a giraffe, reproduced in Plate IV, is of an entirely different character. It was obtained in China by Mr. A. W. Bahr, who kindly placed it at my disposal. The picture is painted on old silk, the surface of which is much disintegrated, measuring 54 x 33½ inches. It is not signed or sealed, or in any way inscribed. The giraffe is of imposing size, and the unknown Chinese artist has with remarkable effect brought out its height in comparison with its two Arab guides. The animal is provided with a green headstall, and the neck is adorned with a tassel of horse-hair dyed red and surmounted by metal-work. This tassel is of Chinese make, and was attached to the animal on its arrival in China. Horses and mules are still decorated with such tassels. The almost regular designs of hexagons covering the body allow the inference that this animal is intended to represent the reticulated species which has been described above with reference to a Persian miniature (p. 39). The two turbaned and bearded Arabs are clad in long, red, girdled gowns and high boots, and are types full of character. Each holds the end of a halter in both his hands. This picture is doubtless a production of the Ming period, and very probably of the fifteenth century.

C. R. Eastman, who in 1917 published this painting in *Nature*, advanced the theory that it had been copied in China from models brought over from Persia, as in his judgment it bears a striking resemblance to the Persian miniature in Plate II. This entire speculation decidedly misses the mark. The two pictures, as every one may convince himself from the reproductions here published, have but one point in common,—the design of hexagons on the skins of the animals. This is simply due to the fact that the Persian and Chinese artists independently endeavored to sketch the same species, a reticulated giraffe. For the rest, their productions in style, composition, and spirit are fundamentally different; the pose and the equipment of the animals are wholly at variance. Mr. Eastman is ignorant of the history of the giraffe in Persia and China, and knows nothing of the numerous importations of live giraffes into both countries. He invents a comfortable theory to suit his convenience, and insinuates to Chinese painters a working method which they never followed. Nothing is known of Persian animal paintings imported into China and copied there, but we know as a fact that the Chinese were always fond of exotic animals and that their artists were in the habit of portraying them, either voluntarily or by imperial command.

It was customary with the Chinese emperors to have unusual animals which were presented by foreign potentates painted or even sculptured by their court artists. To cite only two specific instances which occurred during the Ming period,—a black horse with a white forehead and white feet was offered to the emperor in 1439 by Ulug Beg Mirza, chief of Samarkand and eldest son of Shah Rukh, son of Timur. The emperor ordered a picture of it to be made. In 1490 an envoy from Samarkand, together with an embassy from Turfan, arrived to present a lion and a karakal. When the envoys had reached the province of Kan-su, pictures were taken of these beasts and forwarded by a courier to the emperor. The ministers proposed to decline these presents, but the emperor overruled them and accepted the gift.

For this reason I am convinced also that the Chinese paintings of giraffes of the fifteenth century were done from nature, from study of the live animals sent as gifts to the imperial court. The situation then was exactly the same in China as in contemporaneous Italy. It is indeed a curious coincidence that in the fifteenth century also live giraffes found their way into Italy and engaged the attention of Italian artists, as is set forth in the chapter "The Giraffe in the Age of the Renaissance." Here again there is no mysterious coeval connection between Chinese and Italian or between Italian and Persian artists. The art of all countries creates new forms at all times from the observation of nature. The activity of the Arabs supplied giraffes to Europe as well as to Persia, India, and China, but the interesting fact remains that the fifteenth century was the great age of the giraffe both in the East and West.

It seems that the importations of giraffes into China were restricted just to the fifteenth century and ceased thereafter. During the sixteenth century and under the Manchu dynasty we hear nothing of giraffes being introduced into the country. Through a curious force of circumstances the animal was brought again to the attention of the Chinese in the latter part of the seventeenth century.

This revival is due to the early Jesuit missionaries who endeavored to acquaint their new disciples with the methods and results of European science and who successfully diffused among them knowledge of geography, chronology, mathematics, physics, astronomy, and technology. In the course of the seventeenth and eighteenth centuries these indefatigable workers produced a remarkable literature both in Chinese and Manchu, which exerted no small degree of influence on the thought of Chinese scholarship. He who is eager to understand the intellectual development of Chinese society during that epoch cannot afford to neglect the literary efforts of those humble and enterprising pioneers. One of them, Ferdinand Verbiest (1623-88), who came to China in 1659, published about 1683 a small geographical work in Chinese, entitled *K'un yü t'u shwo,* which among other matters also contains illustrations with brief descriptions of some foreign animals. Eleven of these pictures have been reproduced in the great cyclopaedia *T'u shu tsi ch'eng,* published in 1726, and this series includes the giraffe (Fig. 14). The accompanying text runs thus: "West of Libya there is the country Abyssinia which produces an animal called *u-na-si-yo.* Its head is shaped like that of a horse; its fore feet are as long as those of a big horse, while its hind feet are short. Its neck is long; from the hoofs of the fore feet up to the head it is over twenty-five feet in height. Its skin is variegated in color. It is fed on hay and grass, and is shown in gardens to people as a curiosity. It turns round to show off its beauty to spectators, as though enjoying being looked at."

The source of Verbiest's illustration is Edward Topsell's "Historie of Foure-footed Beastes" (London, 1607). Topsell's picture of the giraffe reproduced in Fig. 18 (p. 68), as stated by himself, was drawn by Melchior Luorigus at Constantinople in the year of salvation 1559, and was afterwards sent to Germany, where it was imprinted at Nuremberg. A comparison of the two figures will show their close interrelation: the animal in outline and pose is identical

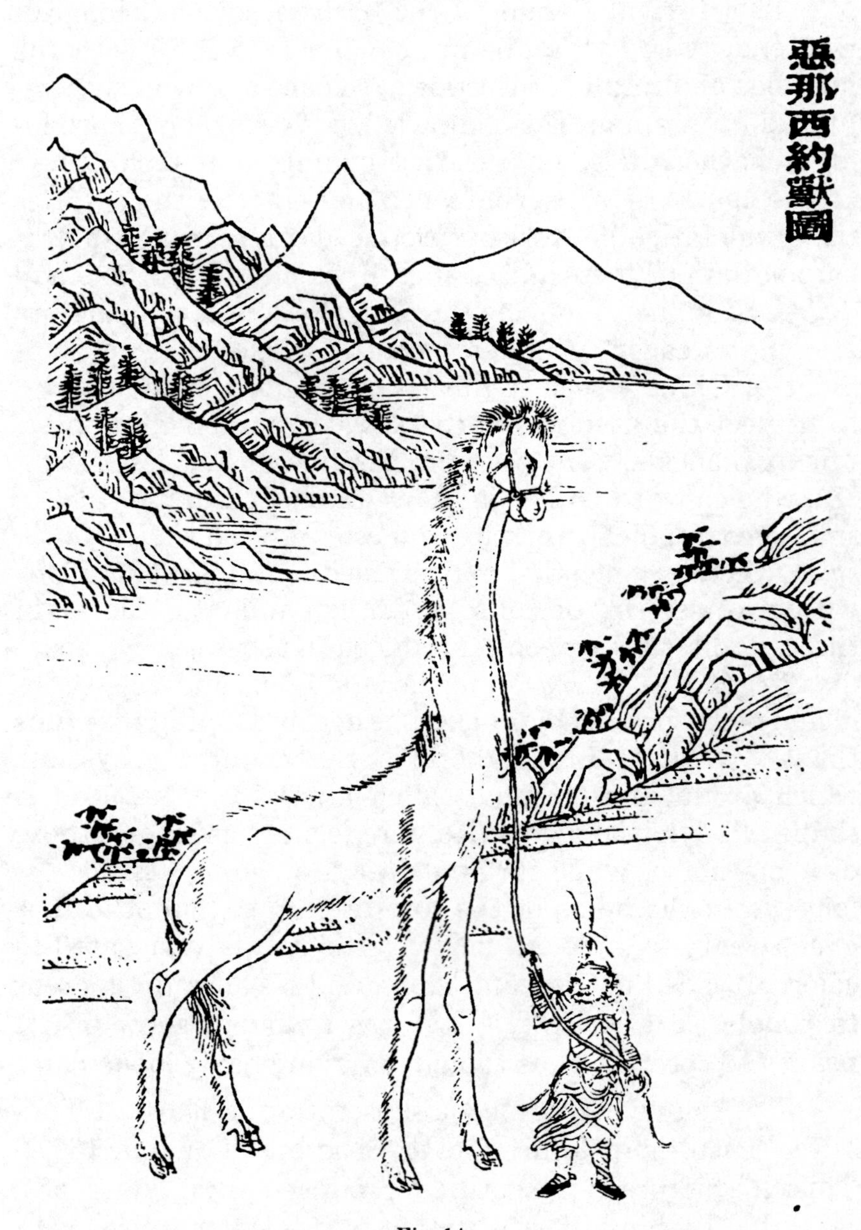

Fig. 14.
Chinese Woodcut of Giraffe Supplied by Ferdinand Verbiest.
From T'u shu tsi ch'eng.

in both, the Arab's head-dress has been changed into a cockade of two feathers in the Chinese engraving, and a landscape of Chinese style has been added to the latter. Verbiest has also drawn on Topsell's description. "When any come to see them, they willingly and of their own accorde, turne themselves round as it were of purpose to shewe their soft haires, and beautifull coulour, being as it were proud to ravish the eies of the beholders." This is the idea expressed by Verbiest in his concluding sentence. A similar observation was made by Vincent de Beauvais (p. 71).

Topsell's influence is also visible in Verbiest's nomenclature, for the curious word *u-na-si-yo* coined by him is not traceable to any African or Oriental language. Topsell, enumerating the Arabic, Chaldaean, Persian, Greek and Latin names of the animal, says that Albertus adds the names Oraflus (hence the older French orafle) and Orasius (cf. p. 72). The latter was chosen by Verbiest and analyzed into *o-ra-si-o*; as there is no equivalent for *ra* in Chinese, he substituted the syllable *na*, and may have felt that he was the more justified in so doing, as Topsell offers an alleged Chaldaean word Ana.

The foreign word *u-na-si-yo*, introduced by Verbiest and only used by him, has never been adopted by the Chinese; but it is noteworthy that the Manchu coined from it a word for giraffe in the form *unasu*. This is contained in the *Ts'ing wen pu hui*, a Manchu-Chinese dictionary compiled in 1786. The Manchu word *unasu* is here explained by a Chinese gloss "*u-na-si-yo*, a strange animal from the country Ya-bi-si (Abyssinia)," briefly characterized with the words of Verbiest. Verbiest's term *u-na-si-yo* has nothing to do with the onager, the wild ass of Central Asia, as has been suggested by Sakharof and Moule.

To cite another example of how Verbiest made use of Topsell's data,—he gives the illustration of a beaver, an animal unknown in China, under the name *pan-ti*, which for a long time was a puzzle to me, as it defies identification with any name for the beaver in Europe and elsewhere. Verbiest's picture is copied again from Topsell, who gives *Canis ponticus* as the beaver's Latin name, so that the Chinese rendering *pan-ti* is doubtless based on *ponticus*. Verbiest's *hu-lo* transcribes Latin gulo, the glutton; his animal *su*, which occurs in Chile in South America, is the Opossum described by Topsell (p. 660) as a "wild beast in the new-found world called Su." This native American name, together with the figure of the animal, was derived by him from A. Thevet's account of Brazil.

The Japanese call the giraffe *hyōda* ("panther-camel") or *kirin* (corresponding to Chinese *k'i-lin*).

GIRAFFE ON A PORTUGUESE COTTON PRINT, EIGHTEENTH CENTURY (p. 87).
In Art Institute, Chicago.

THE GIRAFFE IN INDIA

It has been pointed out in the preceding chapter that, according to Chinese records, giraffes were sent to China in A.D. 1414 by Saifud-din, king of Bengal, and that other African animals like topi and zebra were shipped to China from the kingdom of Malabar as early as the thirteenth century. It is therefore credible that, as H. Schiltberger reports about 1430, giraffes were found at Delhi. He calls them *surnasa* (for *surnafa*) and describes them as being "like a stag, but a tall animal with a long neck, four fathoms in length or longer." These African animals were transported to India by Arabs from the Somali coast by way of the ports of southern Arabia.

India has played a singular role in the historical records of the giraffe. To many ancient and mediaeval writers India was a rather vague notion, and was correlated with Ethiopia or confounded with other countries. Several ancient authors, as mentioned (p. 58), designated India as the home of the giraffe. During the middle ages a distinction was made between India the Greater and India the Lesser (India maior et minor), but there was little concord as to their identity and boundaries, and Abyssinia was termed Middle India. According to a Byzantine chronicle, the emperor Anastasius in A.D. 439 received as a gift from India an elephant and two animals called "camelopardalas." There is no doubt that "India" in this case must be equalized with Ethiopia. Cassianus Bassus, author of a work on agriculture (*Geoponica*, seventh century A.D.), narrates that he saw at Antiochia a camelopard which he says had been brought from India. "India," again, must be understood here as Ethiopia.

André Thevet (Cosmographie universelle, Vol. I, fol. 388b, 1575) was the champion of the strange idea that the habitat of the giraffe was India. He even specifies it "in the high mountains of Cangipu, Plumaticq and Caragan which are in interior India beyond the river Ganges, some five degrees on this side of the tropic of the cancer." From there and several other localities giraffes were brought to an island which he calls Isle Amiadine or Anchédine, and where they were kept by the lords of the country for their pleasure. The Turks found six giraffes there, seized them and forcibly loaded them on their vessels; two of the animals died during the voyage, two others died when embarked at Aden, the two survivors landed safely at Cairo, where Thevet saw them during his three months' stay (compare below, p. 83). There is no doubt that owing to his ignorance of Arabic Thevet misunderstood his informants or interpreters, who he says were "Abyssinians and other Africans." He denies expressly the occurrence of the giraffe in Ethiopia, adding that if it is found there at the courts of the kings and princes, it was transported into that country from India.

Edward Topsell, in his "Historie of Foure-footed Beastes" (1607), defines the distribution of the giraffe thus: "These beastes are plentifull in Ethiopia, India, and the Georgian region, which was once called Media. Likewise in the province of Abasia in India, it is called Surnosa, and in Abasia Surnappa." Abasia, as will be seen (p. 74), is Marco Polo's designation of Abyssinia, and as Abyssinia was comprised under the term Middle India, the confusion with India proper arose in Topsell's mind, or was already contained in the source which he may have consulted.

F. Bernier, who travelled in the Mogul empire during the years 1656-68, reports that he saw at the court of the emperor Aureng-Zeb the skin of a zebra which ambassadors from the king of Ethiopia had brought along. The zebra was alive when it left Africa, but died during the voyage, and the ambassadors had sense enough to preserve its skin. Bernier describes it as "a small species of mule: no tiger is so beautifully marked, and no striped silken stuff is more finely and variously streaked." In view of the fact that India maintained considerable r d with Guendar or Gondar, formerly capital of the Amharic kingdom of Abyssinia, it is quite possible that giraffes also came from there directly to India.

THE GIRAFFE AMONG THE ANCIENTS

The giraffe, being a strictly African animal, remained unknown to the civilizations of Western Asia in ancient times. In the period of the independence of Hellas the Greeks were not acquainted with it. Aristotle, the only great zoologist of antiquity, does not describe it. It has been supposed that the hippardion or pardion mentioned by Aristotle (Historia animalium II, 1) as having "a thin mane extending from the head to the withers," without further particulars, may be the giraffe, but this is highly improbable; at any rate, the evidence for such an identification is insufficient. In the epoch of Hellenism when the geographical horizon had widened and when giraffes were transmitted from Egypt to Rome, we meet the first description of them in late Greek and Roman authors. There is, accordingly, no representation of the animal in Greek art, nor is it found on antique coins or engraved gems.

In 46 B.C. the first giraffe arrived in Rome, and marched in Caesar's triumphal procession; it was subsequently shown in the circus games held by Caesar. This event caused a great sensation, and is referred to by Varro, Horace, Dio Cassius, and Pliny.

Ten giraffes appeared in the circus of Rome in A.D. 247 under the emperor Gordianus III to take part in the celebration of the first millennium that had elapsed since the foundation of Rome. This was the largest number of live giraffes ever brought together at any time. Giraffes were also in the possession of the emperor Aurelianus (A.D. 270–275). In A.D. 274, when he celebrated his triumph over Zenobia, queen of Palmyra, several giraffes appeared in the circus games.

In regard to the habitat of the animal the notions of the ancients were vague. Some authors like Pausanias, Bassus, and others locate it in India; Artemidorus ascribes

GIRAFFE GUIDED BY AFRICAN NATIVE.
Photograph by Courtesy of Carl Hagenbeck.

it to Arabia, Agatharchides to the country of the Troglodytes; Pliny and Heliodorus place its home in Ethiopia.

Agatharchides of Cnidus, a Greek historian and geographer, who lived under Ptolemy Philometor (181-146 B.C.), is the author of a geographical treatise on the Red Sea, which has not been preserved, but extracts of which have been handed down by Diodorus (II, 51) and Photius. "The animals called camelopardalis by the Greeks," Agatharchides relates, "present a mixture of both the animals comprehended in this appellation. In size they are smaller than camels, but shorter in the neck; as to their head and the disposition of their eyes they are somewhat like a pard (*pardalis*). In the curvature of the back again they have some resemblance to the camel, but in color and growth of hair they are like pards (leopards). In like manner, as they have a long tail, they typify the nature of this animal."

Strabo (XVI. 4, 16) describes the giraffe after Artemidorus, a geographer and traveller from Ephesus (about 100 B.C.) as follows:—

"Camelopards are bred in these parts, but they do not in any respect resemble leopards, for their variegated skin is more like the streaked and spotted skin of fallow deer. The hinder quarters are so very much lower than the fore quarters, that it seems as if the animal were sitting upon its rump. It has the height of an ox; the fore legs are as long as those of the camel. The neck rises high and straight up, but the head greatly exceeds in height that of the camel. From this want of proportion, the speed of the animal is not so great, I think, as it is described by Artemidorus, according to whom it cannot be overtaken. It is, however, not a wild animal, but rather like a domesticated beast; for it shows no signs of a savage disposition."

Dio Cassius, in his Roman History (XLII), alludes to the fact that the camelopardalis was introduced into Rome by Caesar for the first time and exhibited to all. He describes the animal "as being like a camel in all respects, except that its legs are not all of the same length, the hind legs being the shorter. Beginning from the rump it grows gradually higher, which gives it the appearance of mounting some elevation; and towering high aloft, it supports the rest of its body on its front legs and lifts its neck in turn to an unusual height. Its skin is spotted like a leopard, and for this reason it bears the joint name of both animals." This plain and clear notice is doubtless based on a personal experience with the giraffe.

In the same manner as the ostrich was believed to resemble the camel (Leaflet 23, p. 24), Pliny (VIII, 27) recognized an affinity of the camel with the giraffe. He describes it under the name cameleopardus and locates it correctly in Ethiopia, where, he says, it is called *nabun*. "It has a neck like that of a horse, feet and legs like those of an ox, a head like that of a camel, and is covered with white spots upon a red ground; hence it has been styled cameleopard. It was first seen at Rome in the circus games held by Caesar, the Dictator. Since that time it has been occasionally seen again. It is more remarkable for the singularity of its appearance than for its fierceness; for this reason it has obtained the name of the wild sheep." Indeed, the giraffe was called in Latin also *ovis fera* ("wild sheep").

Horace (Epistles II, 1) reproaches his fellow-citizens for the pleasure they take in the circus games, and on this occasion paraphrases the name Camelopardalis:—

> Si foret in terris, rideret Democritus, seu
> Diversum confusa genus panthera camelo,
> Sive elephas albus vulgi converteret ora.
>
> "Democritus, if he were still on earth,
> would deride a throng gazing with open
> mouth at a beast half camel, half panther,
> or at a white elephant."

C. Julius Solinus (Collectanea rerum memorabilium, 30, 19) mentions the giraffe, but merely copies Pliny.

The poem *Kynegetika* ("The Hunt"), ascribed to the poet Oppianus (second century A.D.), but written by a poet from Apamea, contains a remarkably good description of the giraffe (III, 461; ed. of P. Boudreaux, p. 119). "Muse! May thy sonorous and harmonious voice sing also of the animals of mixed nature formed by a combination of two different races among which the leopard with speckled back is united with the camel. Father Jupiter, what magnificence shines in thy numerous works! What an abundant variety is revealed in plants, quadrupeds, and marine mammals! How many gifts didst thou bestow on the mortals! Thou whose power has clothed with the leopard's robe this species of camel embellished with the richest colors,—noble and charming animals tamed by man without effort! They have a long neck, their body is sprinkled with various spots; short ears crown their heads devoid of hair in the upper part. Their legs are long, and their feet are large, but these limbs are unequal in size. The fore legs are much more elevated than those behind which are considerably shorter. The lame have such legs. From the middle of the head of these animals issue two horns which are not of the nature of ordinary horns; their soft points surrounded by hair rise on the temples and close to the ears. This species, like deer, has a small mouth slightly split and provided with small teeth as white as silk. Its eyes are vividly lustrous, and its tail, as short as that of a gazelle, is furnished with a tuft of black hair at the end."

Oppianus is the first author who mentions the horns of the giraffe, but curiously enough he does not mention its name.

Heliodorus from Emesa, bishop of Trikka, who lived in the third or fourth century A.D., has given the most detailed description of the animal, which is embodied in his romance The Ethiopics (Aethiopica X, 27). The envoys of the Axiomites of Abyssinia presented a giraffe to the king. "These also presented gifts among which, besides other things, there was a certain species of animal, of nature both extraordinary and wonderful. In size it approached that of a camel, but the surface of its skin was marked with flower-like spots. Its hind parts and the flanks were low, and like those of a lion, but the shoulders, fore legs, and chest were much higher in proportion than the other limbs. His neck was slender, towering up from his large body into a swan-like neck. His head, like that of a camel, was about twice as large as that of a Libyan ostrich. His eyes were very bright and rolled with a fierce expression. His gait also was different from that of every other land or water animal, for his legs were not moved alternately but by pairs, those on the right side being moved together, and then, in like manner, those on the left together, one side at a time being raised before the other, so that in walking he always had one side dangling. For the rest he was so tame and gentle in disposition that his master led him wherever he pleased solely by a small cord fastened around his neck, and he followed him wherever he wanted, as though he were attached to him by means of a very large and strong fetter. At the appearance of this creature the multitude was struck with astonishment, and its form suggesting a name, it received from the populace, from the most prominent features of its body resembling a camel and a leopard, the improvised name of camelopardalis."

When the sacrificial animals at the altars of Helios and Selene (the Sun and Moon) got sight of the odd beast, a stampede ensued; four white horses and a pair of bulls were terrified as if they had beheld some phantom, freed themselves, and galloped wildly away.

Heliodorus' description is picturesque and fairly accurate, save the remark about the fierce glances of the animal, and is apparently based on direct observation. It is noteworthy that he is the first who comments on the amble of the giraffe (see above, p. 8).

A giraffe (reproduced in Fig. 15) is painted as a decoration on the wall of a mortuary vault (columbarium) of the Villa Pamfili at Rome. The animal is conducted by a young guide by means of a long bridle and carries a bell (*tintinnabulum*) around its neck, a symbol of its tameness. On the other side of the man there is an antelope. The original has been destroyed, but a copy of the picture is preserved in Munich.

Two giraffes are represented in a mosaic now preserved in the palace Barberini of Palestrina (the ancient Praeneste, 21 miles from Rome). They are shown grazing and browsing (Fig. 16).

This mosaic was discovered in 1640 and purchased by Cardinal Barberini, who caused a careful drawing to be made of it, and then had it removed to Rome for repairs before having it relaid in his palace at Palestrina. It is said to have formed the pavement of part of the Temple of

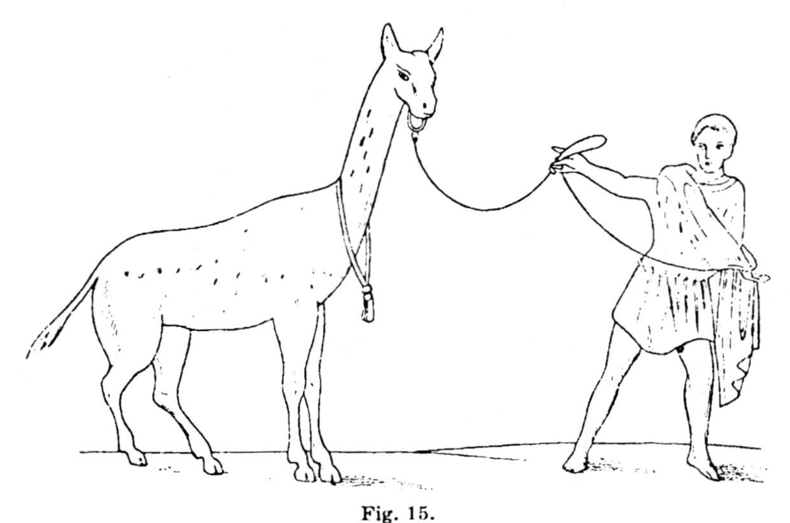

Fig. 15.
Roman Mural Painting of a Giraffe with Guide.
After Daremberg and Saglio.

Fortune at Praeneste, but this view is contested by S. Reinach. The upper portion of the composition illustrates animals of the Egyptian Sudan; they show a striking resemblance to those of the tomb of Marissa.

In the Necropolis of Marissa in Palestine there is in one of the tombs a painted frieze of animals of Graeco-Egyptian style, among these, in the opinion of the discoverers of the tomb, "what is evidently intended for a giraffe" (J. P. Peters and H. Thiersch, Painted Tombs in the Necropolis of Marissa, p. 25. Palestine Exploration Fund,

London, 1905). They describe it as follows: "The neck is very long, but the head, with its rounded ears and large, prominent eye, is much too big. The hind quarters and tail

Fig. 16.
Giraffes in the Mosaic of Palestrina.
After S. Reinach, Répertoire de Peintures.

are those of the deer, the fore legs are as long as the hind legs, and the withers actually lower than the rump. The spotted skin is represented by little black and red spots.

Fig. 17.
Giraffe (?) from Painted Tomb at Marissa, Palestine.
After Peters and Thiersch.

The title above it seems to read: Kamelopardalos." If the latter statement were correct, there would be no doubt of the artist's intention, but in the colored plate (VII) repro-

FOUR GIRAFFES IN HAGENBECK'S KRAAL AT THE FOOT OF THE KILIMANJARO, AFRICA.
Photograph by Courtesy of Carl Hagenbeck.

ducing this portion of the frieze I cannot recognize such a name. Be this as it may, the drawing itself is clumsy and rather represents a deer with a somewhat long neck, without any peculiar characteristics of a giraffe. The animal was probably known to the painter only from hearsay accounts (Fig. 17).

The ancients have not done justice to the giraffe, and have not produced any really artistic representation of it.

THE GIRAFFE AT CONSTANTINOPLE

Menageries were established at Constantinople during the eleventh century when Constantinus IX received an elephant and a giraffe from the Sultan of Egypt. These animals were repeatedly shown in the theatre of Byzance and marvelled at as wonders of nature. The Greeks were passionately fond of circus games and combats of ferocious beasts. The capture of Constantinople through the crusaders in 1203 and the subsequent pillage of the city undoubtedly led to the destruction of the amphitheatre which is no longer mentioned after that date. Notwithstanding, the Byzantine emperors continued to keep exotic animals. In 1257 Michael Paleologus received from the king of Ethiopia a giraffe which he paraded for several days through the streets of the city for the diversion of the Byzantines. This event was regarded as of sufficient importance that Pachymerus, the contemporaneous chronicler of the reign of Michael, took the opportunity of inserting in his work a detailed description of the animal. He emphasizes its gentle disposition and writes that it is so tame that it allows even children to play with it; it lives on grass, but also likes bread and barley no less than a sheep.

Philostorgius (A.D. 364-424), author of an ecclesiastic history (III, 11), speaks of the animals which had come from Ethiopia to Constantinople, and mentions drawings representing giraffes which he had seen at Constantinople himself. He gives a very brief description of the animal, comparing it with a large stag. According to Gyllius, author of a Topography of Constantinople, there were in that city stone statues of giraffes publicly exhibited, together with those of unicorns, tigers, and vultures, but they have since disappeared. It appears from these data that the giraffe must have played a certain role in Byzantine pictorial and plastic art.

The menagerie of Constantinople was visited and described by Pierre Belon in 1546, but no giraffe is mentioned by him. Thirty years later the menagerie was enriched by a giraffe which took part in the festivities occasioned by the circumcision of Mahomet III. Baudier (Histoire générale du Serrail, Lyons, 1659) attended these festivities, and describes a giraffe exhibited on this occasion in the hippodrome. He makes the curious statement that its fore legs are four or five times higher than the hind legs. When conducted through the streets, he says, its head reached into the windows of the houses.

English travellers made the acquaintance of the giraffe at Constantinople. This accounts for the fact that the first English picture of the animal was secured by way of Constantinople.

Fig. 18 is a reproduction of the giraffe inserted in Edward Topsell's "Historie of Foure-footed Beastes," published in London, 1607. In regard to the source of his illustration, Topsell gives the following information: "The latter picture here set down was truely taken by Melchior Luorigus at Constantinople, in the yeare of salvation 1559. By the sight of one of these, sent to the great Turke for a present: which picture and discription, was afterwarde sent into Germany, and was imprinted at Norimberge."

Fynes Moryson, author of the History of Ireland, offers in his "Itinerary" (1597) the following story:—

"Here (at Constantinople) be the ruines of a pallace upon the very wals of the city, called the palace of Constantine, wherein I did see an elephant, called *philo* by the Turkes, and another beast newly brought out of Affricke (the mother of monsters), which beast is altogether unknowne in our parts, and is called *surnapa* by the people of Asia, *astanapa* by others, and *giraffa* by the Italians, the picture whereof I remember to have seene in the mappes of Mercator; and because the beast is very rare, I will describe his forme as well as I can. His haire is red coloured, with many blacke and white spots; I could scarce reach with the points of my fingers to the hinder part of his backe, which grew higher and higher towards his fore-

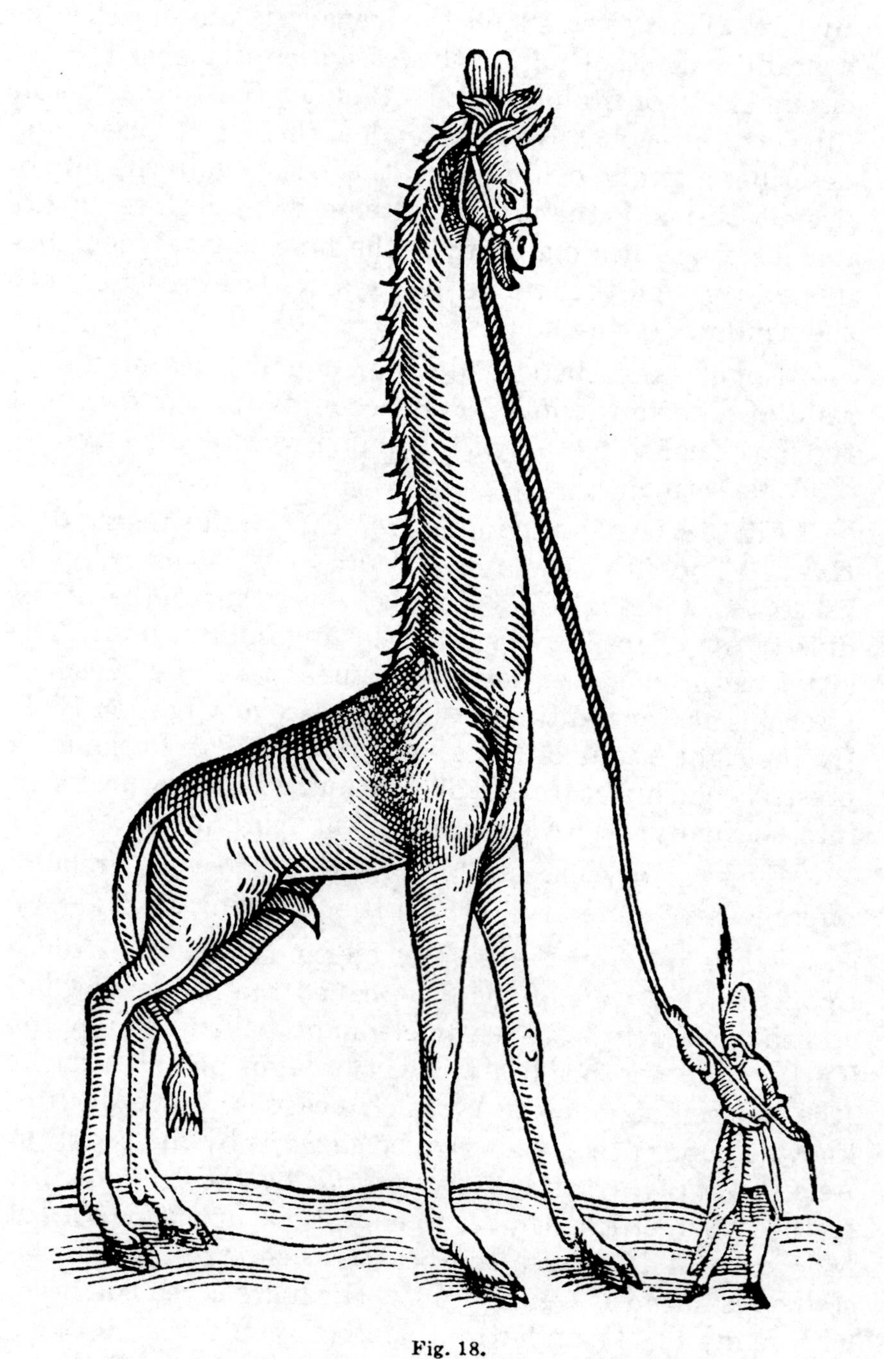

Fig. 18.
Giraffe from E. Topsell's Historie of Foure-footed Beastes (1607).
Drawn in 1559 by Melchior Luorigus at Constantinople.

shoulder, and his necke was thinne and some three els long. So as hee easily turned his head in a moment to any part or corner of the roome wherein he stood, putting it over the beams thereof, being built like a barne, and high for the Turkish building, not unlike the building of Italy, by reason whereof he many times put his nose in my necke, when I thought myselfe furthest distant from him, which familiarity of his I liked not; and howsoever the keepers assured me he would not hurt me yet I avoided these his familiar kisses as much as I could. His body was slender, not greater, but much higher then the body of a stagge or hart, and his head and face was like to that of a stagge, but the head was lesse and the face more beautifull: he had two hornes, but short and scarce halfe a foote long; and in the forehead he had two bunches of flesh, his ears and feete like an ox, and his legges like a stagge."

Of the oriental words given by Moryson, his *philo* for elephant is Turkish, which is derived from Persian *pil* (Aramaic *pil*, Arabic *fil*). His word *surnapa* for the giraffe is Persian *surnāpa* or *zurnāpa*.

John Sanderson, a London merchant, visited Constantinople about the year 1600, and thus relates his impressions at the first sight of a giraffe:—

"The admirablest and fairest beast that ever I saw was a jarraff, as tame as a domesticall deere, and of a reddish deere colour, white brested and cloven footed: he was of a very great height, his fore-legs longer then the hinder, a very long necke, and headed like a camell, except two stumps of horne on his head. This fairest animall was sent out of Ethiopia, to this great Turkes father for a present; two Turkes the keepers of him, would make him kneele, but not before any Christian for any money."

THE GIRAFFE DURING THE MIDDLE AGES

After the fall of the Roman Empire the giraffe remained unknown in most parts of Europe for about a thousand years. Even that small sum of knowledge which the late Greeks and Romans possessed of the animal was lost during that period, and the few mediaeval writers who refer to it are content to quote Solinus; thus Isidorus of Seville (Etymologiarum libri XX, XII, 19, and Origines XII, 2), who wrote about A.D. 636, and who confounds the camelopard with the chameleon and for the rest copies Solinus, and likewise Rabanus Maurus (De universo VIII B), abbot of Fulda and archbishop of Mayence (about A.D. 844).

A new impetus to knowledge was received from the Arabs after their conquest of Spain. The Arabs were fond of animals, and an animal park belonged to the essentials of every Muslim court. When Abderrahman III (A.D. 912–961) in A.D. 936 founded the city Zahra, one mile north of Cordova, in Spain, he established there a garden where rare animals and birds were kept in cages and fenced enclosures. This was the first zoological garden in Europe.

In southern Europe the first great menageries were installed at the court of Frederick II (1212-50), king of the Two Sicilies. This prince, born in Sicily, rather Italian than German, had inherited from his Neapolitan mother a taste for oriental manners and a veritable passion for animals. He made a study of birds, especially those used for the chase, observed them, even dissected them, and wrote a treatise on ornithology. He had an elephant sent to him from India, and he presented to the Sultan of Egypt a white bear in exchange for a giraffe. At Palermo, his usual residence, he created a sort of zoological garden which has been described by Otto von St. Blasio. Frederick was on such good terms with the Muslims that his tolerance gave rise to suspicions of his orthodoxy. He was in correspondence with the Arab philosopher Ibn Sabin. An Arab historian confesses that "the emperor was the most excellent among the kings of the Franks, devoted to science, philosophy, and medicine, and well-disposed toward Muslims."

In 1261 a giraffe was presented to Manfred, a son of Frederick, by the Sultan Beybars (above, p. 36).

It was accordingly the Arabs who acquainted European nations with the live giraffe. This fact is also borne out by our word for the animal, which is derived from the Arabic *zarāfa* or *zurāfa*. The old Spanish form *azorafa* has even preserved the Arabic article *al* (*al-zarāfa*). In modern Spanish and Portuguese it is *girafa*, in French *girafe* (older French *orafle* or *girafle*), Italian *giraffa*. During the middle ages it was sometimes identified with *seraph*: thus E. Topsell (Historie of Four-footed Beastes, 1607) still gives the Arabic name as *Sarapha*, and B. von Breydenbach's picture of the animal is inscribed *seraffa* (p. 76). In Purchas (Pilgrims) the form *ziraph* occurs. Yule thinks it is not impossible that seraph, in its Biblical use, may be radically connected with the giraffe, but this hypothesis is very improbable.

Vincent de Beauvais, author of the Speculum naturale (thirteenth century) refers to the giraffe in three different chapters of his work under three different names, without noticing that these names apply to the same animal. First, he describes it under the name Anabulla (evidently based on Pliny's Ethiopic word *nabun*) as having the neck of a horse, feet and legs of a bull, the head of a camel, and a skin pale red and white in color. Second, he mentions it as camelopardus, copying Solinus or Isidorus. Finally he describes it under the name Orasius, saying that in his time it had been transmitted to the emperor Frederick by the Sultan of the Babylonians. He remarks that the animal seems not to be ignorant of its own beauty, for when it sees people standing around, it turns completely so that it may be admired from every side, for nature has ornamented it with finer colors than all other beasts.

Albertus Magnus (1193-1280), in his work De quadrupedibus (XXII, 2, 1) mentions the giraffe twice, under the name Anabula and again under that of Camelopardulus, without recognizing the identity of the two. He gives Seraph as Arabic and Italian name, and writes that the skin, on account of its decoration, is sold at a high price; he also mentions the giraffe of Frederick II. Neither Vincent nor Albertus alludes to the horns.

The Latinized form *oraflus* (hence older French *orafle*) is distilled from old Spanish *azorafa*, and the form *orasius* occurring in Vincent de Beauvais and Albertus Magnus is due to a misreading of *f* (*ſ*) for *s* (*ſ*), which letters were very similar in ancient manuscripts and printed books.

Fig. 19.
Cameleopardus (Alleged Giraffe).
From the Dialogus Creaturarum Moralisatus (1486).

The climax of all these confusions was finally reached by the creation of a picture of the Camelopardus reconstructed entirely on the basis of mediaeval literary notices and bearing no resemblance whatever to a giraffe. The animal shown in Fig. 19 is reproduced from the Dialogus creaturarum moralisatus, a collection of moralizing animal fables published in Dutch (Gouda, 1480, 1481, 1483, and Antwerp, 1486) and translated into English under the title "The Dialogues of the Creatures Moralized" (London, 1813, with the animal pictures). Our illustration is based on a photograph taken from an original edition of the work in the University Library of Leiden. The text begins,

"Cameleopardus is an animal which has a hoof like a camel, a neck like a horse, feet and legs like a buffalo, and many spots as the animal pardus has on its body." Then follows a conversation of this fictitious creature with Christ, which is not of interest in this connection. A similar fantastic creature accompanies the early editions of Sir John Maundeville's Travels as an illustration of the giraffe (p. 75).

In contrast with this crude ignorance there are a few mediaeval travellers who had occasion to see giraffes and wrote of them somewhat sensibly. Cosmas, a Christian monk from Alexandria, called Indicopleustes ("the Indian Navigator"), in the course of his travels, visited Ethiopia

Fig. 20.
Camelopardalis of Cosmas Indicopleustes.
After J. W. McCrindle, Christian Topography of Cosmas.

about A.D. 525, and in book XI of his "Christian Topography" (written about A.D. 547) gives a brief description of the animals of the country. The giraffe is thus treated by him under the name Camelopardalis: "Camelopards are found only in Ethiopia. They also are wild creatures and undomesticated. In the palace [in the capital Axum] they have one or two that, by command of the king [Elesboas], have been caught when young and tamed to make a show for the king's amusement. When milk or water to drink is set before these creatures in a pan, as is done in the king's presence, they cannot, by reason of the great length of their legs and the height of their chest and neck, stoop down to the earth and drink, unless by straddling with their fore legs. They must therefore, it is plain, in order to drink, stand with their fore legs wide apart. This animal also I have delineated from my personal knowledge of it." Like Herodotus of old, Cosmas was ever athirst after knowledge and possessed of some skill in drawing; he took much delight in covering his manuscript with sketches illustrative of what he had observed, especially types of people and animals. His giraffe, reproduced in Fig. 20, may be designated as a fairly correct outline of the animal.

A giraffe (*orafle*) of crystal as a gift of the Old Man of the Mountain to the king of France is mentioned by Jean Sire de Joinville (Histoire de Saint Louis, written between 1304 and 1309).

Marco Polo alludes to giraffes in three passages of his famous narrative,— for Madagascar, the island of Zanghibar (that is, the country of the Negroes), and for Abyssinia. Polo never visited Madagascar, and his hearsay account of the island contains many errors, among these the giraffe which never occurred in Madagascar and does not occur there. The interesting point, however, is that Polo is the first who recognized a wider geographical distribution of the giraffe and looked for it beyond the limits of Abyssinia to which all former travellers had confined it. With reference to Zanghibar he informs us,—

"They have also many giraffes. This is a beautiful creature, and I must give you a description of it. Its body is short and somewhat sloped to the rear, for its hind legs are short, while the fore legs and the neck are both very long, and thus its head stands about three paces from the ground. The head is small, and the animal is not at all mischievous. Its color is all red and white in round spots, and it is really a beautiful creature."

In the Latin and French versions the animal's name is spelled *graffa*; in Ramusio's Italian version, *giraffa*. Abyssinia is called by Polo Abash (Italian spelling: Abascia;

Latin: Abasia), based on Arabic Habash. He writes that giraffes are produced in the country.

The knight, Wilhelm von Bodensele, whose itinerary was written in 1336 at the request of the Cardinal Talleyrand de Perigord, saw a giraffe at Cairo, calling it *geraffan*.

The earliest notice of the giraffe in English literature occurs in the Travels of Sir John Maundeville of St. Albans (chap. 94), written about the year 1356:—

"In Araby is a kynde of beast that some men call Garsantes [giraffes], that is a fayre beast, and he is hyer than a great courser or a stead [steed], but his neck is nere XX cubytes long, and his crop and his taile lyke a hart and he may loke over a high house." The numerous manuscripts of Maundeville's Travels, owing to the great popularity of the book (scarcely two copies agree to any extent), show many divergences, and in some of them giraffes under the name orafles are ascribed to Chinese Tartary, with the addition, "There also ben many Bestes, that ben clept Orafles. In Arabye, thei ben clept Gerfauntz, that is a Best pomelée or spotted."

As is well known, Maundeville is a fictitious person, and the book going under his name was compiled by a physician of Liège from various sources.

The first printed illustration of a half-way realistic giraffe (Fig. 21) is found in the Peregrinationes in Terram Sanctam ("Peregrinations into the Holy Land") by Bernhard von Breydenbach, dean of Mayence. This work was first published in the same city in 1486, and represents the first illustrated account of a pilgrimage undertaken into the Holy Land in 1483-84, that contains views of places seen en route from Venice to Mount Sinai and drawn by Breydenbach's companion, the painter Erhard Reuwich. The animals sketched by him are the giraffe, inscribed Seraffa, crocodile, rhinoceros, capre de India ("Indian goat"), unicorn (a horse with narwhal's tusk), camel, salamander (gecko), and a great ape of unknown name (Simia sylvanus), accompanied by the statement that "these ani-

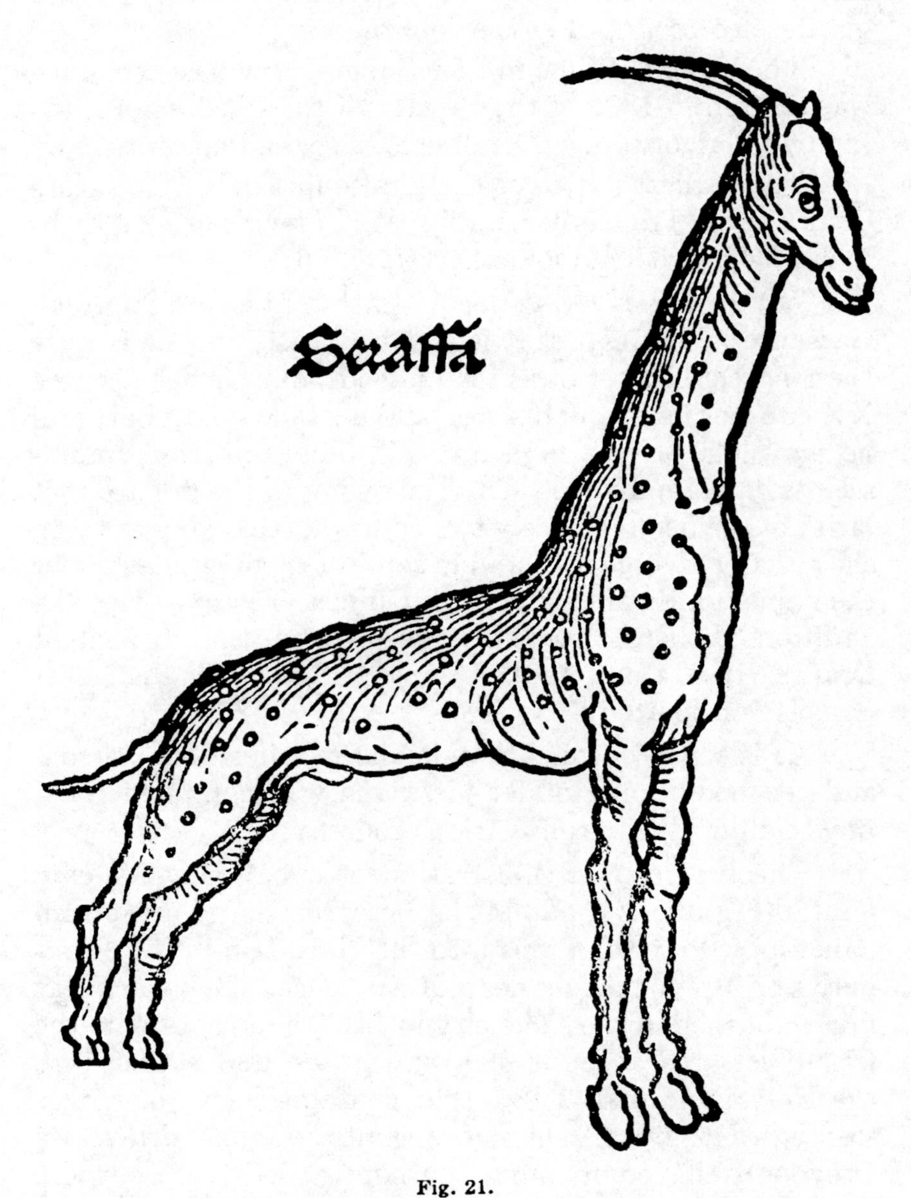

Fig. 21.

Giraffe (Seraffa) by Edward Reuwich.
From B. von Breydenbach's Peregrinationes in Terram Sanctam (1486).

mals are truly depicted, as actually seen by us in the Holy Land"(hec animalia sunt veraciter depicta sicut vidimus in terra sancta). Hugh Wm. Davies, in his Bibliography of Breydenbach (1911), remarks that "this can be believed in regard to the figures of the giraffe and dromedary, which are admirably drawn and probably the earliest printed." I cannot quite approve of this charitable attitude, for the horns of the animal are entirely wrong; in fact, they are not those of a giraffe, but of an antelope or oryx, very like those of *Oryx leucoryx*, the algazel. The tail is also misrepresented; the spots are indicated by small circles. I am inclined to presume that Reuwich drew the picture of the giraffe from memory and that in his effort to remember it visions of the oryx may have crossed his mind; at any rate, some mishap has occurred to him.

Breydenbach's work found a wide distribution: other editions with the woodcuts of the animals are in Flemish (Mainz, 1488), in French (Lyons, 1489), in Latin (Speier, 1490), in Spanish (Zaragoza, 1498), and some later editions, which go to show that in the latter part of the fifteenth century the giraffe was known on paper in most countries of Europe. Not all editions, however, contain the illustrations; thus the Newberry Library of Chicago has a Latin edition printed at Speier, 1486, and a French edition of Paris, 1522, which are minus the woodcuts.

The whole plate of Reuwich's animal pictures was taken over by Nicole le Huen and reproduced in his book "Des sainctes pérégrinations de Jhérusalem et des avirons et des lieux prochains," published at Lyons, 1488. Joly and Lavocat have copied this plate and erroneously assigned the giraffe and other animals to the ingenuity of Nicole le Huen, as Breydenbach's work was not accessible to them.

A tolerably accurate sketch of a giraffe was therefore known in central Europe toward the end of the fifteenth century, but artistic representations of the animal we owe to Italian painters of about the same time, as will be seen in the following chapter.

In his famous edition of Marco Polo's Travels Henry Yule comments that "the giraffe is sometimes wrought in the patterns of mediaeval Saracenic damasks and in Sicilian ones imitated from the former." An inquiry addressed to the Victoria and Albert Museum of London in regard to these designs elicited the following information from Mr. S. L. B. Ashton, in charge of the Department of Textiles: "I am afraid Yule is misleading on this question; the animals on these silks represent some form of deer and could not be taken for giraffes. I imagine that owing to the fact that they are usually represented in confronted pairs with their heads upturned, Yule mistook this length of neck to indicate that they were giraffes."

THE GIRAFFE IN THE AGE OF THE RENAISSANCE

The civilization of the Renaissance in Italy is characterized by the awakening of great interest in natural sciences, particularly in botany and zoology, and by a zeal for collecting curious plants and animals. During the fifteenth century, botanical gardens and animal parks (Italian *serraglio*) were founded in many places in Italy. The joy of exotic beasts led to the importation of live lions, leopards, elephants, camels, giraffes, ostriches, and even crocodiles from the ports of the southern and eastern Mediterranean. Arabs and Turks then were the active purveyors of menagerie animals, in the same manner as the Near East had played this role in the time of the ancient Romans.

One of the chroniclers of Florence relates that in the year 1459, when the Pope Pius II and Maria Sforza were received in that city, bulls, horses, boars, dogs, lions, and a giraffe were enclosed on a public square, but that the lions lay down and refused to attack the other animals. From letters of contemporaries we learn that they observed that lions kept in captivity abandoned their ferocity; and it once happened, as a letter-writer remarks, that a bull drove them back "like sheep into their fold."

Of the collections of exotic animals maintained by the princes of Italy, the most famous was the menagerie of Ferrante, duke of Naples, which contained a giraffe and a zebra,—two animals hitherto not seen in Europe. The duke had received them as a gift from the Caliph of Bagdad, toward the end of the fifteenth century.

Under Lorenzo di Medici the luxury in exotic animals reached its climax at Florence. He had, first of all, leopards trained for hunting whose fame spread into France; moreover, tigers, lions, and bears which he caused to combat with bulls, horses, boars, and greyhounds; elephants which, together with lions, appeared in a triumphal procession, and finally a giraffe presented in 1486 by El-Ashraf Kāit-Bey (1468-96), the Mamluk Sultan of Egypt. This animal was eulogized by the poets Angelo Poliziano and Antonio Costanzo, and was painted in one of the frescoes of the Poggio Cajano Palace in 1521.

Poliziano took matters rather easily, and in his poem confined himself to the remark that he had seen Lorenzo's giraffe; then he proceeds to translate literally the text of Heliodorus cited above (p. 61). Costanzo, however, shows that he really observed the animal, and his data betray the mind of an original thinker. He criticizes Strabo for questioning the animal's fleetness, and reproves Pliny, Solinus, Diodorus, Strabo, Varro, and Albertus Magnus for having suppressed the fact that it is provided with horns. In a Latin epigram addressed by him to Lorenzo the giraffe is introduced as speaking to the latter and lodging a complaint at having thus been deprived of its horns by the writers of the past. Lorenzo's giraffe was so gentle, he says, that it would eat bread, hay, or fruit out of a child's hand, and that when led through the streets, it would take whatever food of this kind was offered to it by spectators.

Lorenzo's giraffe met with a singular fate: it aroused the envy of Anne de Beaujeu, daughter of Louis XI, king of France, who died in 1483. Anne inherited from her father the love for animals, for she purchased a hundred and fifty-six siskins for the large aviary of the castle. She had dreams of owning some day a giraffe, which at that time was the object of curiosity at the Court of Florence and which she alleged Lorenzo di Medici had promised her.

Her letter addressed to him on the 14th of April, 1489, from Plessys du Parc is a document curious enough to be placed here on record. "You know," she wrote, "that formerly you advised me in writing that you would send me the giraffe (*la girafle*), and although I am sure that you

CAPTIVE YOUNG GIRAFFES IN HAGENBECK'S KRAAL AT THE FOOT OF THE KILIMANJARO, AFRICA.
Photograph by Courtesy of Carl Hagenbeck.

will keep your promise, I beg you, nevertheless, to deliver the animal to me and send it this way, so that you may understand the affection which I have for it; for this is the beast of the world that I have the greatest desire to see. And if there is any thing on this side I can do for you, I shall apply myself to it with all my heart. God be with you and guard you." Signed "Anne de France."

The Medicean, however, remained deaf to this prayer and kept his giraffe. It seems that breach of promise suits were not yet instituted at that time.

Giraffes were also kept at other Italian courts; for instance, by Alphonso II, duke of Calabria, in his villa Poggio Reale, and by Duke Hercules I in the Barco Park at Ferrara.

A giraffe is introduced into the background of Gentile Bellini's painting "Preaching of St. Mark at Alexandria," which is in the Brera Gallery of Milan (good photograph in the Ryerson Library of Art Institute, Chicago). G. Bellini (1426-1507) was court painter to the Sultan at Constantinople from 1479 to 1481, and brought back many sketches on his return to Italy, doubtless also the sketch of a giraffe. The painting in question was left unfinished at his death, and was completed by his brother Giovanni. It is an elaborate composition: a throng of monks and turbaned Orientals listening to the sermon of St. Mark on a huge square bordered by Moorish buildings and a cathedral in the background. At the foot of the stairway is planted a solitary and harmless two-horned giraffe, well outlined in its general features.

In 1487 the Sultan of Turkey presented to the Signoria of Florence a giraffe which caused a profound sensation. It was glorified in many painted portraits. Thus a giraffe figures in an "Adoration of the Magi" painted in the school of Pinturricchio (1454-1513) and now in the Pitti Palace of Rome.

Andrea Vannucchi, called Andrea del Sarto (1486-1531), has inserted a giraffe in the procession of the Three

Kings painted by him on a fresco of the Church of the Annunciation (Santissima Annunziata) at Florence (executed about 1510). He did so again in his Tribute to Caesar, dated 1521.

Leo Africanus, an Arabic traveller from Granada (beginning of the sixteenth century), writes, "Of the beast called Giraffa.—This beast is so savage and wilde, that it is a very rare matter to see any of them: for they hide themselves among the deserts and woodes, where no other beasts use to come; and so soone as one of them espieth a man, it flieth foorthwith, though not very swiftly. It is headed like a camell, eared like an oxe, and footed like a... [a word is wanting here in the original]: neither are any taken by hunters, but while they are very yoong."

Pierre Gilles of Albi (or Latinized Gellius) was sent in 1544 to the Orient by command of king François I, in order to "search for and amass ancient books for the king's library." He stopped at Constantinople and Cairo, and in the latter city visited the menagerie of the castle, where the Pasha of Egypt resided. He tells us that he found there three giraffes which he describes thus (in his book De vi et natura animalium XVI, 9):—

"On their foreheads are two horns six inches long, and in the middle of their forehead rises a tubercle to the height of about two inches, which appears like a third horn (in fronte media tuberculum existebat, velut tertium cornu, altum circiter duos digitos). Its neck is seven feet long. This animal is sixteen feet high from the ground, when it holds up its head. It is twenty-two feet long from the tip of the nose to the end of the tail; its fore legs are nearly of an equal height, but the thighs before are so long in comparison to those behind, that its back inclines like the roof of a house. Its whole body is sprinkled with large spots, which are nearly of a square form and of the color of a deer. Its feet are cloven like those of an ox; its upper lip hangs over the under one; its tail is slender, with hair on it to the very point. It ruminates like an ox, and, like cattle, feeds upon herbage and other things. Its mane is like that of a horse and extends from the top of the head to the back. When it walks, it seems to limp, first moving the right feet and then the left ones and simultaneously its sides. When it grazes or drinks, it is obliged to spread its fore legs very widely."

The interesting point is that Gilles is the first who mentions the third horn on the head of the Nubian giraffe.

André Thevet, who introduced tobacco into France (see "Introduction of Tobacco into Europe," Leaflet 19, p. 48), and who accompanied Gilles during part of his travels, likewise noticed the giraffes at Cairo, and gives a sketch of one in his book "Cosmographie de Levant" (Lyons, 1554), reproduced in Fig. 22. He writes, "I do not wish to pass over with silence two giraffes (*girafles*) which I saw there (at Cairo). Their necks are larger than that of a camel; they have on their heads two horns half a foot long, a small one on the front. The two fore legs are large and high, the hind legs are short, as may be seen in the accompanying figure represented as naturally as possible. This beast is the image of the learned and educated men, as Poliziano says; for these, at first sight, seem to be rough, rude, and peeved, although by virtue of the knowledge they have they are far more gracious, human, and affable than the others who have no knowledge whatever of sciences and virtue or who, as is commonly said, have greeted the Muses only at the threshold of the gate." In his "Cosmographie universelle" (Vol. I, fol. 388b, Paris, 1575), Thevet has given a more extensive notice of the giraffe with a very interesting drawing (reproduced in Plate V), but it teems with so many errors and absurdities that it is not worth placing on record. He locates, for instance, the giraffe in India and denies its occurrence in Ethiopia. The giraffe (Plate V) is guided by two Arabs and driven by a third man; another giraffe in the background freely browses under palms. The bodies of the animals are unfortunately misdrawn.

Pierre Belon (1518-64), a prominent French traveller and naturalist, reputed for the exactness of his observations, saw in Cairo the same giraffes as Gilles and Thevet, and has given a more accurate description of them, which is accompanied by the quaint picture of a giraffe drawn by himself from life (Fig. 23). He writes,—

"Formerly the grand lords, whatever barbarians they may have been, rejoiced in having beasts of foreign countries presented to them. In the castle of Cairo we saw

Fig. 22.

Giraffe with Guide.

From André Thevet's Cosmographie de Levant (1554).

several of those which had been brought there from all parts of the world, among these the animal commonly called Zurnapa, by the ancient Romans Camelopardalis. This is a very beautiful beast of the gentlest possible disposition, almost like a lamb, and more amiable or sociable than any other wild animal. Its head is almost similar to that of a stag, save that it is not so large, and bears small, obtuse horns six inches long and covered with hair. There is a distinction between the male and the female inasmuch as the horns of the males are longer; for the rest, both sexes have large ears like a cow, a tongue like an ox and black, and lack teeth in the upper mandible. They have long,

Portraict de la Giraffe.

Fig. 23.
Giraffe.
From Pierre Belon's Observations de Plusieurs Singularitez et Choses Memorables (Anvers, 1555).

straight, and graceful necks and fine, round manes. Their legs are graceful, high in front, and so low behind that the animal seems to stand erect. Its feet are like those of an ox. Its tail hangs down over the hocks, being round and with hair three times coarser than that of a horse. It is slender in the middle of the body. Its hair is white and red. In its gait it resembles the camel. In running, the two front feet go together. It sleeps with the paunch on the ground, and has a callosity on the chest and thighs like a camel. It cannot graze standing without straddling its fore legs, and even then feeds with great difficulty. Therefore it is easily credible that it lives in the fields solely on tree-leaves, its neck being so long that it can reach with its head to the height of a spear."

Aside from exaggerating the proportion of fore and hind legs and the erroneous definition of the gait, Belon's description is fairly exact.

A curious utilization of the hair of the giraffe is mentioned in the Travels of Nicolo dei Conti of the fifteenth century. Conti was a pioneer of European commerce in the East and travelled extensively in Egypt, Arabia, Persia, and India from 1419 to 1444. At his return to Italy he gave an account of his journey to Poggio Bracciolini, secretary of the Pope Eugenius IV. Bracciolini interpolated in his manuscript some information received from emissaries of the Pope to Ethiopia, and the notice of the giraffe emanates from this source. Curiously enough, the animal's name is not given. We read in Conti's Travels, "They informed me that there was also another animal, nine cubits long and six in height, with cloven hoofs like those of an ox, the body not more than a cubit in thickness, with hair very like to that of a leopard and a head resembling that of the camel, with a neck four cubits long and a hairy tail: the hairs are purchased at a high price, and worn by the women suspended from their arms, and ornamented with various sorts of gems."

It is a curious coincidence that a similar allusion to giraffe-tails occurs in the Tractatus pulcherrimus by an unknown author, written in the second half of the fifteenth century and published together with the famous letter of Prester John (see note on p. 97). The giraffe has hitherto not been recognized in this passage, but comparison with Conti's account leaves no doubt of the giraffe being intended. In enumerating the animals of Ethiopia, among these elephant and rhinoceros, this text mentions "another animal in Ethiopia, as they relate, the largest; the hairs of its tail are sold at a great price, and are used by their women as a great ornament." In the same manner as in Conti's notice, the animal is not named, and it is certain that the passage must emanate from the same source,—the Pope's ambassadors to Ethiopia. We remember that giraffe-tails were offered as presents to King Tutenkhamon (above, p. 23), and it is interesting to observe how such old practices have been perpetuated through centuries down to modern times (above, p. 6). The Masai of East Africa still preserve the long hairs of giraffe-tails, and their girls use these hairs as threads to sew the beads on to their clothes. The natives of Kordofan still make bracelets of such hairs, which are traded over the Sudan.

In H. Goebel's "Wandteppiche" (Plate 226) is reproduced a carpet from the beginning of the sixteenth century, doubtfully referred to the manufacture of Oudenarde in Flanders. In this carpet are represented five giraffes equipped with headstalls and collar bands apparently decorated with jewels; one of the animals is provided with three horns. Their necks are straight and too long proportionately; anatomically incorrect and fantastic, they evidently were copied from drawings.

The Art Institute of Chicago owns an interesting print said to be Portuguese and to date from the eighteenth century. It is a gift of Mr. Robert Allerton. A section of it is reproduced in Plate VI. The design, a giraffe guided by an Arab and surrounded by floral patterns, is repeated many times. It is a continuation of the tradition inaugurated by Thevet and Topsell.

THE GIRAFFE IN THE NINETEENTH CENTURY AND AFTER

The first live giraffes received in France and England were gifts of Mohammed Ali, Pasha of Egypt, who also dispatched a live specimen to the Sultan at Constantinople and to the court of Vienna.

In 1826 he presented a giraffe to the king of France who had it placed in the Jardin des Plantes of Paris, which had been established in 1635. This was the first living giraffe who made its appearance in France. Its arrival was a great event and caused a sensation throughout the country. This giraffe was a female, about two years old, eleven feet and six inches in height, originating from Sennaar. She was about six months old when captured by Arabs, and was sold to Muker Bey, governor of Sennaar, who presented her to the Pasha. She was embarked at Alexandria, wearing around her neck a strip of parchment inscribed with several passages from the Koran and purported as an amulet to safeguard her health and welfare. She was accompanied by four Arabs to guide her and by three cows to supply her with milk. She landed at Marseille in November, 1826, sixteen months after leaving Sennaar, and arrived in Paris in June of the following year (1827). She was introduced to the king, Charles X, who then resided in the castle of Saint-Cloud, and was subsequently shown to an ever-increasing multitude of people. Every one was eager to see her, thousands waited in line for hours to catch a glimpse of the animal, the whole press busied itself about her. Articles and poems (*chansons*) were devoted to her, and she became so popular that she penetrated into the realm of fashion which seized her forms and colors, creating dresses à la girafe, hats and neckties à la girafe, and combs à la girafe. At Nevers she was modeled in faience, at Epinal she was glorified in colored pictures. She even entered the sanctum of politics, and a bronze medal was cast, showing a giraffe who addresses these words to the country: "There is nothing that has changed in France, there is only another beast here." This giraffe gladdened the hearts of Parisians for nearly twenty years. It may now be seen stuffed in the Natural History Museum of the Jardin des Plantes. It is a curious coincidence that it is just a hundred years since this first live giraffe arrived in Paris, and an Associated Press dispatch from Paris of July 30, 1927, announces that this centenary will be duly celebrated. In 1843 a giraffe was presented by Clot Bey to the menagerie of the same museum in Paris.

In 1827 Mohammed Ali, Pasha of Egypt, presented a Nubian giraffe to George IV, king of England. This was the first giraffe received alive in Britain. Unfortunately, it survived but a few months at Windsor. The animal, in its surroundings at Windsor, was painted by James Laurent Agasse; this picture is preserved in the Royal Collection and reproduced by Lydekker (in *Proceedings of the Zoological Society of London*, 1904, Vol. II, p. 340). A portrait of Mr. Cross, the animal-dealer, together with two Arabs, is introduced into the scenery. Owing to the immature condition of the animal, the frontal horn was not fully developed; the animal, as shown in the painting, displays all the characteristics of the typical Nubian race of *Giraffa camelopardalis*, such as the net-like style of the markings, the white "stockings," and the comparatively large size of the spots on the upper part of the legs.

Another painting in the Royal Collection, representing a group of giraffes, is by R. B. Davis, a well-known painter, and is dated "September, 1827." It is described as "two giraffes belonging to George IV," and on the back it is titled "portrait of the Giraffe belonging to his Majesty." According to Lydekker, this species is intended for the Southern or Cape form, as the old bull has no frontal horn, while the markings are of the blotched, instead of the netted, type, and the lower parts of the legs are spotted, although not quite so fully as they ought to be. Lydekker thinks that Davis might have taken Paterson's specimen of a Cape giraffe in the British Museum as his model; if this conclusion be correct, the painting is of very considerable interest, as that race now appears to be extinct.

Lieutenant W. Paterson (Narrative of Four Journeys into the Country of the Hottentots and Caffraria in 1777-79, p. 127, London, 1790), who was commissioned by Lady Strathmore to botanize in the then unknown region of Caffraria, offers an excellent copper-plate representing a "Camelopardalis" shot by him in South Africa and describes it as follows: "The color of these animals is in general reddish, or dark brown and white, and some of them are black and white; they are cloven footed; have four teats; their tail resembles that of a bullock; but the hair of the tail is much stronger, and in general black; they have eight fore teeth below, but none above; and six grinders, or double teeth, on each side above and below; the tongue is rather pointed and rough; they have no foot-lock hoofs; they are not swift; but can continue a long chase before they stop; which may be the reason that few of them are shot. The ground is so sharp that a horse is in general lame before he can get within shot of them, which was the case with our horses, otherwise I should have preserved two perfect specimens of a male and female. It is difficult to distinguish them at a distance, from the shortness of their body, which, together with the length of their neck, gives them the appearance of a decayed tree." Paterson sent home an immature male specimen of a Southern giraffe which he had shot and which was presented by Lady Strathmore to John Hunter, the distinguished surgeon. The animal's skull with some of the bones is still preserved in the Museum of the Royal College of Surgeons. The giraffe itself was finally acquired by the British Museum, where it was still extant in 1843, though in bad condition.

In 1836, four young giraffes from Kordofan, about two years old, were safely received at the London Zoological Gardens. The animals—three males and a female—flourished, and became the progenitors of a long line of English-bred giraffes, the first calf being born in June, 1839. It was followed by two others, the old female dying at the age of eighteen years. The animals continued to breed, and during the period between 1836 and the death of the last of the old stock in 1892, no less than thirty individuals were exhibited in the Regent's Park menagerie, seventeen of which had been born there. A pair of young animals, presented by Col. Mahon and likewise obtained from Kordofan, arrived in London in the summer of 1902.

The first living example of the Southern giraffe was imported into Europe in 1895 for the Zoological Garden of London at the price of £500. It had been captured on the Sabi River in Portuguese territory and brought down to Pretoria, whence it was conveyed to Delagoa Bay and shipped to Southampton.

In 1863 Lorenzo Casanova, an adventurous traveller and animal collector, returned from the Egyptian Sudan to Europe with a transport of six giraffes, the first African elephants, and many other rare mammals. In 1864 he entered with the firm Carl Hagenbeck into a contract according to which all animals to be secured on his future expeditions to Africa should be ceded to the latter. In 1870 the largest consignment of wild animals that ever reached Europe arrived at Trieste, consisting of fourteen giraffes, ninety other mammals, and twenty-six ostriches. The giraffes were distributed over the zoological gardens of Vienna, Dresden, Berlin, and Hamburg. About that time the itinerant menagerie-owners and showmen also began to keep giraffes; thus Carl Kaufmann, famous animal-trainer and disciple of Gottlieb Kreutzberg, who always endeavored to gather novel and interesting beasts, had a superb collection of trained lions, tigers, elephants, hippopotamus, rhinoceros, and giraffes. Renz, the celebrated circus-director, utilized giraffes, antelopes, buffalo, and many other creatures for the equipment of his pantomime "The Festival of the Queen of Abyssinia."

An inquiry addressed to the firm Carl Hagenbeck at Stellingen near Hamburg elicited the information that during the period 1873-1914 this firm imported a total of a hundred and fifty giraffes in four species,—*Giraffa camelopardalis* of Lower Nubia and Abyssinia, *G. capensis* of the Cape territory, *G. hagenbecki* from Gallaland, and *G. tippelskirchi* from former German East Africa. The largest specimen imported by Hagenbeck, about eleven and a half feet in height, came from the Galla country, and was transmitted to the Zoological Garden of Rome. Prior to 1914 Hagenbeck maintained at the foot of the Kilimanjaro in Africa a station for captive animals, where the captured young giraffes moved freely in a larger kraal, as shown in Plates VIII-IX made from photographs due to the courtesy of the firm Carl Hagenbeck. In its wonderful park at Stellingen the giraffes occupy a large stretch of land with a fine building of Arabic style. Like other animals, giraffes can be perfectly acclimatized almost everywhere, and do not suffer from the inclemencies of the European winter. Among the numerous interesting observations recorded by Carl Hagenbeck in his memoirs we read also that the hairs of the giraffes adapt themselves to the new conditions of life and that toward the end of the winter their hairs were found to be one and a half times longer than they usually are.

Only young animals of about eight feet in height are captured. They are hunted and lassoed by horsemen. This is comparatively easy, but the task of accustoming them to their new life, caring for them and rearing them, above all, their transportation presents difficult problems. On their way to the coast the animals must run. A strap is placed around the base of their neck, and they are governed by means of two halters, one in front and one behind. On board ship or train they are stowed in large boxes which in size must correspond to the height of the animal with its neck outstretched. The average price for a young giraffe before the war was about $1500-2000. At present when giraffes but very seldom are offered on the market, prices are arbitrary and fluctuating, and vary between $5000 and $7500.

The Zoological Society of Philadelphia keeps records of all the animals that have arrived there for the zoological garden which is the oldest in the United States. The earliest record there relating to the arrival of giraffes is an entry under August 11, 1874, when five males and one female were purchased.

The zoological garden in Lincoln Park, Chicago, received two giraffes, a male and a female, two years old, in October 1913, as a gift from Mrs. Mollie Netcher Newberger. The female died in December, 1915; the male, in May, 1919. Both were mounted, and are now on exhibition at the Boston Store. A giraffe in the Bronx Zoological Garden, New York, according to newspaper reports, is said to have given life to three young ones.

The London Zoological Garden now has only two giraffes—Maudie and Maggie. Maudie is a Nubian giraffe from the Sudan; and Maggie, a Kordofan giraffe, born in the menagerie, who has weathered twenty years of captivity.

In modern applied and commercial art the giraffe has not been entirely forgotten. It is familiar to our newspaper cartoonists. The advertisement of a well-known throat remedy is accompanied by a giraffe's head and neck. The British Uganda Railway displays a poster with a very effective colored picture of a giraffe. In the London Illustrated News of May 29, 1926 appeared a series of eleven comical sketches of giraffes from the hand of J. A. Shepherd under the title "Humours of the Zoo: Studies of Animal Life, No. XV." As to art-crafts, I have noticed metal figures of giraffes as radiator caps on automobiles. Yet, a wider application might be made of this motif; for instance, in pen-racks and lamp-holders, an electric bulb being carried between the horns. Carl F. Gronemann, who has drawn the giraffe-heads for the cover and vignette of this leaflet, has thereby furnished excellent examples of how such animal designs may be employed in the graphic arts, for book-ornaments, bindings, or book-plates. Our sculptors and artists in oil have almost neglected this subject. While we have excellent photographs of both wild and tame giraffes, a really artistic painting or statuette of them remains to be done, and the inspiration coming from the works of the ancient Egyptians and Chinese may be helpful to the modern artist.

A very artistic picture of four giraffes browsing among acacias, by the American artist, Robert Winthrop Chanler, is now in the Musée du Luxembourg, Paris; it is reproduced in *The American Magazine of Art*, 1922, No. 12, p. 535.

NOTES

In regard to the role of the giraffe in Hottentot folk-lore (p. 29) compare the stories recorded by L. Schultze, Aus Namaland und Kalahari (Jena, 1907), pp. 405, 417, 489, 531. The Masai of East Africa have a good story of the Dorobo and the Giraffe (A. C. Hollis, The Masai, Their Language and Folk-lore, 1905, p. 235).

Page 35. Quatremère (Histoire des Sultans Mamlouks de l'Egypte, Vol. I, 1840, pp. 106-108) has extracted from Arabic manuscripts quite a number of records referring to presentations of giraffes. Only those which are of importance on account of their historical associations have been mentioned by me. In regard to al-Mahdi (p. 35), see T. K. Hitti, Origins of the Islamic State, Vol. I, p. 381 (Columbia University Press, 1916). The essential point is to recognize that the Muslim rulers of mediaeval Egypt were exceedingly active in sending giraffes as gifts into many parts of the world. The Abbassid Caliphs had an animal park at Baghdad which has been described by a Greek embassy in A.D. 917 (see G. Le Strange, *Journal Royal Asiatic Society*, 1897, p. 41).—The giraffe occurs also among Egyptian shadow-play figures of Cairo. One of these is illustrated by P. Kahle, *Der Islam*, Vol. II, p. 173 (possibly a giraffe in Fig. 34, Vol. I, p. 294).—In regard to the derivation of the Arabic word *zarāfa* from the Ethiopic and the relations of these words to Egyptian, compare F. Hommel, Die Namen der Säugetiere bei den südsemitischen Völkern (1879), p. 230.—Masudi is not the first Arabic author who wrote about the giraffe. There is an earlier lengthy account by Al-Jahiz (who died in A.D. 869) in his Kitāb al-hayawān ("Book of Animals"), Vol. VII, p. 76 of the edition published at Cairo, 1907; but the text is partially corrupt and very abstruse, and as its essential points are all contained in the authors cited above, I have not reproduced it.—The Persian story of the young giraffe (p. 39) meets with a curious parallel to what the Arabs say about the young rhinoceros: the period of gestation of the mother rhino is four years, the young one stretches its head out of the mother's womb and browses at the trees around; at the lapse of four years it leaves the womb and runs away with lightning speed, for fear that its mother might lick it with her tongue which is so rough that once it licks an animal, the latter's flesh will separate from the bones in a moment (compare G. Ferrand in *Journal asiatique*, 1925, Oct.-Dec., p. 267).

As Prof. Sprengling kindly informs me, one of the earliest Arabic references to the giraffe occurs in Bashshār Ibn Burd, the blind, deformed poet of the late Omayyad and early Abbassid period, who died in A.D. 783. In a satire on the early Mutagilite Wāsil Ibn Atā, named Abu Hudhaifa, nicknamed al-Ghazzāl, the weaver (because he frequented the weavers to observe the chastity of their women), when the latter made a derogatory exclamation about the poet's neck, he says:—

Why should I be bothered by a weaver, who, if he turns his back, has a neck
Like an ostrich of the desert; and if he faces you,
The neck of the giraffe? What have I to do with you?

Some Arabic philologists regard *zarafa* as a purely Arabic word and derive it from the Arabic root *zrf*, which means "assembly." Hence Sibawaih, the great grammarian of the Arabs, who died in A.D. 793 or 796, writes, "God created the giraffe with its fore legs longer than its hind legs. It is named with the name of the assembly, because it is in the form of an assembly of animals. Ibn Doraid writes it *zurafa* and doubts that it is an Arabic word." Ibn Doraid, of course, is justified in his doubt; he was a celebrated philologist of Basra and lived from A. D. 837 to 934.

The giraffe in Chinese records (p. 42) was first pointed out by H. Kopsch (*China Review*, Vol. VI, 1878, p. 277), who translated the description of a Kilin with reference to Aden from a Chinese biography of Mohammed. This text, however, has no independent value, but is literally copied from Ma Huan's account. This brief notice induced De Groot to contribute to the same journal (Vol. VII, p. 72) an article on "The Giraffe and The Kilin," in which he tries to show that the Kilin of ancient Chinese tradition may be identical with the giraffe. This, of course, is a reversion of logic. It is impossible to assume that the ancient Chinese were acquainted with the giraffe, which in the present geological period did not anywhere occur in Asia; nor do the ancient descriptions of the Kilin, as assumed by De Groot, fit the giraffe. The climax of sinological romance is reached by A. Forke (Mu Wang und die Königin von Saba, p. 141), according to whom the Chinese were acquainted with the giraffe in the earlier Chou period through the travels of King Mu to the west. The giraffe, on the other hand, was not recognized by Bretschneider (*China Review*, Vol. V, 1876, p. 172) in the Kilin of Arabia purchased by a Chinese envoy in 1430. O. Münsterberg (Chinesische Kunstgeschichte, Vol. II, p. 65) sees a "wounded giraffe" on a Han bas-relief of Teng-fung, Ho-nan. The animal in question is simply a deer. The alleged "giraffe-like Kilin" on a bronze basin of the Han period (cf. A. C. Moule in the article cited in the Bibliography) is the so-called spotted deer (*Cervus mandarinus*), called by the Chinese *mei hua lu* ("plum-blossom stag"). Its spots are represented either by small circles or even by plum-blossoms of realistic style.

The reader interested in the relations of the Chinese with the east coast of Africa may consult F. Hirth, Early Chinese Notices of East African Territories, *Journal American Oriental Society*, Vol. XXX, 1909, pp. 46-57.

The animal *a-t'a-pi* (p. 43) is referred to by W. W. Rockhill (*T'oung Pao*, 1914, p. 441) with the remark, "I have no means of determining what animal is meant." *Damaliscus jimila*, according to Roosevelt, extends from Mount Elgon and the northern highlands of Uganda southward over the Man Escarpment and Victoria Nyanza drainage to what formerly was central German East Africa; westward as far as the Edward Nyanza and Lake Kivu; also near the coast from the Sakaki and Tana Rivers northward as far as the Juba River. The topi is one of the most conspicuously colored of all antelopes, being inversely countershaded. The body coloration is a bright cinnamon-rufous overlaid everywhere by a silvery sheen which gives the coat a resplendent effect. The red color is deepest on the head, throat, and sides and lightest on the rump, hind quarters, and tail, where it fades to pure cinnamon. The shoulders are marked by a broad black patch which extends down on the fore legs as far as the knees and completely circles the upper part of the leg. The hind quarters are marked by a much larger black patch which extends down on the limbs as far as the hocks above which it forms a complete band around the leg.

Ma Huan's account of Aden containing the description of the giraffe (p. 45) was first translated by G. Phillips in *Journal Royal Asiatic Society*, 1896, pp. 348-351, and subsequently by A. C. Moule in the article cited in the Bibliography.

In regard to the opossum (p. 54) cf. C. R. Eastman, Early Figures of the Opossum, *Nature*, Vol. 95, 1915, p. 89.

Page 58. The learned S. Bochart, in his famous Hierozoicon (Vol. I, col. 908, 1675) rejected the opinion that Aristotle was acquainted with the giraffe, but subsequently Pallas, Allamand, G. Schneider in his translation of Aristotle's History of Animals, as well as Joly and Lavocat, have championed the opposite view, which, however, is untenable. O. Keller (Die antike Tierwelt) offers little on the giraffe; he does not place the accounts of the ancients on record, nor does he discuss them. H. Rommel (Die naturwissenschaftlich-paradoxographischen Exkurse bei Philostratos, Heliodoros und Tatios, 1923, p. 61) gives a brief critical evaluation of the texts.

An interesting essay on the former statues in Constantinople (p. 66) was written by R. M. Dawkins, Ancient Statues in Mediaeval Constantinople, *Folk-lore*, Vol. XXXV, 1924, pp. 209-248.

The text of Jean de Joinville (p. 74) is as follows: "Entre les autres joiaus que il envoia au roy, li envoia un oliphant de cristal mount bien fait, et une beste que l'on appelle orafle, de cristal aussi, pommes de diverses manières de cristal, et jeuz de tables et de eschiez; et toutes ces choses estoient fleuretées de ambre, et estoit li ambres liez sur le cristal à beles vignetes de bon or fin."—Natalis de Wailly, Histoire de Saint Louis par Jean Sire de Joinville (1878), p. 163.

The complete title of this curious little work (p. 86) is Tractatus pulcherrimus de situ et dispositione regionum et insularum tocius Indiae, nec non de rerum mirabilium ac gentium diversitate. A critical editon of the text is given by F. Zarncke (Der Priester Johannes II, pp. 174-179).

B. LAUFER.

BIBLIOGRAPHY

BRYDEN, H. A.—On the Present Distribution of the Giraffe South of the Zambesi. Proceedings Zoological Society of London, 1891, pp. 445-447.

Great and Small Game of Africa. London, 1899. Giraffe: pp. 488-510.

BURCKHARDT, J.—Die Kultur der Renaissance in Italien. 2 vols. 12th ed. Leipzig, 1919.

EASTMAN, C. R.—Early Representations of the Giraffe. Nature, Vol. 94, 1915, pp. 672-673.

Illustration of the giraffe of Pamfili (incomplete) after O. Keller and an Egyptian design from Thebes after Ehrenberg.

More Early Animal Figures. Nature, Vol. 95, 1915, p. 589.

Two Egyptian figures, one after Wilkinson, another from Hierakonpolis after Quibell.

Chinese and Persian Giraffe Paintings. Nature, Vol. 99, 1917, p. 344.

Chinese painting of A. W. Bahr representing giraffe and accompanied by erroneous conclusions (see above, p. 49).

Giraffe and Sea Horse in Ancient Art. American Museum Journal, Vol. XVII, 1917, p. 489.

Same matter as preceding article.

FERRAND, G.—Le nom de la giraffe dans le Ying yai cheng lan. Journal Asiatique, July-August, 1918, pp. 155-158.

In this very interesting article G. Ferrand makes the point that the Chinese name *k'i-lin* for the giraffe is based on Somali *giri* or *geri*. This ingenious supposition is not entirely convincing for several reasons. First, a direct contact of the Chinese with the Somali is unproved. Second, the old Chinese pronunciation *gi-lin* holds good only for the T'ang period, not for the fifteenth century when the Chinese actually made the acquaintance of the giraffe and when the word was articulated *k'i-lin* as at present. Third, the name *k'i-lin* was applied to the animal in China when it arrived there as early as 1414, the Chinese naturally believing that it virtually was the k'i-lin of their ancient lore. Ferrand insists that Ma Huan heard the Somali word *giri* at Aden, but Ma Huan himself did not visit Aden; his account of Aden is based on the report of the eunuch Li who was at Aden in 1422, but at least eight years earlier the giraffe was designated *k'i-lin* on Chinese soil. For these reasons the Somali hypothesis appears to me unnecessary. The question is merely of an adaptation of an old name to a novel animal, not of an attempt at transcribing a foreign word. The Somali name was not transmitted anywhere; it was the Arabic name *zurafa* which was conveyed both to China and to Europe.

GRABHAM, G. W.—An Original Representation of the Giraffe. Nature Vol. 96, 1915, pp. 59-60.

Reproduction from G. A. Hoskins, Travels in Ethiopia (1835), of a giraffe from an Egyptian monument, with reference to Eastman's articles in Nature.

HAGENBECK, C.—Von Tieren und Menschen, Erlebnisse und Erfahrungen. Berlin, 1908.

JOLY, N., and LAVOCAT, A.—Recherches historiques, zoologiques, anatomiques et paléontologiques sur la girafe. Mémoires de la Société des sciences naturelles de Strasbourg, Vol. III, 1846, pp. 1-124 in quarto. 17 plates.

This is the most extensive monograph on the giraffe ever published and particularly good in the historical section. The authors give the complete texts of Greek, Latin, Byzantine and mediaeval writers on the giraffe, but English authors are neglected, and Oriental lore was unknown at that time.

LOISEL, G.—Historie des ménageries de l'antiquité à nos jours. 3 vols. Paris, 1912.

LYDEKKER, R.—On Old Pictures of Giraffes and Zebras. Proceedings of the Zoological Society of London, 1904, Vol. II, pp. 339-345. Refers to the English paintings of giraffes mentioned on p. 89.

MAXWELL, W.—Stalking Big Game with a Camera in Equatorial Africa. New York, 1924. Chap. VI: Camera Incidents with the Masai Giraffe.

MOULE, C. A.—Some Foreign Birds and Beasts in Chinese Books. Journal of the Royal Asiatic Society, 1925, pp. 247-261.

The value of this article rests on the fact that for the first time illustrations of animals from a Chinese book of the fifteenth century are given, but the data are not critically digested.

PHIPSON, EMMA.—The Animal-lore of Shakespeare's Time, pp. 130-133. London, 1883.

RENSHAW, G.—Natural History Essays. London, 1904. The Northern Giraffe, pp. 99-113; 5 illustrations.

ROOSEVELT, T. and HELLER, E.—Life-histories of African Game Animals. 2 vols. New York, 1914. Chap. XI: The Reticulated and Common Giraffes.

SALZE.—Observations faites sur la girafe envoyée au roi par le Pacha d'Egypte. Mémoires du Muséum d'histoire naturelle, Paris, Vol. XIV, 1827, pp. 68-84.

This is the first description of the giraffe in France based on a live specimen and enriched by information given by the Arab guides of the animal.

WINTON, W. E. de.—Remarks on the Existing Forms of Giraffe. Proceedings Zoological Society of London, 1897, pp. 273-283.

YULE, H.—Hobson-Jobson. London, 1903. "Giraffe": pp. 377-378.

The quotations given are mere extracts and not complete; the translations from Greek authors are very inexact.

182

中国的斗蟋蟀

Cricket Champions of China

Passion of the Chinese for Insect Musicians and Fighters Is Age-old and Unceasing

KEEPING animals or birds in captivity may seem to some people infinitely heartless and cruel, but the keeping of crickets in China in order to enjoy their musical efforts and fighting ability cannot be so considered because of the loving care bestowed on these pets. The Chinese have made such a patient study of their insect-musicians and fighters and are so fond of them that their health, comfort, and proper feeding always come first in the thoughts of their keepers.

Cricket fights are not as cruel as cock and quail fights in which the Chinese also indulge, nor are the three combined as cruel as the bull fights of Spain and Latin America.

China, venerable "Wise Man of the East," is shown in this article prepared from material in a monograph by Dr. Berthold Laufer, Curator of Anthropology at Field Museum of Natural History, Chicago, as a sentimental, loving student of the lowly cricket which, in many lands, is merely considered an omen of good fortune.

—The Editor.

OF the many insects that are capable of producing sound in various ways, the best known and the most expert musicians are the crickets, who during the latter part of summer and in the autumn fill the air with a continuous concert. They are well known on account of their abundance, their wide distribution, their characteristic chirping, and the habit many of them have of seeking shelter in human habitations.

Of crickets there are three distinct groups— known as mole-crickets, true crickets, and tree-crickets. The first-named are so called because they burrow in the ground like moles; they are pre-eminently burrowers. The mole-crickets feed upon the tender roots of various plants. The true crickets are common everywhere, living in fields, and some species even in our houses. They usually live on plants, but are not strictly vegetarians; sometimes they are predaceous and feed mercilessly upon other insects. The eggs are laid in the autumn, usually in the ground, and are hatched in the following summer. The greater number of the old insects die on the approach of winter; a few, however, survive the cold season. The tree-crickets principally inhabit trees, but they occur also on shrubs, or even on high herbs and tall grass.

Like their near relatives, crickets have biting mouth parts, and, like the grasshoppers and katydids, rather long hind legs which enable them to jump. Although many of them have wings when full grown, they move about mainly by jumping or hopping. When the young cricket emerges from the egg, it strongly resembles the adult, but it lacks wings and wing-covers, which gradually appear as the insect grows older and larger. The final development of wings and wing-covers furnishes the means whereby the male cricket can produce his familiar chirping sound. It is only the adult male that sings; the young and the females cannot chirp.

All photos courtesy Field Museum of Natural History

BOY CRICKET CATCHERS

Twelfth century Chinese painting showing boys playing with crickets. One is shown with trap and jar

On examining the base of the fore wings or wing-covers of the male cricket, it will be noticed that the veins at the base are fewer, thicker, and more irregular than those on the hind or lower wings. On the underside of some of these thick veins will also be seen fine, transverse ridges like those on a file. The wing-covers of the female have uniform, parallel veins, without a trace of ridges. The male cricket produces his chirping sound by raisin[g] his wing-covers above his body an[d] then rubbing their bases together, s[o] that the file-like veins of the under su[r]face of the one wing-cover scrape th[e] upper surface of the lower.

The sound made by crickets is, o[f] course, not a true song, but a mechan[i]cal production, as are all of the soun[ds] produced by insects. The object of th[e] chirping or stridulating is somewha[t] conjectural. It may be a love-son[g,] mating-call, or an expression of som[e] other emotion. The fact that th[e] crickets are able to sing onl[y] when they are full grown and cap[a]ble of mating would seem to sug[]gest that their chirping is a lov[e] song.

This commonly held view, how[]ever, is contested by Frank [E.] Lutz in a recent article on "In[]sect Sounds" published in *Natur[al] History.* Dr. Lutz starts from th[e] opinion that not everything [in] nature has a practical or util[i]tarian purpose and that man[y] striking characters and chara[c]teristics of animals and plants a[re] of no use to their possessors or t[o] any other creature; they seem t[o] him to be much like the figur[es] in a kaleidoscope, definite an[d] doubtless due to some intern[al] mechanism, but not serving an[y] special purpose.

THE Chinese, perhaps, hav[e] made a not uninterestin[g] contribution to this problem. O[f] the many species of crickets use[d] by them, the females are kep[t] only of one—the black tree cricket, called by them *kin chun[g]* ("Golden Bell," with reference t[o] its sounds), as they assert tha[t] this is the only kind of cricke[t] that requires the presence of th[e] female to sing. The females of al[l] other species are not kept by th[e] Chinese. As soon as the insect[s] are old enough for their sex to b[e] determined, the females are fed to bird[s] or sold to bird-fanciers.

But whether captive insects are in[]structive examples for the study of th[e] origin and motives of their chirps i[s] another question. Our canaries an[d] other birds in confinement likewis[e] sing without females. Whatever th[e] biological origin of insect sound ma[y] be it seems reasonable to infer that th[e] endless repetition of such sounds ha[s] the tendency to develop into a purel[y] mechanical practice which the insect in dulges as a pastime for its own diversion.

The relation of the Chinese to crickets and other insects presents one of their most striking characteristics and one of the most curious chapters of culture-historical development. In the primitive stages of life man took a keen interest in the animal world. First of all, he closely observed and studied large mammals, and next to these, birds and fishes. A curious exception to this almost universal rule is presented by the ancient Chinese.

THEY were more interested in the class of insects than in all other groups of animals combined; while mammals, least of all, attracted their attention. Their love of insects led them to observations and discoveries which still elicit our admiration. The curious life-history of the cicada was known to them in early times, and only a nation which had an innate sympathy with the smallest creatures of nature was able to penetrate into the mysterious habits of the silkworm and present the world with the discovery of silk. The cicada as an emblem of resurrection, the praying-mantis as a symbol of bravery, and many other insects play a prominent role in early religious and poetical conceptions as well as in art, as shown by their effigies in jade.

In regard to mammals, birds, and fishes, Chinese terminology does not rise above the ordinary, but their nomenclature of insects is richer and more colorful than that of most languages. Not only do they have a distinct word or even several terms for every species found in their country, but also numerous poetic and local names for the many varieties of each species for which words are lacking in English and other tongues.

In their relations to crickets the Chinese have passed through three distinct periods: during the first period running from the times of early antiquity down to the T'ang dynasty, they merely appreciated the cricket's powerful tunes; under the T'ang (A.D. 618-906) they began to keep crickets as interned prisoners in cages to be able to enjoy their concert at any time; finally, under the Sung (A.D. 960-1278) they developed the sportof cricket-fights and a regular cult of the cricket.

The praise of the cricket is sung in the odes of the *Shi king*, the earliest collection of Chinese popular songs. People then enjoyed listening to its chirping sounds, while it moved about in their houses or under their beds. It was regarded as a creature of good omen, and wealth was predicted for the families which had many crickets on their hearths. When their voices were heard in the autumn, it was a signal for the weavers to commence their work.

CRICKET TYPES

1a and 1b show yellowish tree cricket, male and female; 2a and 2b, black tree cricket; 3 shows katydid

The sounds produced by the mitred cricket recall to the Chinese the click of a weaver's shuttle. One of its names therefore is *tsu-chi*, which means literally "one who stimulates spinning." "Chicken of the weaver's shuttle" is a term of endearment for it.

As it happened in China so frequently, a certain custom first originated in the palace, became fashionable, and then gradually spread among all classes of the populace. The women enshrined in the imperial seraglio evidently found solace and diversion in the company of crickets during their lonesome nights. Instead of golden cages, the people availed themselves of small bamboo or wooden cages which they carried in their bosom or suspended from their girdles.

There are various methods of catching crickets. They are usually captured at evening. In the north of China a lighted candle is placed near the entrance of their hole, and a trap box is held in readiness. Attracted by the light, the insects hop out of their retreats, and are finally caught in the traps made of bamboo or ivory rods. Some of these ivory traps are veritable works of art: they are surmounted by carvings of dragons, and the trap doors shut very accurately.

CRICKET GOURDS

Note the exquisite molded design of one in lower left corner. Covers are of carved ivory or white jade

IN the south, men avail themselves of what is called a fire-basket (*fo lam*) which is made of iron rods and in which a charcoal fire is kept burning. This fire drives the insects out of their dens. Sometimes the cricket-hunters reach their object by pouring water into the holes where the insects hide. Sometimes they endeavor to entice them from the nest by placing at its entrance the fruit of *Nephelium longana* (*lung yen*, "dragon's eyes").

Many people rear hundreds of crickets in their homes, and have several rooms stacked with the jars which shelter the insects. The rich employ experts to look after theirs. As soon as you enter a house like this, you are greeted by a deafening noise which a Chinese is able to stand for any length of time.

During the summer the insects are kept in circular pottery jars made of a common burnt clay and covered with a flat lid, which is sometimes perforated. Many potters made a special business of these cricket houses, and impressed on them a seal with their names; for example, Chao Tse-yü, who lived in the first part of the nineteeth century and

CRICKET KEEPERS' ACCESSORIES

Bamboo and ivory traps, feeding dish nippers, pot cleaning brush, frame for use while cages are cleaned

whose productions still enjoy a special reputation. There are old pots said to go back as far as the Ming dynasty (1368-1643), and these are highly prized.

The crickets keep cool in these jars, which are often shaped in the form of a gourd, as the heat does not penetrate the thick clay walls. Tiny porcelain dishes decorated in blue and white or small bits of clay contain food and water for the insects, and they are also provided with beds or sleeping boxes of clay. Jars of somewhat larger size serve for holding the cricket-fights.

During the winter months the crickets are transferred to specially prepared gourds which are provided with loose covers wrought in open work so as to admit fresh air. This is said to be a special variety of the common gourd (*Lagenaria vulgaris*), the cultivation of which was known to a single family of Peking.

THE gourds used as cricket habitations are all artificially shaped. They are raised in earthen moulds, the flowers are forced into the moulds, and as they grow they assume the shape of and the designs fashioned in the moulds. There is accordingly an infinite variety of forms: there are slender and graceful, round and double, cylindrical and jar-like ones. Those formerly made for the palace are decorated with figures and scenes in high relief fashioned in the clay mould. The technique employed in these ancient pieces is now lost; at least they are no longer made, although there are poor modern imitations in which the surfaces are carved, not moulded.

The covers of the gourd, flat or tall, are made of jade, elephant or walrus ivory, coconut shell, and sandalwood, all elaborately decorated, partly in high relief, partly in open work, or in the two methods combined, with floral designs, dragons, lions and other animals. Gourd vines with flowers and fruits belong to the most favorite designs carved in the flat ivory covers; gourd and cricket appear to be inseparable companions.

A kind of cement which is a mixture of lime and sandy loam is smeared over the bottom of the gourd to provide a comfortable resting-place for the tenant. The owner of the cricket may carry the gourd in his bosom wherever he goes, and in passing men in the street you may hear the shrill sound of the insect from its warm and safe place of refuge. The gourds keep the insects warm, and on a cold night they receive a cotton padding to sleep upon.

In the summer the insects are generally fed on fresh cucumber, lettuce, and other greens. During their confinement in autumn and winter masticated chestnuts and yellow beans are given them. In the south they are also fed on chopped fish and various kinds of insects, and even receive honey as a tonic.

IT is quite a common sight to see the idlers congregated in the tea-houses and laying their crickets out on the tables. Their masters wash the gourds with hot tea and chew chestnuts and beans to feed them. Then they listen to their songs and boast of their grinding powers. The Chinese cricket books give many elaborate rules for proper feeding which vary with the different species and with every month.

The fighting crickets receive particular attention and nourishment, a dish consisting of a bit of rice mixed with fresh cucumbers, boiled chestnuts, lotus seeds, and mosquitoes. When the time for the fight draws near, they get a tonic in the form of a bouillon made from the root of a certain flower. Some fanciers allow themselves to be stung by mosquitoes, and when these are full of blood, they are given their favorite pupils. In order to stir their ferocity prior to a bout, they are sometimes also compelled to fast. As soon as it is recognized from their slow movements that they are sick, they are fed on small red insects gathered in water.

A tickler is used for stirring the crickets to incite them to sing. In Peking fine hair from hare or rat whiskers inserted in a reed or bone handle is utilized for this purpose; in Shanghai, a fine blade of crab or finger grass. The ticklers are kept in bamboo or wooden tubes, and the rich indulge in the luxury of having an elegant ivory tube surmounted by the carving of a lion.

THE tympanum of good singers is coated with a bit of wax to increase or strengthen the volume of sound. A small needle about three inches long with blunt end, about the size of a darning needle, is heated over a candle and lightly dipped in the wax. The insect is held between the thumb and forefinger of the closed hand, and the wax is applied to the wing-covers.

In the course of many generations, the Chinese through long experience and practice, have accomplished what we may call a natural selection of fighting crickets. The good fighters are believed to be incarnations of great heroes of the past, and are treated in every respect like soldiers. Kia Se-Tao, the first author who wrote on the subject, says that "rearing crickets is like rearing soldiers." The strongest and bravest of these who are most appreciated at Peking and Tientsin come from the southern province of Kwang-tung. These fighters are dubbed "generals" or "marshals," and seven varieties of them are distinguished, each with a special name.

CARVED WALNUT SHELL

For keeping singing crickets—usually carried in the girdle. Note intricate design

Those with black heads and gray hair in their bodies are considered best. Next in appreciation come those with yellow heads and gray hair, then those with white heads and gray hair, then those with golden wings covered with red hair, those of yellow color with blood-red hair, who are said to have two tails in form of sheep's horns, finally those yellow in color with pointed head and long abdomen and those supposed to be dressed in embroidered silk, gray in color and covered with red spots like fish-scales. The good fighters, according to Chinese experts, are recognized by their loud chirping, their big heads and necks, long legs, and broad bodies and backs.

THE "Generals," as stated, receive a special diet before the contest, and are attended to with utmost care and great competence. Observations made for many centuries have developed a set of standard rules which are conscientiously followed. The trainers, for instance, are aware of the fact that extremes of temperature are injurious to the crickets. When they observe that the insects droop their tiny mustaches, they know that they are too warm, and endeavor to maintain for them an even temperature and exclude all draughts from them. Smoke is supposed to be detrimental to their health, and the rooms in which they are kept must be perfectly free from it.

CRICKET POTS

The summer habitations of the insects. The one in lower left-hand corner contains clay beds in which they sleep

The experts also have a thorough understanding of their diseases, and have prescriptions at hand for their treatment and cure. If the crickets are sick from overeating, they are fed on a kind of red insect. If sickness arises from cold, they get mosquitoes; if from heat, shoots of the green pea are given them. A kind of butterfly known as "bamboo butterfly" is administered for difficulty in breathing. In a word, they are cared for like pet babies.

The tournaments take place in an open space, on a public square, or in a special house termed Autumn Amusements. There are heavy-weight, middle and light-weight champions. The wranglers are always matched on equal terms according to size, weight, and color, and are carefully weighed on a pair of tiny scales at the opening of each contest. A silk cover is spread over a table on which are placed the pottery jars containing the warring crickets. The jar is the arena in which the prize fight is staged.

As a rule, the two adversaries facing each other will first endeavor to flee, but the thick walls of the bowl or jar are set up as an invincible barrier to this attempt at desertion. Now the referee who is called "Army Commander" or "Director of the Battle" intercedes, announcing the contestants and reciting the history of their past performances, and spurs the two parties on to combat. For this purpose he avails himself of the tickler described above, and first stirs their heads and the ends of their tails, finally their large hind legs.

The two opponents thus excited stretch out their antennae which the Chinese not inaptly designate "tweezers," and jump at each other's heads. The antennae or tentacles are their chief weapons. One of the belligerents will soon lose one of its horns, while the other may retort by tearing off one of the enemy's legs. The two combatants become more and more ferocious and fight each other mercilessly. The struggle usually ends in the death of one of them, and it occurs not infrequently that the more agile or stronger one pounces with its whole weight upon the body of its opponent, severing its head completely.

Cricket-fights in China have developed into a veritable passion. Bets are concluded, and large sums are wagered on the prospective champions. The stakes are in some cases very large, and at single matches held in Canton, are said to have sometimes aggregated $100,000. It happens quite frequently that too ardent amateurs are completely ruined in the game. Gambling is forbidden by law in China as elsewhere, but such laws are usually winked at, and the official theory in this case is that stakes consist of presents of sweet cakes.

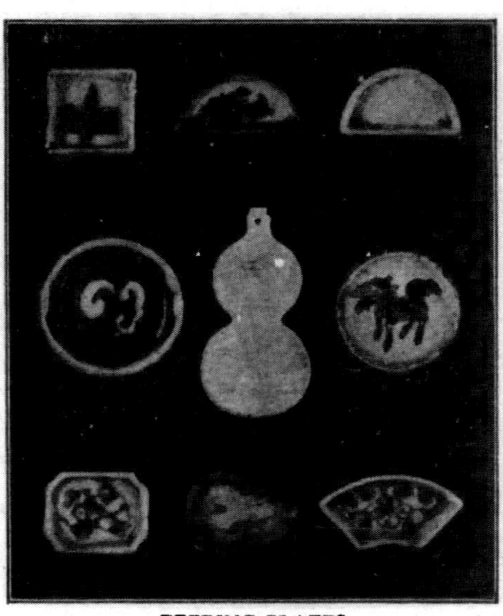

FEEDING PLATES

Of delicate porcelain. In center is an ivory, gourd-shaped tablet on which names of champions are inscribed

Choice champions fetch prices up to $100, the value of a good horse in China, and owners of famous crickets travel long distances to meet their competitors and congregate with them in order to match their champions. Some amateurs delight in raising them by the hundreds in the hope of producing the champion of the champions of the season, who is honored with the attribute of Grand Marshal. These men are by no means low-brows, but highly cultured men, and those in responsible government positions are found in this class.

TWO localities near Canton, Fa-ti and Cha-pi, not far from Whampoa, enjoy a special reputation for cricket-fighting. At these places extensive mat sheds are erected and divided into several compartments. In each section a contest goes on, the pot which forms the arena being placed on a table. In order to acquaint prospective betters with the merits of the crickets matched against each other, a placard is posted on the sides of the building, setting forth the various stakes previously won by each cricket. Great excitement is manifested at these matches, and considerable sums of money change hands.

The sum of money staked on the contest is lodged with a committee who retain ten percent to cover expenses and hand over the balance to the owner of the winning cricket. The lucky winner is also presented with a roast pig, a piece of silk, and a gilded ornament resembling a bouquet of flowers. This decoration is deposited by him either on the ancestral altar of his house to inform his ancestors of his good luck and to thank them for their protection, or on a shrine in honor of Kwan-ti, a deified hero, who is the personification of all manly virtues and a model of gentlemanly conduct.

The names of the victorious champions are inscribed on an ivory tablet carved in the shape of a gourd, these tablets like diplomas are ı iously kept in the houses of the fo nate owners. Sometimes the characte of the inscription are laid out in gold. The victory is occasion for great rejoicing and jollification. Music is performed, gongs are clanged, flags displayed, flowers scattered, and the tablet of victory is triumphantly marched in front, the jubilant victor struts in the procession of his overjoyed compatriots, carrying his victorious cricket home. The sunshine of his glory falls on the whole community in which he lives, and his village will gain as much publicity and notoriety as an American town which has produced a golf or baseball champion.

In southern China, a cricket which has won many victories is honored with the title "conquering or victorious cricket" (*shou lip*); on its death it is placed in a small silver coffin, and is solemnly buried. The owner of the champion believes that the honorable interment will bring him good luck and that excellent fighting crickets will be found in the following year in the neighborhood of the place where his favorite cricket lies buried.

ALL these ideas emanate from the belief that able cricket champions are incarnations of great warriors and heroes of the past from whom they have inherited a soul imbued with prowess and fighting qualities. Dickens says, "For all the Cricket Tribe are potent Spirits, even though the people who hold converse with them do not know it (which is frequently the case)."

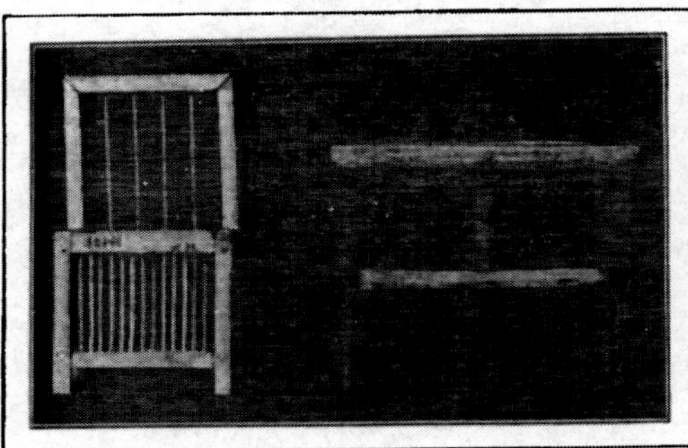

CAGES

In Shanghai and Hangchow, grasshoppers are also held captive, usually being kept in wooden cages rather fancifully shaped like tables, chairs, or similar articles of furniture. Cicadas were formerly also kept in small cages which were suspended from the eaves of the house or from branches of trees, but this custom is no longer practised. However, boys sometimes cage them

A proverbial saying with reference to a man who failed or has been defeated is, "A defeated cricket—he gives up his mouth," which means as much as "throwing up the sponge."

The following Chinese stories may give an insight into the cricket rage.

Kia Se-tao, a minister of state and general who lived in the thirteenth century, and who wrote, as mentioned, an authoritative treatise on the subject, is one of the cricket fanciers famous in history. He was completely obsessed with an all-absorbing passion for the cricket cult. The story goes that one day, during a war of the Mongols against the imperial house of Sung, an important city fell into the hands of the foe. When Kia Se-tao received news ' the disaster, he was found kneeling . . the grass of a lawn and taking part in a cricket fight. "In this manner you look out for the interests of the nation!" he was reprimanded. He was not in the least disturbed, however, and kept his attention concentrated on the game.

AN anecdote of tragical character is told with reference to an official of Peking, who held the post of director of the rice-granaries of the capital. He once found a cricket of choice quality and exceptional value. In order to secure this treasure, he exchanged his best horse for it and resolved to present this fine specimen to the emperor. He placed it cautiously in a box and took it home. During his absence his prying wife craved to see the insect which had been bought so dearly. She opened the box, and fate ordained that the cricket made its escape. A rooster happened to be around and swallowed the cricket. The poor woman, frightened by the consequences of her act, strangled herself with a rope. At his return the husband learned of the double loss he had suffered and, seized by despair, committed suicide. The Chinese narrator of the story concludes, "Who would have imagined that the graceful singer of the fields might provoke such a tragedy as this?"

During the period 1426-36, a magistrate, wishing to befriend the governor, presented him with a cricket which proved to be a remarkable fighter. Thereupon the governor ordered the beadles of his district to keep him supplied with crickets. Their price rose very high, and when the beadles sent for a single one it was enough to ruin several families. One man found a handsome cricket which he proceeded to feed carefully in a bowl. His son, curious to see the specimen, accidentally let the cricket escape. The son attempted suicide and when revived, was found to have lost his mind. The father that evening found a small cricket which, when set to fight, vanquished all others. This he presented to the governor and the governor, in turn, showered him with honors. After six months, the son recovered and said his spirit had gone into the cricket and he had fought skillfully to amend the loss he had caused his father.

AS far as is known, China is the only nation that has developed cricket-fights. The Japanese, although fond of chirping insects, which they keep as pets in little cages, do not use them for fighting purposes. Kipling writes in his Jungle-book, "The herd-children of India sleep and wake and sleep again, and weave little baskets of dried grass and put grasshoppers in them; or catch two praying-mantises and make them fight." This may be an occasional occurrence in India, but it has not developed into a sport or a national pastime. In regard to Japan the reader may be referred to Lafcadio Hearn's essays "Insect-Musicians," inserted in his "Exotics and Retrospectives," and "Semi" [Cicada] in his "Shadowings."

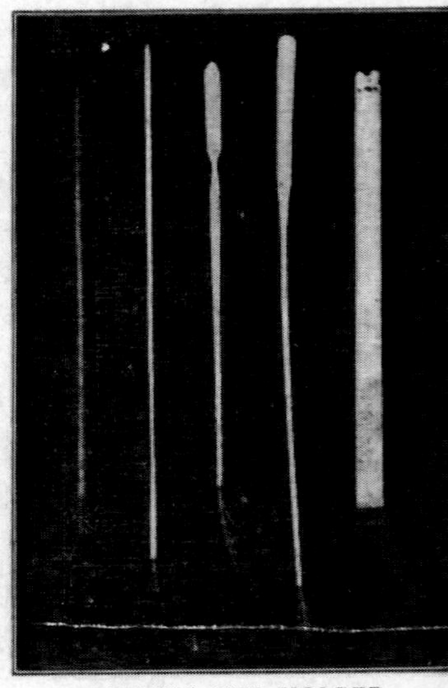

TICKLERS AND HOLDER

For inciting crickets to sing or fight: of rat or hare whiskers in reed, bone, or ivory

183

竞技场上的蟋蟀

THE CRICKET IN THE ARENA.

CRICKET-FIGHTS AS A NATIONAL SPORT IN CHINA, AND CRICKETS AS "MUSICAL" PETS.

Abridged from an Article in the "Scientific American," based on a Monograph by Dr. BERTHOLD LAUFER, Curator of Anthropology, Field Museum of Natural History, Chicago.

KEEPING animals or birds in captivity may seem to some people heartless and cruel, but the keeping of crickets in China in order to enjoy their musical efforts and fighting ability cannot be so considered because of the loving care bestowed on these pets. Cricket-fights are not as cruel as the cock and quail fights in which the Chinese also indulge, nor are the three combined as cruel as the bull-fights of Spain and Latin America.

It is only the adult male cricket that sings; the young and the females cannot chirp. The male

"GOOD FIGHTERS ARE BELIEVED TO BE INCARNATIONS OF GREAT HEROES": TYPES OF CRICKETS KEPT IN CHINA AS PETS OR FOR FIGHTING PURPOSES.

The upper pair (Nos. 1a and 1b) are the male and female of a yellowish tree-cricket, and the lower pair (Nos. 2a and 2b) those of a black tree-cricket—the only kind of which the female is preserved, as necessary to make the male "sing." The middle one (No. 3) is a katydid.

produces his chirping sound by raising his wing-covers above his body and then rubbing their bases together, so that the file-like veins of the under surface of the one wing-cover scrape the upper surface of the lower.

Of the many species of crickets used by the Chinese, the females are kept only of one—the black tree-cricket, called by them *kin chung* ("Golden Bell," with reference to its sounds), as they assert that this is the only kind of cricket that requires the presence of the female to sing.

The ancient Chinese were more interested in insects than in all other groups of animals combined. The women in the imperial seraglio evidently found solace and diversion in the company of crickets during their lonesome nights. Instead of golden cages, the people had small bamboo or wooden cages which they carried in their bosom or suspended from their girdles.

There are various methods of catching crickets. They are usually captured at evening. In the north of China a lighted candle is placed near the entrance of their hole, and a trap box is held in readiness. Attracted by the light, the insects hop out of their retreats, and are caught in the traps made of bamboo or ivory rods. Some of these ivory traps are veritable works of art. Many people rear hundreds of crickets in their homes, and have several rooms stacked with the jars which shelter the insects. As soon as you enter a house like this, you are greeted by a deafening noise.

During the summer the insects are kept in circular pottery jars made of a common burnt clay and covered with a flat lid, sometimes perforated. Many potters made a special business of these cricket-houses, and impressed on them a seal with their names. There are old pots said to go back as far as the Ming dynasty (1368-1643), and these are highly prized. The crickets keep cool in these jars. Tiny porcelain dishes decorated in blue and white or small bits of clay contain food and water for the insects, and they are also provided with beds or sleeping boxes of clay. Jars of somewhat larger size serve for holding the cricket-fights.

During winter the crickets are transferred to specially prepared gourds with loose covers wrought in open-work to admit fresh air. The gourds used as cricket habitations are all artificially shaped. They are raised in earthen moulds; the flowers are forced into the moulds and as they grow they assume the shape of designs.

The covers of the gourd are made of jade, elephant or walrus ivory, coconut shell, and sandalwood, all elaborately decorated. A kind of cement is smeared over the bottom of the gourd to provide a comfortable resting-place for the tenant. The owner of the cricket may carry the gourd in his bosom wherever he goes. The gourds keep the insects warm, and on a cold night they receive a cotton padding to sleep upon. In summer the insects are generally fed on fresh cucumber, lettuce, and other greens. During autumn and winter masticated chestnuts and yellow beans are given them. In the south they are also fed on chopped fish and various kinds of insects, and even receive honey as a tonic.

The fighting crickets receive particular attention and nourishment. When the time for the fight draws near, they get a tonic in the form of a bouillon made from the root of a certain flower. Some fanciers allow themselves to be stung by mosquitoes, and, when these are full of blood, they are given their favourite pupils. In order to stir their ferocity, they are sometimes also compelled to fast. A tickler is used for stirring the crickets to incite them to sing. In Peking fine hair from hare or rat whiskers inserted in a reed or bone handle is utilised for this purpose; in Shanghai, a fine blade of crab or finger grass. The ticklers are kept in bamboo or wooden tubes, the rich having an elegant ivory tube surmounted by the carving of a lion.

In the course of many generations, the Chinese have accomplished what we may call a natural selection of fighting crickets. The good fighters are treated in every respect like soldiers. The strongest and bravest, most appreciated at Peking and Tientsin, come from the southern province of Kwang-tung. These fighters are dubbed "generals" or "marshals," and seven varieties are distinguished. Those with black heads and grey hair in their bodies are considered best. Next in appreciation come those with yellow heads and grey hair, then those with white heads and grey hair, then those with golden wings covered with red hair, those of yellow colour with blood-red hair; finally, those yellow with pointed head and long abdomen and those supposed to be dressed in embroidered silk, grey, and covered with red spots. The good fighters are recognised by their loud chirping, big heads and necks, long legs, and broad bodies and backs.

The tournaments take place in an open space, on a public square, or in a special house termed "Autumn Amusements." There are heavy-weight, middle- and light-weight champions. The wranglers are carefully weighed on a pair of tiny scales at the opening of each contest. A silk cover is spread over a table on which are placed the pottery jars containing the warring crickets. The jar is the arena in which the prize fight is staged. As a rule, the two adversaries will first endeavour to flee, but the referee, who is called "Army Commander" or "Director of the Battle," intercedes, announcing the contestants and reciting the history of their past performances, and spurs the two parties on to combat with the tickler, first stirring their heads and the ends of their tails, finally their large hind-legs.

The two opponents, thus excited, stretch out their antennæ and jump at each other's heads. The antennæ or tentacles are their chief weapons. One of the belligerents will soon lose one of its horns, while the other may retort by tearing off one of the enemy's legs. The struggle usually ends in the death of one of them, and not infrequently the more agile or stronger one pounces with its whole weight upon its opponent, severing its head completely. Cricket-fights in China have developed into a veritable passion. Large sums are wagered. The stakes are in some cases very large, and at single matches held in Canton are said to have sometimes aggregated 100,000 dollars. Frequently too ardent amateurs are completely ruined. Choice champions fetch prices up to 100 dollars, the value of a good horse in China. Some amateurs delight in raising them by the hundreds in the hope of producing the champion of the champions of the season, who is honoured with the attribute of Grand Marshal.

Two localities near Canton, Fa-ti and Cha-pi not far from Whampoa, enjoy a special reputation for cricket-fighting. At these places extensive mat sheds are erected and divided into several compartments. In each section a contest goes on, the pot which forms the arena being placed on a table. The lucky winner is presented with a roast pig, a piece of silk, and a gilded ornament resembling a bouquet of flowers. This decoration is deposited by him either on the ancestral altar of his house, to inform his ancestors of his good luck and to thank them for their protection, or on a shrine in honour of Kwan-ti, a deified hero who is the personification of all manly virtues and a model of gentlemanly conduct.

The names of the victorious champions are inscribed on an ivory tablet carved in the shape of a gourd, and these tablets, like diplomas, are religiously kept. The victory is occasion for great rejoicing. The jubilant victor struts in the procession of his overjoyed compatriots, carrying his victorious cricket home. His glory falls on the whole community, and his village will gain as much publicity as an American town which has produced a golf or baseball champion. In southern China, a cricket which has won many victories bears the title "conquering or victorious cricket" (*shou lip*); on its death it is placed in a small silver coffin, and is solemnly buried.

All these ideas emanate from the belief that able cricket champions are incarnations of great warriors. Dickens says, "For all the Cricket Tribe are potent Spirits, even though the people who hold converse with them do not know it (which is frequently the case)."

As far as is known, China is the only nation that has developed cricket-fights. The Japanese, although

SPORT DATING FROM THE SUNG ERA (960—1279): BOY CRICKET-CATCHERS, ONE WITH TRAP AND JAR—A TWELFTH-CENTURY CHINESE PAINTING.

Photographs by Courtesy of the Field Museum of Natural History, Chicago.

fond of chirping insects, which they keep as pets in little cages, do not use them for fighting purposes. Kipling writes in his "Jungle-Book": "The herd-children of India sleep and wake and sleep again, and weave little baskets of dried grass and put grass-hoppers in them; or catch two praying-mantises and make them fight." This may be an occasional occurrence in India, but it has not developed into a sport or a rational pastime

184

飞机制造的历史背景

FIELD MUSEUM OF NATURAL HISTORY
FOUNDED BY MARSHALL FIELD, 1893

PUBLICATION 253

ANTHROPOLOGICAL SERIES VOLUME XVIII, NO. 1

THE PREHISTORY OF AVIATION

BY

BERTHOLD LAUFER
Curator of Anthropology

---

12 Plates in Photogravure

CHICAGO
1928

PRINTED IN THE UNITED STATES OF AMERICA
BY FIELD MUSEUM PRESS

CONTENTS

List of Plates 5
Introduction 7
The Romance of Flying in Ancient China 14
Kites as Precursors of Aeroplanes 31
The Dawn of Airships in Ancient India 44
From Babylon and Persia to the Greeks and the Arabs . . 58
The Air Mail of Ancient Times 71
Notes 88
Bibliographical References 94
Index 95

LIST OF PLATES

I. Winged Deity Attended by Bird-men. Stone bas-relief of Han period, A.D. 147, Shan-tung, China.

II. Aerial Contest of Dragon-chariot and Dragon-riders. Stone bas-relief of Han period, A.D. 147, Shan-tung, China.

III. Aerial Contest of Dragon-chariot and Dragon-riders. Continuation of the panel shown in Plate II.

IV. Ki-kung's Flying Chariot. Chinese Woodcut from T'u shu tsi ch'eng.

V. Francesco Lana's Flying Boat. From Lana's Prodromo, 1670.

VI. Flying Taoist Saint. Chinese landscape in ink from General Munthe Collection now in Los Angeles Museum.

VII. The Goddess Si Wang Mu Flying Astride a Crane. Scene from an embroidered Chinese screen of the K'ang-hi period (1662-1722) in Blackstone Chinese Collection of Field Museum.

VIII. Boys Flying a Kite. Scene from a Chinese painted roll by Su Han-ch'en of the twelfth century in collections of Field Museum.

IX. Earliest English Illustration of a Kite. From John Bate's The Mysteries of Nature and Art, 1634.

X. Two Apsarases or Heavenly Nymphs Flying downward and Surrounding the Buddha Amitābha. Marble sculpture with votive inscription yielding date A.D. 677. Blackstone Chinese Collection of Field Museum.

XI. Kai Kawus' Flight to Heaven. From a Persian illustrated manuscript of the Shahnameh, dated 1587-88. Courtesy of Metropolitan Museum of Art, New York.

XII. The Aerial Voyage of Domingo Gonsales. From F. Godwin's Man in the Moone, 1638.

THE PREHISTORY OF AVIATION

INTRODUCTION

A French miniature of the fourteenth century depicts the Spirit or Angel of Youth, who is never fatigued and whose course nothing can arrest. He is arrayed with wings on his feet, soaring over the sea. The wings are tinted green, the color of hope. Youth has fair hair and a blue robe. He carries on his shoulders a pilgrim who is in the vigor of age, and while crossing the water, addresses to him these lines:—

I am called Youth, the nimble,
The tumbler and the runner,
The grasshopper, the dasher,
Who cares not a glove for danger.
I see, I come, I bound, I fly,
I sport and caracole.
My feet they bear me whither I will,
They've wings; your eyes may see them.
Give here thine hand, with thee I'll fly
And carry thee over the sea.

On May 20-21, 1927, Colonel Charles A. Lindbergh accomplished his solitary transoceanic flight from New York to Paris and stirred the entire world. We experienced the same thrill as in our boyhood days when we were first reading about the campaigns of Alexander the Great or Columbus' voyages of discovery.

Yet, the desire to fly is as old as mankind. "Oh that I had wings like a dove! for then would I fly away, and be at rest," sings the royal psalmist (Psalms 55, 6). In all ages man's imagination was fired by the sight of soaring birds and was seized by the ambition to migrate and to sail upon the wind like one of them. Many daring men tried nobly and less nobly to emulate the ways of the eagle in the air. It is a long record of ventures, experiments, and failures, but remains the most fascinating romance in the history of mankind.

In his excellent "History of Aeronautics in Great Britain," J. E. Hodgson divides the history of the subject into four eras covering very unequal periods of time. His first he titles the legendary and prehistoric era with its tale of mythological and fabulous stories of flight, verging gradually into the historic, and extending to about the end of the fifteenth century. As examples of this prehistoric period he cites Daedalus and Icarus from Greek mythology and Bladud, the flying king of Britain in 800 B.C. However, China, India, and the rest of Asia remain out of consideration in this scheme. Moreover, the ideas registered by Hodgson under his three historical periods of Europe, like aspiration, speculation, endeavor, romance, limited achievement, are no less conspicuous in the Orient. His second or first historical period in Europe, which dates from the sixteenth to beyond the latter half of the eighteenth century, is characterized thus: "The practicability of flight was a matter of speculation and discussion, became the subject of imaginative romance, and was made the object of theoretical projects and not a few practical attempts."

It will be demonstrated on the following pages that all these features were in evidence among Oriental nations in early times, many centuries before their dawn in Europe, and that the fundamental ideas underlying the principles of our present aviation take their root in the Orient. My conception of the so-called mythological and legendary period in the history of aviation differs widely from that of my predecessors. It is a comfortable method and no more than a conventional form of thinking to stamp early traditions as mythological or legendary. This is a scholastic phrase from which little is gained, no tangible significance accrues. An inquisitive mind is intent on unravelling the fabric of a myth, on seeking an interpretation of its origin. If myth it is, how did the myth spring into existence? As there is a logic of human reasoning, so there is a logic of human imagination. The imaginative faculty of the human mind cannot conceive things that have absolutely no reality in existence; the product of our imagination is always elicited by something that exists or that we have reason to believe to exist.

H. G. Wells, who, being a novelist and artist, is possessed of the insight, vision, and intuition which most scholars lack, is on the right track when in his "Outline of History" he comments on the Daedalus story as follows: "Greek legend has it that it was in Crete that Daedalus attempted to make the first flying-machine. Daedalus ('cunning artificer') was a sort of personified summary of mechanical skill. It is curious to speculate what germ of fact lies behind him and those waxen wings that, according to the legend, melted and plunged his son Icarus in the sea." In his manner of reasoning Wells certainly is superior to the majority of schoolmasters who pitifully dismiss Daedalus as a myth.

The prehistory of mechanical science is shrouded in mystery, because primitive man was unable to render an intelligent account of it. In the same manner as natural phenomena were regarded by him as wonders or miracles wrought by supernatural agencies, so any mechanical devices were interpreted as the outcome of witchcraft: the skilled artificer and every investigator and experimenter of prehistoric and early historic days has gone down in history or tradition as a sorcerer, enchanter, wizard, or magician, who made a pact with demoniacal powers. Many of these so-called magicians were simply clever mechanics whose work was beyond their contemporaries' comprehension or whose achievements were so singular and awe-inspiring that supernatural forces were believed to have inspired their genius. This is the reason why those who made attempts at aerial flights are usually associated with magic and necromantic art or why in our middle ages solely devils and witches are endowed with the faculty of flying. John Wilkins, in 1648, wrote seriously, "Witches are commonly related to passe unto their usual meetings in some remote place; and as they doe sell windes unto mariners, so likewise are they sometimes hired to carry men speedily through the open air. Acosta affirms that such kind of passages are usuall amongst divers sorcerers with Indians at this day. So Kepler in his Astronomicall dream doth fancy a witch to be conveyed to the moon by her Familian."

The ancient traditions regarding mechanical wonders must therefore be divested of their legendary garb and exposed in their historical nucleus. On the other hand, it is always the marvelous and romantic that lingers in the memory of man. The dry and bare bones of historical events are apt to be relegated to the wastebasket of oblivion. We do not retain in our minds the dates of wars and battles or the chronological tables of dynasties we had to memorize in school, but we remember many heroes by anecdotes and bons mots which the stern historian will frown at as unauthentic. No historian's pen has preserved a record of the Trojan War, but Homer has sung it in the form of epic poetry which has been enjoyed by a hundred generations and which has been more often read than any accurate report of a war published by the competent staff of any war ministry. Alexander the Great is not remembered by Oriental nations as their conqueror, but as a deified hero of marvelous exploits, as he appears in the Greek Romance going under his name. Therefore it is just man's ingrained love for the fabulous and fanciful, for the wondrous and extraordinary to which we are indebted for the preservation of ancient records of flight.

In the same manner as astrology was the precursor of astronomy and alchemy evolved into the science of chemistry, so there is an abundance of primitive lore which godfathers the history of aviation. To distinguish that primeval stage from aviation as an accomplished fact of the present time, we might coin for the former the new term "aviology" in imitation of astrology, but the public mind is sufficiently alarmed by an exuberance of ologies, and it is therefore preferable to speak simply of the prehistory of aviation. It must not be imagined that the latter is set apart as a thing in itself, fundamentally distinct from the history of aviation. The two, in fact, are closely allied and interwoven, inseparable, merging into each other, and the recent historical development is unintelligible without a knowledge of its prehistoric setting and background. Thus, it will be seen, our aeroplanes are pedigreed from kites which have their origin in China. Our modern progress in aviation is not solely due to efforts of the present generation, stupendous and admirable as they may be, but presents the process of a gradual evolution of ideas which have grown out of the imagination, endeavors, experiments, triumphs, and failures of many past ages. Stress must be laid on the word "imagination," for there is no field of human exertions in which imagination and romantic dreams have played a greater role and have proved more fertile than in the development of aviation. Intuition, romance, and adventure are its leading motives; for man, from the very moment he had grown into a full-fledged human being, never lived on bread and love alone. We have conquered the air in this age of science and unprecedented progress of mechanics, but in the last instance this conquest goes back to the trend of man's mind toward the romantic and adventurous. Describing merely the gradual perfection of mechanical devices does not make a complete history of aviation. It is the spirit and the idea behind the devices that count, the idea itself means everything. The will to fly is the will to conquer, and this will has pervaded the hearts of men in the earliest stages of the great civilizations of Asia.

Many visions and reveries of the Orient have been brought true by modern inventions, but the Orient merits credit for the genesis of the idea. The notion of Roentgen rays, for instance, was anticipated both in ancient China and India. The Chinese have several accounts concerning metal mirrors which would light up the interior organs of the human body. The emperor Ts'in Shi (259-210 B.C.) is credited with the possession of such a mirror which was styled "the precious mirror that would illuminate the bones of the body," or "the mirror illuminating the gall." This mirror was discovered in the palace of the Ts'in emperors at Hien-yang in Shen-si Province by the founder of the Han dynasty in 206 B.C., and is described as follows: "It was a rectangular mirror four feet wide, five feet and nine inches high, brilliant both on its outer and inner sides. When a man stood straight before it to see his reflection, his image appeared reversed. When some one placed his hands on his heart, he observed his five viscera placed side by side and not impeded by any obstacle. When a man had a hidden malady within his organs, he could recognize the seat of his complaint by looking into this mirror and laying his hands on his heart. Moreover, when a woman had perverse sentiments, her gall would swell and her heart palpitate. The emperor Ts'in Shi therefore constantly availed himself of this mirror to test the women of his seraglio: those whose gall would swell and whose heart would be agitated, he ordered to be killed."

Jīvaka, a celebrated physician of ancient India and contemporary of Gautama Buddha, called the king of doctors, at least had the idea that it was necessary to illuminate the organs of the body for the purpose of making a diagnosis and perform surgical operations. He practised trephining, and this appeared to his contemporaries so wondrous that it was interwoven with many legends. Jīvaka is said to have discovered in a load of fagots a marvelous gem possessed of the virtue that "when placed before an invalid, it illuminated his body as a lamp lights up all objects in a house, and so revealed the nature of his malady." He laid this gem on the head of a sick man, and found that there was a centipede inside of his head (probably a brain tumor); he opened his skull with an instrument and pulled the centipede out with a pair of heated pincers, whereupon the patient recovered. According to another version, it was a piece of wood from a tree, called "the king of physicians," which enabled Jīvaka to see plainly the five viscera, the intestines, and the stomach; and he availed himself of a golden knife in opening the skull.

True it is that the first actual bombardments from the air took place but recently during the World War, but the idea itself is not novel. It was forestalled in the seventeenth century by Francesco Lana (below, p. 22), and the first air-bombardier was the giant bird Rukh when he hurled huge bowlders at Sindbad's ship.

The story of a flying Uganda warrior who engaged in efficient bombardments from the air was graphically recorded in 1871 by the famous explorer, Henry M. Stanley, in his work "Through the Dark Continent":—

"One of the heroes of Nakivingi [one of the ancient kings of Uganda,—the Charlemagne of Uganda, as Stanley calls him] was a warrior named Kibaga, who possessed the power of flying. When the king warred with the Wanyoro, he sent Kibaga into the air to ascertain the whereabouts of the foe, who, when discovered by this extraordinary being, were attacked on land in their hiding-places by Nakivingi, and from above by the active and faithful Kibaga, who showered great rocks on them and by these means slew a vast number. It happened that among the captives of Unyoro Kibaga saw a beautiful woman, who was solicited by the king in marriage. As Nakivingi was greatly indebted to Kibaga for his unique services, he gave her to Kibaga as wife, with a warning, however, not to impart the knowledge of his power to her, lest she should betray him. For a long time after the marriage his wife knew nothing of his power, but suspecting something strange in him from his repeated sudden absences and reappearances at his home, she set herself to watch him, and one morning as he left his hut, she was surprised to see him suddenly mount into the air with a burden of rocks slung on his back. On seeing this she remembered that Wanyoro complaining that more of their people were killed by some means from above than by the spears of Nakivingi, and Delilah-like, loving her race and her people more than she loved her husband, she hastened to her people's camp, and communicated, to the surprise of the Wanyoro, what she had that day learned. To avenge themselves on Kibaga, the Wanyoro set archers in ambush on the summits of each lofty hill, with instructions to confine themselves to watching the air and listening for the brushing of his wings, and to shoot their arrows in the direction of the sound, whether anything was seen or not. By this means on a certain day, as Nakivingi marched to the battle, Kibaga was wounded to the death by an arrow, and upon the road large drops of blood were seen falling, and on coming to a tall tree the king detected a dead body entangled in its branches. When the tree was cut down, Nakivingi saw to his infinite sorrow that it was the body of his faithful flying warrior Kibaga."

If this tradition had been recorded in recent years, we should be inclined to trace it to the influence of World-War stories spreading to Africa, but it was recorded by Stanley in 1871 when there were no Zeppelins and aeroplanes in sight.

In the seventeenth century Joseph Glanvill predicted that to future ages it might become "as ordinary to buy a pair of wings to fly into remotest regions, as it then was to buy a pair of boots."

John Logan, a Scotch poet of the eighteenth century, has the lines:—

Oh could I fly, I'd fly with thee!
We'd make with joyful wing
Our annual visit o'er the globe,
Companions of the spring.

Erasmus Darwin (1731-1802), grandfather of Charles Darwin, in his *The Botanic Garden* (1789), utters the prophetic words:—

Soon shall thy arm, unconquer'd steam! afar
Drag the slow barge, or drive the rapid car;
Or on wide-waving wings expanded bear
The flying chariot through the field of air.

In 1907 Dr. Alexander Graham Bell wrote, "It has long been recognized by a growing school of thinkers that an aerial vehicle, in order to cope with the wind, should be specifically heavier than the air through which it moves. This position is supported by the fact that all of nature's flying models, from the smallest insect to the largest bird, are specifically heavier than air in which they fly, most of them many hundreds of times heavier, and that none of them adopts the balloon principle in flight... It is certainly the case that the tendency of aerial research is to-day reverting more and more to the old lines of investigation that were pursued for hundreds of years before the invention of the balloon diverted attention from the subject. The old devices have been re-invented; the old experiments have been tried once more. Again, the birds are recognized as the true models of flight, and again men have put on wings, but this time with more promise of success."

THE ROMANCE OF FLYING IN ANCIENT CHINA

Among the many singular coincidences of events that loom up in ancient books of the East and the West, none perhaps is more captivating than that an imperial flyer appears at the threshold of the earliest recorded history of China and that a royal flyer opens the chapter of the early history of Great Britain.

Bladud, the legendary tenth king of Britain, father of King Lear and founder of Bath, is said to have made wings of feathers by means of which he attempted an aerial flight that unfortunately resulted in his death in 852 before our era. This story is recorded by Geoffrey of Monmouth (A.D. 1100-54) in his Historia Regum Britanniae, written in or about the year 1147 (first printed in 1508). Naturally, Bladud is made by tradition a necromancer and performer of magical tricks, in the same manner as attempts at flying were connected with magic in China and elsewhere.

The Chinese emperor Shun, who lived in the third millennium before our era (traditional date 2258-2208 B.C.), is not only the first flyer recorded in history, but also the very first who made a successful descent in a parachute,—an experiment first made or repeated in the midst of our civilization as late as A.D. 1783.

Shun's early life teemed with thrilling adventures. His mother died when he was quite young. His father, Ku Sou, himself of imperial descent, took a second wife by whom he had a son. He grew very fond of his offspring from this new union, but gradually conceived a dislike for Shun, which resulted in several conspiracies against the poor youngster's life. In each case, however, he was miraculously rescued, and in spite of severe persecution continued in exemplary and dutiful conduct toward his father and stepmother. By his filial piety he attracted the attention of the wise and worthy emperor Yao whose name is suggestive of China's golden age. Yao had two gifted daughters, Nü Ying and O Huang, who instructed Shun in the "art of flying like a bird." In the commentary to the Annals of the Bamboo Books (that is, records inscribed on tablets of bamboo), an authentic ancient historical book, Shun is indeed described as a flyer. There it is written, "Shun's parents detested him. They made him plaster a granary and set fire to it at its foundation. Shun donned the work-clothes of a bird, and flying made his escape." Then his parents caused him to descend a deep well and heaped stones on top of it. Shun donned the work-clothes of a dragon and crawled out of the well from the side. He was endowed with a dragon's countenance and thus shared the dragon's natural ability to fly and to crawl. Shun was not a notoriety seeker; he did not fly for the sake of glory or establishing a record; he flew because sheer necessity compelled him to fly to save his soul. For this purpose he availed himself of a flying apparatus based on the principle of bird-flight.

But this is not all. Se-ma Ts'ien, the father of history, as he is justly called, has preserved the following tradition. Ku Sou bade his son Shun build a granary and ascend it, and thereupon set the structure on fire. Shun, who stood on top of the tower, spread out two large reed hats which he used as a parachute in making his descent, and landed on the ground unscathed. Considering the fact that Chinese reed hats are umbrella-shaped, circular, and very large in diameter (some such hats in the Museum's collection from Korea measure two feet three inches to three feet in diameter), this feat would not seem impossible. In the use of the parachute the Chinese have forestalled us a considerable span of time; for Leonardo da Vinci (1452-1519), the great artist, scientist, and mechanician, was the first in our midst who conceived the idea of the parachute. Leonardo writes, "If a man have a tent roof of calked linen twelve yards broad and as many yards high, he will be able to let himself fall from any great height without danger to himself." In one of his manuscripts he has also given the figure of a man descending with this kind of parachute. About 1595 Fausto Veranzio, a Venetian, published a modified design, doubtless inspired by Leonardo's sketch, in which a sort of square sail extended by four rods of equal size is used. There is, however, one great difference between Leonardo and Shun: the former was merely a theorist who never used a parachute, while the latter really performed the trick. The first real descent in a parachute in Europe was not made till 1783 when Lenormand carried out a successful experiment from an observatory at Montpellier.

To complete Shun's story,—he married the two sisters, his teachers in the art of flying; and Yao, his father-in-law, gave him a share in the government of the empire. On the latter's death he succeeded to the throne and ruled as the model of a good and wise sovereign.

Chinese writers fable about a country of Flying Folks (Yü min), located in an island in the south-eastern ocean, living on high peaks near the sea-shore, and described as people with long jaws, bird's-beaks, red eyes, white heads, covered with hair and feathers, able to fly, but not over a long distance; they are said to resemble human beings, but to be born from eggs. The conception of bird-men is quite familiar to ancient Chinese mythology. A deity with outspread wings, seated on a pedestal, is shown on the pediment of a gravestone of the Han period (Plate I), from about the middle of the second century of our era. Winged attendants fly above him, and others approach him, holding gifts or offerings. On the left are two kneeling figures, holding tablets in the attitude of adoration, the first with a horse's-head, the second with a bird's-head, both winged, but for the rest human and clad in wide, long gowns. Behind this pair appears a walking bird with long tail-feathers, but with human head, holding the leaf of a plant. This picture represents the abode of the aerial spirits. In Assyrian-Babylonian monuments winged figures, man-headed or bird-headed, are frequent, but they are always represented standing or walking, never flying, which makes for a net distinction from the Chinese flying bird-men. As in the sculptures of Mesopotamia winged bulls, lions, griffins, and horses appear, so we meet also in China statues and relief representations of winged monsters, tigers, lions, and horses.

Lei Kung, god of thunder and lightning, has wings attached to his shoulders, usually wings of a bat, and by means of these appendages he directs his course through the air to wherever he desires to produce a thunder-storm.

Tung Yung, a legendary personage, who is supposed to have lived in the second century of our era, was rewarded for his filial piety by the Spinning Damsel, an astral deity, from whom he received two boys whom she had deposed under an elm-tree. One of these had under his arm-pits fleshy excrescences in the shape of wings, and his face displayed a bird's-beak. When grown up, he excelled in muscular strength and supported his father and younger brother with the fruit of his manual labor. His employer complained of his insatiable appetite, but was pleased, as he performed the work of two men. One day the bird-man announced that his mother had appeared to him the previous night, inviting them to rejoin her; then he unfolded his wings, shouldered his father, and sailed up skyward.

"Ascending to heaven by means of flight" is expressed in Chinese "by means of feathers he was transformed and ascended as an immortal"; and "feather scholar" or "feather guest" is a term for a Taoist priest.

Winged flight, however, appears but seldom as a real attempt. The emperor Shun is practically the sole example, and seems to have found few imitators, quite in distinction from Daedalus, whose feat has stimulated so many until recent times. Another instance of winged flight known to me is one that occurs in a dream.

T'ao K'an, a celebrated Chinese statesman (A.D. 209-334), once had a dream which led to his advancement. He dreamt that he scaled the heights of heaven with the aid of eight wings, and passed through eight of the celestial doors, but was driven back from the ninth by the warder, who cast him down to earth. When he landed there, the wings on his left side were broken. Subsequently he entered public life, and was appointed governor of eight provinces, which was interpreted as a realization of his dream.

The first description of an air-journey is found in the celebrated poem Li Sao ("Fallen into Sorrow") by K'ü Yüan (332-295 B.C.), a loyal statesman, who enjoyed the confidence of his sovereign until impeached through the intrigues of rivals. Despondent over his disgrace and concious of his own integrity, he found solace in composing a poem, which is an allegorical picture of his search after a prince who would listen to good counsels in government. The poet kneels at the grave of the emporer Shun, and is then carried up into the air in a chariot built in the form of a phoenix to which are yoked four dragons smooth as jade. In this vehicle, through dust and wind, he suddenly ascends on high toward the K'un-lun range of mountains. Wang-shu, the charioteer of the moon, is his precursor, and Fei-lien, god of winds, follows him as attendant.

> I ordered the phoenix to fly aloft,
> And continue its flight day and night.
> But a whirlwind brought together my opponents,—
> Clouds and rainbow were led to meet and oppose me.
> In multitudes they assembled, now dividing, now collecting.
> In confusion they separated, some going above, others beneath.

In his search he surveys the earth to its four extreme points, travels all over the sky, and then descends to the earth. Again he undertakes a journey into the air above the Kun-lun Mountains, in a chariot adorned with jade and ivory drawn by a team of eight flying dragons.

> I turned my course to K'un-lun;
> Long was the way, and far and wide did I wander,
> Amidst the dark shade were displayed the rainbows in the clouds,
> While the jade bells about the chariot tinkled.
> I started in the morning from the Ford of the Sky,
> And in the evening I arrived at the extreme west.

The idea of a flying chariot or airship, usually drawn by dragons, is not alien to ancient Chinese art. An aerial contest of winged beings astride scaly and horned dragons (Plates II-III) is skilfully represented on a grave-stone of the Han period (second century A.D.). This picture is animated by life and motion: an exalted winged god is enthroned in a flying chariot set in motion by fleet dragons and floating over clouds. The pilot of the airship is leaning forward, tapping a dragon's tail as though eager to spur him on. Two winged standard-bearers mounted on swiftly moving dragons follow the car as escorts. Four dragon-riders precede it, and the procession moves on toward a winged flag-bearer standing on a cloud, while another person kneels in front of him. The clouds are represented as birds on the wing, their bodies consisting of spirals which are symbolic of clouds.

Huang Ti, one of the ancient legendary emperors, attained immortality by mounting a fantastic creature with the body of a horse and wings of a dragon (called *tse-huang* or *ch'eng-huang*). According to another version of the legend, he made his ascent on a long-bearded dragon strong enough to transport also his wives and ministers,—more than seventy persons. The officials of lower rank who were not able to find a seat on the dragon's back (not unlike the strap-hangers in our street-cars) clung to the hairs of the dragon's beard; these, however, gave way, the passengers plunged to the ground and also dropped the emperor's bow. The multitude of spectators reverentially watched the apotheosis from a distance, and when Huang Ti had reached his destination in heaven, they picked up his bow and the dragon's hairs. This story is a cheery example of the Chinese sense of humor: other myth-framers would have been prone to push their principles to extremes and, endowing the dragon's beard with divine strength, would have conveyed the strap-hangers straight heavenward.

Under the reign of the emperor Ts'in Shi (259-210 B.C.), Mao Mong, great grandfather of Mao Ying, styled "the true man of sublime origin," ascended Mount Hua, mounted the clouds and bestrode a dragon which was hidden in the clouds, rising into the azure spaces of heaven in broad daylight.

In a hymn to the god of Heaven composed under the emperor Wu (140-87 B.C.) of the Han dynasty, the deity appears amid dark clouds in a chariot drawn by winged flying dragons and adorned with many feathered streamers; the rapidity with which the deity descends is compared with that of the horses of the wind.

Pei Ti, god (literally, "emperor") of the north, much worshipped at Canton under the name Pak Tai, was after a long career of holiness elected to the office of chief minister of the gods. Angelic messengers descended from heaven, presenting him with silk robes, red shoes, flying swords, and a chariot of nine colors in which he ascended to the celestial abode at the time of the reign of Huang Ti.

When the mind of a nation is filled with the romance of the air, when the air surrounding it is populated with winged genii and flying chariots, and when such subjects are glorified by art and adorn the stone walls of the grave chambers, it is the logical step that imagination thus impregnated leads one or the other to attempt the construction of some kind of an airship.

The *Ti wang shi ki* ("History of the Ancient Emperors"), written by Huang-fu Mi (A.D. 215-282), contains this notice:—

"Ki-kung-shi was able to make a flying chariot which driven by a fair wind travelled a great distance. At the time of the emperor Ch'eng T'ang (1766-54 B.C., founder of the Shang dynasty) the west wind blew Ki-kung's chariot as far as Yü-chou (Ho-nan). The emperor ordered this chariot to be destroyed that it should not become known to the people. Ten years later when the east wind blew, the emperor caused another chariot of this kind to be built by Ki-kung and sent him back in it."

The term "flying chariot" (*fei ch'o*) used in this passage is now current in China for the designation of an aeroplane.

A similar account is contained in the *Po wu chi*, written by Chang Hua in the third century of our era, but with the difference that the invention of the flying chariot is ascribed to the Ki-kung *nation.* A tribe of this name is mentioned in the *Shan hai king* ("Book of Mountains and Seas"), an ancient collection of (partially absurd) geographical fables, where the Ki-kung are characterized as single-armed (the very name means "one upper arm"), three-eyed, hermaphrodites, and riding on striped horses. Shen Yo (A.D. 411-513), the commentator of the Bamboo Annals, speaks of the Ki-kung or their chief as having arrived in a chariot at the court of the emperor Ch'eng T'ang in 1766 B.C., but he says nothing of a flying chariot, nor does the *Shan hai king*, which attributes to them horses as means of conveyance. At the outset it is hardly probable that single-armed hermaphrodites should have a special talent for aviation. It is therefore obvious that in the above notice of the *Po wu chi* two distinct traditions are contaminated: there was, in my opinion, an individual who lived in times of antiquity, Ki-kung by name, who invented an airship or endeavored at least to construct one; and there was also a tradition current about a fabulous tribe accidentally bearing the same name, which had nothing to do with aviation; because, however, the Ki-kung people arrived at the imperial court in chariots, it was easy to confound or identify these chariots with the flying chariot made by the mechanic, Ki-kung by name. On the other hand, it is suspicious that the latter also is supposed to have come to the court of Ch'eng T'ang, and this date is surely the outcome of an afterthought and devoid of historical value. Be this as it may, the interesting point to be retained is that the Chinese possess an apparently old tradition regarding an airship driven by the force of the winds.

A wood engraving of what in the estimation of Chinese draughtsmen this airship looked like is on record and reproduced in Plate IV. Here we see two men standing in a square, box-like affair, with flags flying, comfortably sailing in the clouds; the car is set in motion by two curious wheels. It will be noted that the two men are just human, having two eyes and two arms; and it should be borne in mind that this illustration is not contemporaneous with the story, nor is it handed down from ancient times, but that it is of comparatively recent origin and merely reconstructed upon the slender fabric of the ancient tradition. It has as much value for the reconstruction of the airship in question as, for instance, Doré's illustrations of the Bible have for the reconstruction of ancient Hebrew life and archaeology. Professor Giles, who first called attention to this drawing in his article "Traces of Aviation in Ancient China," has also published an earlier woodcut of the same subject, taken from a rare book in the Cambridge University Library, that was published in China in the latter part of the fourteenth century; the Ki-kung car illustrated in this book, aside from minor details, is practically identical with the later production aforementioned. Professor Giles adds this interesting comment: "It is noticeable at once that the occupants of the car, especially in the later illustration, are not one-armed. Also, that the wheels fore and aft are at right angles to the direction in which the car is flying through rolling clouds; and further, what is most curious of all, that the wheels appear to be constructed on the screw system, like the propeller of a steamer. Now, in the published description of Latham's flying-machine, we read, 'For the cross-Channel flight a fifty horsepower Antoinette motor has been mounted. This drives a screw which, placed in front of the machine, cleaves a way through the air, pulling the machine after it. It is called a tractor screw.'"

In the attempt to reconstruct the flying chariot, the Chinese draughtsman stressed the second part of the compound and produced the picture of a two-wheeled cart. In this point he is decidedly wrong, for a vehicle of this type could never rise into the air. We have to fall back on the words of the account itself, in order to form some idea of what this airship might have been. The sole indication of a motive power given in the text is the wind: the vehicle in question depended upon favorable winds, and was propelled by the east wind if it wanted to go east, and by the west wind when it was to return west. For this reason it cannot be presumed that a car or chariot, in the strict sense of the word, is involved. The word *ch'o*, which means a "car," refers also to machines, engines, or contrivances which do not move; thus, for instance, *hua ch'o* ("smooth car") signifies a "pulley"; *shan ch'o* ("fan car"), a "winnowing mill." Now, as far as ancient China is concerned, there were only two devices known as capable of setting a vehicle in motion,—a sail and a kite. As to sails, the Chinese very efficiently applied them (and presumably still apply them) to wheelbarrows, as I repeatedly noticed myself on my travels in Shan-tung Province; but a sail alone cannot lift any vehicle from the ground. This, however, may be accomplished by several powerful kites. The Chinese were the inventors of the flying-kite, as will be set forth in the following chapter, and were in possession of kites at an early date, assuredly in the third century of our era, the date of the *Ti wang shi ki*. I imagine, therefore, that Ki-kung's "flying chariot" was built on the aerostatic principle, being driven by a combination of sails and kites, and was very much like the kite-chariot constructed by George Pocock in 1826 and discussed in detail below (p. 41). The "chariot" part of Ki-kung's machine may have been a very simple affair: all he needed was a seat for himself, which may have been made of light wood, bamboo, or basketry.

The famous boat-shaped aerial car, theoretically conceived by the Jesuit Francesco Lana (1631-87), is reproduced in Plate V, not for its own sake, but because it exhibits some affinity with Ki-kung's machine and, mutatis mutandis, may help us visualize it to better advantage. It was Lana's idea of lifting his ship into the air by means of four large, hollow globes of very thin sheets of copper, from which the air had been wholly extracted, thereby causing them to weigh less than the surrounding atmosphere, and enabling them to rise and support the weight of the ship in the air; propulsion and direction were to be obtained by sails and oars. The question of the practicability of this proposal does not concern us here; what I wish to point out is merely this, that if in Lana's sketch the four copper globes are replaced by four powerful paper kites, we may realize what the Chinese aerostat might have been.

The Chinese emperor, in the above story, caused the airship to be destroyed, as he did not wish his own people to see it. He evidently was anxious to remain intrenched on his throne and to steer clear of innovations that might menace the safety of his realm. Francesco Lana, in his Prodromo (1670, p. 61), gives us the best explanation of the reasons which may have prompted that autocrat to his action. Having developed his plan of an airship with sail and oars, as pointed out above, the Jesuit author winds up thus: "I do not see any other difficulty that could prevail against this invention, save one, which seems to me weightier than all others, and this is that God will never permit such a machine to be constructed, in order to preclude the numerous consequences which might disturb the civil and political government among men. For who sees not that no city would be secure from surprise attacks, as the airship might appear at any hour directly over its market-square and would land there its crew? The same would happen to the courtyards of private houses and to ships crossing the sea, for the airship would only have to descend out of the air down to the sails of the sea-going vessels and lop their cables. Even without descending, it could hurl iron pieces which would capsize the vessels and kill men, and the ships might be burnt with artificial fire, balls, and bombs. This might be done not only to ships, but also to houses, castles, and cities, with perfect safety for those who throw such missiles down from an enormous height."

The first author in Europe who discussed the possibility of a flying chariot was John Wilkins (1614-72), bishop of Chester from 1668 and subsequently Master at Trinity College, Cambridge, one of the founders of the Royal Society, to which he acted as first secretary. His writings, particularly his "Mathematicall Magick" (1648), contributed much toward arousing public interest in the problem of flight. He distinguishes (p. 199) "four several ways whereby this flying in the air hath been or may be attempted. Two of them by the strength of other things, and two of them by our own strength: 1. By spirits or angels. 2. By the help of fowls. 3. By wings fastened immediately to the body. 4. By a flying chariot." This fourth and last way seems to him altogether probable and much more useful than any of the rest. "And that is by a flying chariot, which may be so contrived as to carry a man within it; and though the strength of a spring might perhaps be serviceable for the motion of this engine, yet it were better to have it assisted by the labour of some intelligent mover as the heavenly orbs are supposed to be turned. And therefore if it were made big enough to carry sundry persons together, then each of them in their severall turns might successively labour in the causing of this motion; which thereby would be much more constant and lasting, then it could otherwise be, if it did wholly depend on the strength of the same person. This contrivance being as much to be preferred before any of the other, as swimming in a ship before swimming in water."

Kung-shu Tse, also called Lu Pan, "the mechanician of Lu," because he was a native of the state of Lu in Shan-tung Province, was a contemporary of Confucius and a clever mechanician. In the work going under the name of the philosopher Mo Ti (chap. 49) who lived in the fifth century before our era, he is said to have carved a magpie from bamboo and wood; when completed, he caused this artificial bird to fly, and only after three days it came down to earth. According to another tradition, Kung-shu himself made an ascent riding on a wooden kite in order to spy on a city which he desired to capture. Other Chinese writers ascribe the manufacture of a wooden kite to Mo Ti, or to the collaboration of both Kung-shu Tse and Mo Ti, saying that it could fly for three days without resting. Han Fei, a philosopher of the third century before our era, relates that Mo Ti worked for three years at a wooden kite, but that after flying for a single day it was smashed. It is obvious that in these various accounts there is a confusion of "three days" and "three years," while no clear idea is conveyed of the construction and mechanism of the artifact. Some Chinese authors regard this wooden kite as the beginning and forerunner of the later toy, the paper kite; but this view seems erroneous, as the bird is described as being carved from wood, and as paper was unknown during the period in question. It appears to have been rather an automatic, mechanical contrivance that was capable of rising to some extent into the air,—a sort of affinity to the dove of Archytas (p. 64). Certain it is that it was not Mo Ti, as asserted by Han Fei and Lie-tse, who made the flying kite: in the first place, Mo Ti was a philosopher of ethical and social tendencies who did not engage in manual labor; second, Mo Ti himself saw it fit to condemn the invention of the flying kite as an idle and useless plaything.

Kung-shu Tse is credited with several other inventions,—two kinds of a grinding mill and a scaling-ladder used in besieging cities, known as "cloud-ladder" (*yün t'i*). He is also said to have made wooden horses which moved by means of springs and could draw carriages; or, according to another version, he made for his mother a wooden coachman who drove an automobile. At present Kung-shu Tse is worshipped as the patron-saint of carpenters.

There is a curious incident on record in the Book of Rites (*Li ki*), which illustrates the fact that even in his youth his thoughts were concentrated on problems of engineering and that his contemporaries were averse to his innovations. A certain individual's mother had died, and Kung-shu asked leave to lower the coffin into the grave by means of a new mechanical contrivance invented by him. Its application was objected to by one present at the funeral on the ground that the ancient practices of the principality of Lu ought to be upheld, and it was ironically suggested to the inventor that he should test his ingenuity rather on his own mother than on that of another man.

It is small wonder that later legends have grossly exaggerated Lu Pan's mechanical skill. Thus a story is current that he made a wooden kite which was mounted by his father, and the old man flew as far as Wu-hui, a town in the prefecture of Su-chou, Kiang-su Province, the ancient kingdom of Wu. The people there took the landing flyer for a devil and slew him. Lu Pan, infuriated at this detestable crime, carved a wooden effigy of some evil spirit, whose hand pointed in the direction of Wu and caused a drought there for a period of three years. On consulting the oracle, the inhabitants of Wu recognized that this calamity was brought about by Lu Pan, and appeased his wrath with supplications and presents, whereupon he chopped off the hand of the statue, and rain fell abundantly in the kingdom. According to another legend of comparatively recent date, Lu Pan made a wooden kite; all it was necessary to do was to rap at the door-post three times, and the kite flew off, carrying away the person who was mounted on it. There is reason to believe that Kung-shu Tse and Lu Pan are two distinct individuals and that the two were merged into one by subsequent traditions, but this question does not concern us here; we are interested in the mechanical contrivance itself, and for this purpose the texts of the early philosophers only merit consideration.

Wang Ch'ung (A. D. 27-97), philosopher, critic, and sceptic, who poked fun at the literati, discredits Lu Pan's invention in the following discourse:—

"From wood he carved a kite capable of flying for three days without descending. It is possible that he made a wooden kite and was able to fly it; but the report that it did not alight for three days is exaggerated. Carving it from wood, he gave it the shape of a bird;

how, then, could it fly without resting? If it could soar, why just for three days? In case it was equipped with a mechanism by which it was set in motion and continued to fly, it might not have descended. In this case it should be said that it flew continually, not for three days. There is a report that through his own skill Lu Pan lost his mother. Being a skilled mechanic, he had constructed for her a wooden carriage and horses with a wooden charioteer. When the apparatus was completed, he set his mother in the carriage which sped away and never returned. Thus he lost his mother. Provided the mechanism of the wooden kite was well arranged, it must have been like that of the wooden carriage and horses; in this case it would have continued to fly without stopping. On the other hand, a mechanism functions but a short while, and for this reason the kite could not have kept up its motion for more than three days. This also holds good for the wooden carriage, which should have come to a stop after three days on the road, instead of going on so that his mother was lost. Apparently the two stories are untrustworthy."

It is obvious that as early as the first century of our era real knowledge of this contrivance was lost.

Aside from the dove of Archytas to which reference has been made, Lu Pan's wooden magpie or kite meets with another curious parallel in the West. The astronomer Regiomontanus, who lived at Nuremberg in the fifteenth century, is said to have constructed an eagle which, on the emperor's (Charles V) approach to the city, he sent out high in the air a long way to meet him, and which accompanied him to the city gates. I. B. Hart furnishes this comment, "Shorn of all the inevitable additions of credulous narrators, the probability is that Regiomontanus, who was of a mechanical turn of mind, fashioned a clockwork contrivance which, more by luck possibly than by design, acted as a glider when released." Regiomontanus is also credited with having had an automaton in perpetual motion in his workshop and with having made a fly which, taking its flight from his hand, would fly around the room, and at last, as if weary, would return to his master's hand. Francesco Lana, in his "Prodromo" (1670, pp. 50-51), has given directions as to how to make birds which fly through the air. Considering the fact that such like contrivances are reported from different parts of the world and at widely varying times, we cannot refrain from concluding that a grain of truth must underlie these accounts and that Lu Pan's wooden kite also, even granted that like other inventions it has been magnified, to some extent was an object of reality and had a foundation in fact.

Perhaps it was a primitive form of glider, perhaps it was connected with and raised by a flying kite.

Starting from realistic means of flight, Chinese efforts did not continue in this direction. Strangely enough, from realism they developed into mysticism and magic. From the second century B.C. alchemical lore coming from the West began to infiltrate Chinese thought; quest of the elixir of life and the desire to transmute base metals into gold allied themselves with ancient native conceptions of formulas for securing longevity and immortality in a better land. The notion of flight was a link of paramount importance in this chain of mystic dreams which held the minds of the people enthralled for many centuries.

Liu An, commonly known as Huai-nan Tse (second century before our era) was much given to alchemistic studies and to search for the elixir of life on which he published several treatises. Tradition credits him with the discovery of an elixir which he finally drank, with the effect that he rose to heaven in broad daylight. The vessel which contained the beverage of immortality he dropped into his courtyard, and when the dogs and poultry sipped the dregs, they immediately sailed up to heaven after him.

Li Shao-kün, an adept of alchemy and the magic arts under the emperor Wu of the Han dynasty (140-87 B.C.), over whom he gained great influence, made an elixir of life and pretended to be able to transmute cinnabar into gold. He described his magic powers in this strain, "I know how to harden snow and change it into white silver. I know how cinnabar transforms its nature and passes into yellow gold. I can rein the flying dragon and visit the extremities of the earth. I can bestride the hoary crane and soar above the nine degrees of heaven."

Indeed, the riding conveyance favorite with Taoist saints for taking passage into the beyond is the crane, a bird famed in Chinese lore and endowed with many supernatural attributes. He is said to reach a fabulous age, and when six hundred years old, to dispense with solid food, but to continue to drink water. Of the four kinds of crane,—the black, the yellow, the white, and the blue ones,—the black one is the longest-lived. He is hence reputed as the patriarch of the feathered tribe, and manifests a particular interest in human affairs. Men have repeatedly been transformed into the shape of a crane, and he transports to the regions of eternal bliss those who have attained the degree of sainthood in this life, as he also serves as a vehicle to the goddes Si Wang Mu (Plate VI). This picture is a small section from a large embroidered screen in twelve panels in the Museum's collection, depicting the celebration of the goddess' birthday when the Eight Immortals appear to offer congratulations and rich gifts. The goddess surrounded by attendants alights from her celestial quarters on the back of a flying crane.

Wang Tse-k'iao, who was the eldest son of king Ling of the Chou dynasty and lived in the sixth century before our era, studied the black art for thirty years under a magician named Fou-k'iu Kung. One day he sent a message to his kin, saying that he would appear to them on the seventh day of the seventh moon on the summit of a mountain; and indeed, on the appointed day, he was seen riding through the air on a white crane, waving his farewell to the world and ascending to heaven to join the ranks of the immortals.

Ting-ling Wei (second century of our era), a student of the black art, was transformed into a crane a thousand years after his death, that he might revisit earth and his old home, when he bewailed the changes that time had wrought upon men and their hearts.

Wang K'iao, who lived in the first century of our era, used to report regularly at court; but as he had no chariot or horses, the emperor Ming of the Han dynasty was curious to learn how he managed to travel such a long distance, and instructed the Grand Astrologer to find out. The imperial messenger was amazed to discover that Wang did the trick by riding upon a pair of wild ducks, which bore him swiftly through the air. Hence he lay in wait and threw a net over the birds; but when he went to seize them, he found only a pair of official shoes which had been presented to Wang by the emperor.

Less frequently a tiger is made the aerial courser. This was the climax of the life of Madame Ts'ai Luan, who lived in the fourth and fifth centuries of our era and made a study of the black art. These efforts, however, did not have any riches in store for her; for she remained poor and eked out a meagre livelihood by making copies of a dictionary of rhymes, which she sold to scholars. Her reward came ten years later when she and her husband went up to heaven on a pair of white tigers.

Some accomplished the ascent to heaven even without the medium of a riding animal. Thus Ma Tse-jan, reputed for his wide knowledge of simples and in great demand as a physician, studied the doctrines and practices of Taoism, and was ultimately taken up to heaven alive.

Plate VII illustrates a Taoist saint comfortably flying in the air from cliff to cliff, simply driven by the wind, while two wanderers on the mountain path gaze at him in bewilderment. This picture is a landscape drawn in ink, probably of the Ming period.

"Shoes which enable one to ascend the clouds" (*teng yün li*) are ascribed by tradition to Sun Pin; they were made of fish-skin and enabled their wearer to walk on water and to tread onc louds. "Flying cloud shoes" (*fei yün li*) are attributed to the famous poet Po Kü-i (A.D. 772-846): when he was engaged in preparing an elixir on Mount Lu in Kiang-si Province, one of the haunted grottoes of the Taoists, he made these shoes of fine black damask, cutting a cluster of clouds out of plain raw silk which he dyed with four choice aromatics. When he moved around in these shoes, he looked like smoky mist, as though clouds were rising from beneath his feet and as though he would before long ascend to the celestial palace. These magic shoes are an echo of the thousand-league boots of our folk-lore of which more will be said in the chapter on India.

In the long history of this struggle for the conquest of the air, two singular ideas finally come to the fore—levitation by means of starvation and application of remedies taken internally. The slogan of this school was: Live on air to conquer the air!

These Taoist ideas may partially be traceable to India. The Buddhist saints of the Tantra school also had the notion of obtaining supernatural powers which would enable them to transmute their bodies and to assume any shape at will, as well as to traverse space with the most rapid possible motion.

Leading a natural life in the seclusion of mountains in close contact with nature was believed to be conducive to obtaining longevity and immortality. The highest ambition of many Taoist hermits, then, was to reduce their weight, to lighten their bodies, to release their souls, and thus to obtain the ability to fly toward heaven. Chang Tao-ling (A.D. 34-156), known as the first Taoist pope, retired to the mountains and devoted himself to the study of alchemy and to cultivating the virtues of purity and mental abstraction. The white tiger presiding over the west and the green dragon ruling the quarter of the east appeared in the air above his habitation, and finally he reached his goal and found the elixir of immortality. He swallowed a dose of it, and the sixty year old man was suddenly transformed into a vigorous youth. Soon after when he made a pilgrimage to a sacred mountain in the proximity of the city of Ho-nan, he met on the road a man who disclosed to him the location of a cave; it concealed, he intimated, occult writings whose study would enable him to obtain the power of flying skyward. After fasting and purifying himself he found the books in question, and by studying these he attained the gift of ubiquity, and was capable of assuming simultaneously various shapes. After years of meditation and efforts spent on exorcism of demons he was deemed worthy of appearing before Lao-tse, and ascended as an immortal to the heavens with his wife and two favorite disciples.

An-k'i Sheng, a legendary magician who lived on the Isles of the Blest in the Eastern Ocean and possessed the power of rendering himself visible or invisible at will, visited the Lo-fou Mountains in Kwang-tung Province, where he subsisted only on the stalks of water-rushes; by virtue of this diet he finally became emancipated from the dross of earth, and ascending the summit of the White Cloud Mountain, rose to heaven before the eyes of his companion. The recipe "living on air" was tried, for instance, by Chang Liang, who died in 187 before our era. He began to eliminate food in the hope of gaining levitation of the body and finally immortality, but failed, because he once yielded to the solicitations of the empress and ate a bit of rice.

A good example of this sort of hunger-strike apostle is presented by Li Pi (A.D. 722-789), who as a youth was keenly devoted to the study of Taoism and would roam in the mountains, pondering upon the secret of immortality. He declined to marry, abstained from all substantial food except fruit and berries, and practised the art of breathing, which is believed by the Taoists to conduce to immortality. He finally became reduced to a skeleton, and received the nickname "the Collar-bone Immortal of Ye." Chang Cho also, a scholar of the ninth century, trained himself to get along without food. He was able to cut butterflies out of paper, which would flutter about and return to his hands. Lu Ts'ang-yung, son of an official, flunked in the civil service examinations and retired with his brother to the mountains, where they lived as hermits and studied the art of getting along without food (which in our own times many students of art and science have involuntarily imitated).

The climax of this movement was reached in the preparation of a nostrum for promoting the art of flying.

T'ao Hung-king (A.D. 452-536), a distinguished physician and a celebrated adept in the mysteries of Taoism, compounded what is known as the "flying elixir" (*fei tan*): it did not contain any medicinal drugs, but was a mixture of gold, cinnabar, azurite, and sulphur,—ingredients which had been contributed by the emperor. This compound is said to have had the color of hoarfrost and snow and to have been bitter of taste. When swallowed, it was believed to produce levitation of the body. The emperor tasted and tested it, found it beneficial, and conferred honors on the manufacturer. I think this is the only example in the history of the world in the way of teaching to fly by means of a medicine taken internally; but from the viewpoint of Chinese alchemical and religious lore it is quite intelligible how this notion could spring into existence.

In speaking of the Yogins of India, Marco Polo writes, "They are extremely long-lived, every man of them living to a hundred and fifty or two hundred years. They eat very little, but what they do eat is good; rice and milk chiefly. And these people make use of a very strange beverage; for they make a potion of sulphur and quicksilver mixed together, and this they drink twice every month. This, they say, gives them long life; and it is a potion they are used to take from their childhood." The alchemists of both Asia and Europe regarded sulphur and mercury, combined under different conditions and various proportions, as the origin of all metals. Mercury was called the mother of metals; sulphur, the father.

The desire to obtain eternal youth focused on the elixir of immortality, the fountain of youth, or the rejuvenescent water of life, has haunted mankind through all ages. It was this theme that occupied Nathaniel Hawthorne's mind throughout his life and that culminated in his unfinished romance which rested upon his coffin.

KITES AS PRECURSORS OF AEROPLANES

A flying-kite may be defined as an aeroplane which cannot be manned, and an aeroplane may be defined as a kite which can be manned. This definition implies the interrelation of the two mechanical devices. How this development was brought about will be demonstrated on the following pages. Kites were invented and first put to a practical test in ancient China; hence the Chinese must be credited with a substantial contribution to the advance of aeronautics. In January, 1894, O. Chanute wrote in Chicago, "It would not at all be surprising to find, should a stable aeroplane be hereafter produced, that it has its prototype in a Chinese kite." And history proves him right.

It must not be imagined that the Chinese kite is anything like the flimsy, cross-shaped structure of wood covered with paper of a diamond-shaped surface that we used to fly in our boyhood days. This toy is a poor degenerate orphan put to blush in comparison with the ingenious creations of the Chinese, which are wonders of both technique and art. The ordinary Chinese kites are made of a light, elastic framework of bamboo over which is spread a sheet of strong paper painted in brilliant hues with human or animal figures. They generously display that love of art and that whole gamut of decorative design which runs through the life of the Chinese nation. Favorite subjects are mythological figures and monsters, dragons, actors, and heroes of popular plays, beautiful women, animals of all descriptions, birds of prey, serpents, frogs and fishes, flies and butterflies as well as centipedes, also flower-baskets and boats. In the bird-kites the thin paper attached to the wings is moved by the wind and simulates the flapping of the wings. It goes without saying that, as indicated by the common term "paper kite" (*chi yüan*) for the device, kites are a favorite pattern frequently in evidence. Not only the great variety of quaint shapes and designs is amazing, but also the correct calculation or premeditated evaluation of the distance effect; viewed in close proximity the kite pictures may seem disproportionately large or exaggerated or even distorted, while naturally they are designed for a distant vista and in fact, when towering high in the air, appear most beautiful and so life-like that they may be taken for real birds. Again, the kite in the air is hardly ever stationary, but constantly on the move, hovering and soaring, and as it moves on, appears more and more as a legitimate denizen of the atmosphere.

And then the stupendous skill of the hands which manipulate the flying monsters! A long coil of tough cord is wound over a reel, and it is the reel, not the cord, which is held in the hand and is continually turned as the paper plane rises.

The most complicated and ingenious of these flying-machines is the centipede kite. One which I obtained at Peking in 1901 for the American Museum of Natural History in New York (together with a collection of some seventy kites, all of different types) measures forty feet in length, and is made to fold up accordion-like. The fierce head of the creature with huge eyes and gaping jaws is surmounted by long, protuding horns. The body consists of a series of some twenty-five disks, about a foot in diameter, formed of a bamboo frame covered with paper. These are painted with concentric zones in black, yellow, and white. The disks are connected with one another by two cords which keep them equidistant, and are fastened to a transverse bamboo rod from which sticks run crosswise to the centre of the disks. The latter revolve when the kite is being flown. The rear disk is provided with streamers that form the tail. It requires great skill to raise this kite, and cords are attached to three or more points of the body to keep it under control. In a strong wind several men are required to hold the reel. Seen in the air with its gigantic proportions, its huge glaring eyes swiftly twirling in their sockets, its weird, wriggling, serpentine motions, it conveys the impression of some fossil monster of bygone ages having suddenly come back to life. A centipede kite of smaller dimensions is also made in Hawaii.

O. Chanute justly calls attention to the fact that this device resembles in arrangement the multiple disk kites suggested and designed for life-saving in shipwrecks by E. J. Cordner, an Irish Catholic priest, in 1859.

Mechanically kites are constructed on the principle underlying the behavior of a soaring bird which performs its movements with peculiarly curved and warped surfaces.

The ninth day of the ninth month in the autumn is devoted to the festival called Ch'ung-yang which is celebrated by ascending hills. Friends and acquaintances join for a picnic on some eminence in the neighborhood of their town and set kites adrift, as the autumnal breezes favor the sport. This also is the great day for holding kite contests. Any kite, no matter to whom it belongs, may be cut down by another. For this purpose the cord near the kite is stiffened with crushed glass or porcelain smeared on with fish-glue. The kite-flyer manoeuvres to get his kite to windward of that of his rival, allows his cord to drift against that of his opponent and by a sudden jerk to cut it through, so that the hostile kite is brought down.

A musical kite was first invented by Li Ye of the tenth century of our era, an expert kite-maker, who was purveyor of kites for his imperial majesty. He made an ordinary paper kite with a string attached to it and fastened to the kite's head a bamboo flute. He flew this kite in such a manner that when the wind struck the holes of the flute and produced sounds like those of a harpsichord (*cheng*), which originally had twelve, at present, however, has thirteen, brass strings. Hence a new term for kites came into vogue—"wind harpsichord" (*fung cheng*), which is now used indiscriminately for any kite.

Such flutes are still occasionally used in connection with kites. They consist of a short bamboo tube closed at the ends and provided with three apertures,—one in the centre and one at either extremity. When the kite is flying, the air, in rushing into the holes of the instrument, produces a somewhat intense and plaintive sound, which can be heard at a great distance. Sometimes three or four of these bamboo tubes are placed one above another over the kite, and in this case a very pronounced deep sound is produced. Imagine that hundreds of such kites may be released at a time and are hovering in the air, and there is a veritable aerial orchestra at play. This music has a beneficial effect, for it is thought to scour evil spirits from the atmosphere. To strengthen this benevolent influence, a captive kite, during the prevalence of winds, is often affixed to the roof of a house when during the whole night it will emit plaintive murmurs. Still more frequently, at least in Peking in the age of the Manchu dynasty, there was attached to the top of a kite a musical bow of light willow-wood or bamboo strung with a silken cord. When struck by the wind, the instrument would produce humming sounds like an Aeolian harp. Thus the Chinese were the first who knew how to produce "music on the air." At night paper lanterns with a lighted candle inside were sometimes suspended from small kites in the shape of butterflies, and these were again attached to the main or pilot kite of much larger dimensions. Ear and eye are thus treated to a feast, but this is not all. To make the performance still more spectacular, messengers consisting of bamboo frames with fire-crackers attached are sent up the strings from which the kite is governed, and the crackers are timed to explode on reaching the top.

Archdeacon J. H. Gray, in his excellent book "China" (1878), says, "In the centre of Chinese kites, four or five metallic strings are fixed on the principle of the Aeolian harp. When they are flying, 'slow-lisping notes as of the Aeolian lyre' are distinctly heard."

He then records from oral tradition the following story in explanation of this musical apparatus: "During the reign of the emperor Liu Pang, the founder of the Han dynasty, a general who was much attached to the dynasty which had been obliged to give way before the more powerful house of Han, resolved to make a last vigorous effort to drive Liu Pang from the throne he had recently usurped. A battle, however, resulted in the army of the general being hemmed in and threatened with annihilation. At his wit's end to devise a method of escape, he at last conceived the ingenious idea of frightening the enemy by flying kites, fitted with Aeolian strings, over their camp in the dead of night. The wind was favorable, and when all was wrapt in darkness and silence, the forces of Liu Pang heard sounds in the air resembling *Fu Han!* Beware of Han! It was their guardian angels, they declared, who were warning them of impending danger, and they precipitately fled, hotly pursued by the general and his army."

Kites were originally used in China for military signalling and for such purposes only, but in the beginning they were not connected with any religious practices, as is erroneously stated by several authors. Thus Hodgson (History of Aeronautics, p. 368) writes, "It cannot be doubted that the kite, though of uncertain, is nevertheless of very ancient origin... Though in wide-spread use as a pastime among the Chinese, Japanese, Maoris, and other peoples, its origin is usually ascribed to religion." Originally it was not a toy either; this is a later development which set in from the time of the Sung dynasty.

There is no Chinese document dealing with kites that contains a word about religious observances in connection with them. The idea that a kite functions as a scapegoat charming away the owner's sins and misfortunes is a recent development to be found locally and sporadically, but it is not general; it is more developed in Korea than in China, but at any rate it bears no relation to the origin of kites.

The beginning of kites in China cannot be clearly traced. It is a curious fact that just in those things which are characteristically Chinese their records fail us—partially perhaps because these things seem trivial, partially because Chinese scholars are of the bookish type and poor observers of real life. A paltry object like a kite was somewhat below their dignity; nevertheless kites have been made the subject of a score of poetical compositions. In times of early antiquity kites did not exist: they are not mentioned, for instance, as it might be expected, in the treatise on the Art of War which Sun Wu wrote in the sixth century before our era and of which we have an excellent English translation by Lionel Giles.

It has been pointed out in the preceding chapter that Kung-shu's wooden bird was not a flying-kite. The earliest notion of this device looms up in the life of Han Sin, who died in 196 before our era. He is known as one of the Three Heroes who assisted Liu Pang in ascending the throne as first emperor of the Han dynasty. He was desirous of digging a tunnel into the Wei-yang Palace, and is said to have flown a paper kite for the purpose of measuring the distance to the palace. Some explain that he did so by measuring the length of the cord fastened to the kite; others with a bolder grip of imagination pretend that he himself ascended on the kite to gain a free outlook on the palace. It is more probable that Han Sin introduced kites into warfare, using them in trigonometrical calculations of the distance from the hostile army. Be this as it may, the story is not well authenticated; it is not contained in contemporaneous records, but only in comparatively late sources. If for no other reason, it is suspicious that Han Sin's kite is said to have been made of paper, while paper was invented only three hundred years later.

Chinese authors are wont to speak of "paper kites," but rag-paper was invented by Ts'ai Lun only in A.D. 105. Ever since paper has come into use, kites have been made of this material, and no other is employed for them. Nevertheless it is not reasonable to argue that prior to the invention of rag-paper kites could not have been made; the framework might have been covered with silk, hemp, or some other light fabric as well,—only Chinese records are reticent as to this point. The Polynesians enlist bark-cloth (tapa) for their kites, and as will be seen below, the first kite made in England was of linen, while Benjamin Franklin's famous kite was of silk.

In A.D. 549 Hou King (502-552) besieged the city of T'ai in which Kien-wen, subsequently the emperor Wu of the Liang dynasty, was bottled up. Unable to communicate with the outside world, Kien-wen had a paper kite made with a message attached to it and sent it up into the air that his friends might be advised of his perilous plight. One of Hou King's officers, Wang Wei by name, saw the kite rise and ordered his best archers to take a shot at it (first example of anti-aircraft practice). The kite dropped, but, as tradition has it, was transformed into a bird that escaped into the clouds, no one knowing where it went—which probably means that the kite, after all, had not been hit. This story is on record in the *Tu i chi* written by Li Yu of the T'ang dynasty. In A.D. 781 when Chang P'ei, a loyal general, defended the city Lin-ming against T'ien Yüe who had revolted against the reigning house of T'ang, Chang released a kite to inform Ma Sui of the predicament of the garrison which was exposed to starvation. Again, in this case, the kite was espied by the hostile camp, and T'ien commanded his archers to bring it down, but in this attempt they failed. The garrison held out until Ma Sui came to its relief when a crushing defeat was inflicted on the besiegers.

Many European authors who are only too prone to accentuate the topsy-turvy-dom of the Chinese assert wrongly that kite-flying is exclusively pursued in China by adults, not by boys. It is certainly true that the men are passionate and expert kite-flyers, and it is equally true that many kites, owing to their enormous size and weight, can be manipulated by grown-ups only. But how should the man acquire his skill had he not gained his practice from early boyhood days? Boys assuredly play with kites, and have done so from the days of the Sung dynasty (the end of the tenth century). From that period onward there are many records and pictures testifying to the kite-flying of youngsters, and they are even encouraged to indulge in this wholesome pastime for the reason that "it makes them throw their heads back and open their mouths, thus getting rid of internal heat." Plate VIII illustrates an outdoor scene: boys sporting with kites from a Sung painting depicting the games and entertainments of a hundred boys, by Su Han-ch'en, a renowned artist of the twelfth century.

From China kites were diffused to all other nations of eastern Asia who experienced the influence of Chinese civilization—Korea, Japan, Annam, Camboja, Siam, Malaysia, inclusive of the Philippines and Borneo. As in China, kite-flying has developed into a national pastime in Korea and Japan which received their culture from the mother-country.

In some parts of Indonesia, Micronesia, and Melanesia kites are turned to a practical purpose for catching fish. A fishing-line to the end of which is fastened a baited hook or noose is attached to a kite which is flown from the end of a canoe over the water; the bait is made to play over the surface of the sea by the movements of the kite in the wind. When the fish bites, the kite goes down. In Polynesia kite-flying is pursued for amusement only, chiefly in New Zealand, the Cook group, Tahiti, Hervey Islands, the Marquesas, Tuamotu, Easter Island, and Hawaii; kites are unknown in Samoa and Tonga.

Kites were introduced into India from China either through the Malays or Chinese immigrants or both. Kite-flying is a popular amusement in India during the spring. Matches are often made for considerable stakes. As in China, the strings are coated with crushed glass smeared on with glue, and each player seeks to manoeuvre his kite so as to cut his rival's string. Respectable elderly gentlemen also take keen interest in the game.

In Siam, kite-flying was a state ceremony as well as a public festivity. Large paper kites were sent up into the air with the object of promoting the seasonal wind by the fluttering noise made by them. The festival was obviously connected with agriculture and the appearance of the north-east monsoon.

The ancients were not acquainted with the flying-kite. Archytas' wooden dove, as pointed out on p. 64, is not a kite. There is no reference to a kite in the writings of any Greek or Roman author. The fact remains that kites were introduced into Europe from the East not earlier than the end of the sixteenth century. The Chinese were the inventors of it, and all data at our disposal go to prove that the kite spread from the Far East westward to the Near East and finally to Europe, and that it makes its debut in Europe as a Chinese contrivance, not as a heritage of classical antiquity.

Muṣailima, the false prophet, a contemporary of Mohammed, is said to have employed at night paper kites with musical bows in order to convey the impression to his adherents that he was communicating with angels. Al-Jāhiz, who died in A.D. 869, in his Book of Animals (Kitāb al-hayawān), speaks of "flags of the boys which were made of Chinese carton and paper; to these tails and wings were attached, little bells were tied to their fronts, and on breezy days they were released into the air from long and firm threads."

Kite-flying is well known to the Turks as a sport both for children and adults. The kite is called in Osmanli *kartal* ("eagle"), in Jagatai and Cumanian *sar* ("sparrow-hawk").

In European literature kites are first described by the Italian Giovanni Batista in his book on natural magic (Magia naturalis, 1589) and by J. J. Wecker (De secretis, 1592). The Jesuit Athanasius Kircher (Ars magna lucis, 1646) was well acquainted with kites. As is well known, he also wrote a book on China which is based on information received from the members of his order working in China.

Francesco Lana (1670, p. 50) informs us that in his time kites were manipulated by the children of Italy. He calls them *drago* ("dragon"), while the Italians now designate them *aquilone* ("north wind"), *cometa* ("comet"), or *cervo-volante* ("flying-stag") in accordance with French *cerf-volant*. German *drache* is doubtless based on the Italian appellation, and the Russians speak of a serpent (*zmäi* or *zmäika*). The Spaniards style a kite *pajaro* ("bird, sparrow") or *papagayo* ("parrot"). Curiously enough, the correct Chinese term "kite" (*yüan*) is preserved in our English word, and this seems to hint at the fact that paper kites were directly imported from China into England with the correct label attached.

J. Strutt, in his classical book "The Sports and Pastimes of the People of England," informs us that the kite probably received its name from having originally been made in the shape of the bird called a kite and that in a short French and English Dictionary published by Miege in 1690 the word *cerf-volant* is said to denote a paper kite,—the first registration of the word he found. "I have been told," he winds up, "that in China the flying of paper kites is a very ancient pastime, and practised much more generally by the children there than it is in England. From that country perhaps it was brought to us, but the time of its introduction is unknown to me; however, I do not find any reason to conclude that it existed here much more than a century back" (Strutt wrote in 1801). Certainly kites were used in England a considerable time prior to 1690.

In the middle of the seventeenth century kites were commonly employed in England for the purpose of letting off fireworks. John Bate, in his "Mysteries of Nature and Art" (1634), describes the making of a kite to this end, though he avoids the word itself. "You must take a piece of linnen cloth of a yard or more in length," he writes, "it must be cut after the form of a pane of glass; fasten two light sticks cross the same, to make it stand at breadth; then smear it over with linseed oil and liquid varnish tempered together, or else wet it with oil of Peter; and unto the longest corner fasten a match prepared with saltpeter water upon which you may fasten divers crackers, or saucissons; betwixt every of which bind a knot of paper-shavings, which will make it flie the better; then tie a small rope of length sufficient to raise it unto what height you shall desire, and to guite it withall; then fire the match, and raise it against the wind in an open field, and as the match burneth, it will fire the crackers and saucissons, which will give divers blows in the ayer." Bate's kite is reproduced in Plate IX. It has the shape of a lozenge, and is equipped with a tail to afford stability. This type of kite was commonly used in England down to the latter half of the nineteenth century.

S. Butler (*Hudibras*, 1664) alludes to the kite in scoffing at the prophecies based on the appearance of comets:—

> It happen'd as a boy, one night,
> Did fly his tarcel of a kite;
> His train was six yards long, milk-white,
> At th' end of which, there hung a light,
> Inclosed in lanthorn made of paper...

A tarcel is a young hawk. The Oxford English Dictionary contains an interesting quotation from Marvell (1672): He may make a great paper-kite of his own letter of 850 pages.

In Europe, finally, kites were employed for scientific purposes, for the first time, it is said, by Alexander Wilson, professor of astronomy at Glasgow University, who claimed in 1778 that he used kites attached to wire for electrical experiments long before Franklin in 1749 and that with four or five paper kites strung one above another he raised thermometers to an altitude of three thousand feet, in order to determine the temperature in the clouds.

In 1752 Benjamin Franklin made his experiment of collecting atmospheric electricity through the medium of a kite covered with silk and fitted at the top with a metal point. This experiment demonstrated the identity of lightning with electricity. Franklin's kite consisted of a framework in the shape of a cross made of two light strips of cedar. Over this frame was stretched a silk handkerchief tied to the four ends. From the top of the upright stick of the cross extended a sharp-pointed wire the length of a foot. A silk ribbon was tied to the end of the string which held the kite, the end next the hand, and a key suspended at the junction of the twine and silk. The kite was raised by Franklin during a thunderstorm in June, 1752, and almost immediately he experienced a spark on applying his knuckles to the key. When the cord was moistened by a passing shower, the electricity grew abundant. A Leyden jar was charged at the key, and by the spark thus obtained spirits were ignited, and other experiments performed.

Franklin's memorable experiments established definitely the service of the kite for scientific purposes. It was adapted especially to meteorological work, self-registering thermometers being sent up with it. Thus it is an efficient means of obtaining observations in the free air at moderate elevations. For all greater heights a balloon is used; the kite, however, can be used in stormy weather when the balloon is not serviceable; but the special advantage of the kite lies in the fact that the self-recording apparatus is thoroughly ventilated by the wind, and therefore gives the temperature and moisture of the free air with the least possible error introduced by solar heat or instrumental radiation.

Both in China and Japan there are stories current about men riding on kites through the air. There is a tradition alive in Japan that Yui-no Shosetsu, who tried to overthrow the Tokugawa government in the seventeenth century, made a large kite on which he ascended in order to spy on the Shogun's palace of Yedo. The Shogun and his court were taken aback, and the construction of large kites was forthwith forbidden under penalty of death. Shosetsu was subsequently seized and compelled to commit harakiri. A famous brigand of the seventeenth century, Ishikawa Goyemon, is said to have attempted to steal the gold from the huge golden fish or dolphin on the tower of the Castle of Nagoya by mounting on a kite. He succeeded in abstracting three golden fins. Since that hime large kites have been prohibited in Owari. Tametomo, of the Minamoto family, a hero of the twelfth century, who lived in exile on Oshima Island, is said to have sent his young son from there to Kamakura by means of an enormous kite. There are other stories of valiant cavaliers who used kites as airplanes by flying on them over the enemy's camp for purposes of reconnaissance.

On September 24, 1927, the Associated Press reported by cable from Constanza, Rumania, that while Robert M. Patterson, American Chargé d'affaires in Rumania, was motoring along the beach on the Black Sea, he heard cries for help from a small naked boy flying a huge kite which threatened to fly away with him or to pull him into the sea. The frightened boy turned out to be five year old King Michael, who, the dispatch said, despite his elevation to the throne, cares more about kites than kingdoms. Mr. Patterson, who knew Michael from babyhood, stopped his motor and ran to his rescue. Taking the thick cord from the boy's blistered hands, he pulled in the kite which was twice the size of the young king. It required all his strength. "Don't tell my mother," whispered the anxious monarch, "she will kill me, she doesn't know I'm out." Mr. Patterson placed him in his car, and driving to Princess Helene's residence, deposited the little king safely in the hands of his English nurse.

Now if this or a similar story were found in a Chinese record, all the gray-haired sinologues would shake their wise heads and denounce it as an anecdote without a foundation of reality; but such things will happen, and even more than that.

Athanasius Kircher was well posted on the subject of kites, and in his work "Ars magna lucis" (Rome, 1646, p. 826) mentions the fact that in his time kites were made of such dimensions that they were capable of lifting a man.

The fact that it is not impossible to lift a person into the air by means of two or several powerful kites combined may be inferred from experiments made in England and America during the nineteenth century. About the year 1826 the principle of the kite was turned to a practical purpose by George Pocock, a schoolmaster of Bristol. Interested in kite-flying from the days of his boyhood, Pocock found through various experiments that by attaching several kites, one beneath another, they could be elevated above the clouds. Then he attached to the cord of the kite a board which was dragged along rapidly like a sledge, and next a car with a full load of passengers was drawn easily over the turf. The first person who soared aloft in the air by this invention was a lady, who, seated in an arm-chair, was raised by the kite to a height of a hundred yards. Several years later he developed this "aeropleustic art" (a term invented and used by him alone) by constructing a four-wheeled carriage which he termed "char-volant" (flying car). It was set in motion by two or more large kites made to fold up and controlled as to angle and obliquity by four lines. He demonstrated that two large kites with a surface of a hundred square feet sent up in a gentle breeze had a draught power of three hundredweight or nine hundredweight in a brisk gale. On January 8, 1827, Pocock claimed to have covered several miles between Bristol and Marlborough at twenty miles an hour—a speed which he remarks need not be thought dangerous—and that on this occasion the London mail-coach was easily overtaken. Pocock proposed also to apply kites to marine purposes for towing boats or life-saving from shipwrecks on a lea shore, and suggested their military use for elevating a man in reconnaissances and signalling—which the Chinese had done centuries ago. Pocock's kite-chariot, of course, was not a practicable scheme, as it depended on the winds, but it has a decidedly Chinese flavor. In 1876 Joseph Simmons, it is said, was drawn into the air to a height of six hundred feet or more by means of two superposed kites, and then adjusting his weight by means of guy lines glided down to earth. He filed a patent for his invention (No. 2428, 1875) as "improved means and apparatus for conveying or carrying human beings or objects into mid-air." In

1868 Biot, a French engineer and a lifelong experimenter with kites, was lifted from the ground by a large apparatus of this kind. In 1894 Captain B. F. S. Baden-Powell, of the Scots Guards, constructed a kite 36 feet high consisting of four or five superposed kites, with which he successfully lifted a man on different occasions to a height of a hundred feet. In 1897 Lieutenant H. D. Wise made experiments in the United States with large kites of the Hargrave type and succeeded in lifting a man forty feet above the ground.

It was Laurence Hargrave, an Australian, who then gave a fresh impetus to scientific kite-flying. He realized that the structure best adapted for a good kite would also be suitable as a basis for the structure of a flying-machine. He introduced a new principle and invented what is known as the "cellular construction of kites." This is a kite composed of two rectangular cells separated by a considerable space—known as "the Hargrave box kite" (figured in *National Geographic Magazine*, 1903, p. 221). This type of kite, which surpassed in stability all previous examples, formed the starting-point of Alexander Graham Bell's epoch-making researches and his constructions of triangular and tetrahedral kites. In 1903 Dr. Bell wrote, "I have had the feeling that a properly constructed flying-machine should be capable of being flown as a kite; and, conversely, that a properly constructed kite should be capable of use as a flying-machine when driven by its own propellers."

In December, 1907 Dr. Bell experimented with a gigantic man-lifting kite, the Cygnet, more than forty feet long, which was sent up both with and without a man. Lieutenant Selfridge, of the United States Army, ascended with this kite to a height of 168 feet and remained in the air for over seven minutes. Illustrations of these highly interesting experiments may be viewed in the *National Geographic Magazine* for 1908.

Dr. Bell's prophetic words uttered in 1903 have at present been fulfilled. The wings of the modern biplane are closely patterned after the Hargrave box-kite on which Dr. Bell inaugurated his experiments. The man-lifting kite has developed into an airplane. The speed plane of our times is but a first cousin to the kite.

Another Chinese apparatus deserves mention here, as it served as a source of inspiration to Sir George Cayley (1774-1857), one of the great pioneers of modern aviation. His interest in aeronautics was aroused in boyhood by the invention of the balloon in 1783. He himself tells us that his first experiment in such matters was made as early as 1796 with a Chinese or aerial top, which served at once to illustrate the principle of the helicopter and the air-screw. Though but a toy of a few inches in length, its capacity to demonstrate certain elementary, but important principles in aeronautics made a lasting impression on Cayley's youthful mind, and only three years before his death he sent to Dupuis Delcourt a drawing of one which he had had made, the best he had ever seen, and capable of rising ninety feet in the air. This drawing is reproduced in Hodgson's book (Fig. 135). The original of one of these aerial tops is still in the possession of Mrs. Thompson, a grand-daughter of Cayley.

THE DAWN OF AIRSHIPS IN ANCIENT INDIA

Although the Aryan Indians of the Vedic period had numerous aerial deities, like the Gandharvas, elfs haunting the "fathomless spaces of air," no allusion is made in the Rigveda to their manner of locomotion, and none is described as possessed of wings. Among the divine steeds there is one named Dadhikrā, praised for his swiftness, speeding like the wind, and equipped with bird's wings; he is likened to a swooping eagle and even directly called an eagle. The Vedic gods did not fly, but preferred driving in luminous cars usually drawn by fleet horses, in some cases by cows, goats, and spotted deer. Indra, the favorite national god of the Vedic Indians, primarily a storm and thunder-god, conqueror of the demons of drought and darkness, and also a god of battle who aided the advancing Aryans in their struggle against the Dasyus, the aboriginal inhabitants of India, is borne on a golden chariot which is swifter than thought. This vehicle is drawn by two or more tawny, sun-eyed chargers with flowing golden manes and hair like peacocks' feathers. Snorting and neighing, they rapidly traverse vast distances, and Indra is transported by them "as an eagle is borne by its wings." His weapon is the thunderbolt (*vajra*), which personifies the lightning stroke and with which he slays his foes. A myth of post-Vedic times tells of quaking mountains provided with wings and gifted with the power of flight: they flew around like birds, alighted wherever they pleased, and with their incessant motion made the earth unsteady. With his thunderbolt Indra clipped the wings of the restive mountains and settled them permanently in their place; their wings were transformed into thunder clouds.

The Açvins ("horsemen"), twin deities, presumably symbolizing the dawn and the morning star, travel in a sun-like chariot all parts of which are golden and whose construction is based on the number three, having three seats, triple wheels, and triple fellies. It moves lightly, is swifter than "thought or the twinkling of an eye," and is drawn by horses, but more commonly by birds, swans, eagles, or eagle steeds. It touches the ends of heaven and extends over the five countries, it races round heaven, and traverses heaven and earth in a single day. They also have a car which without draught animals traverses space. The Açvins are domiciled in heaven or in the air, and appear at the time of the early dawn when they yoke their chariot to descend to earth and receive the offering of worshippers. Ushas, the maiden goddess of dawn, the most poetical figure of the Vedic pantheon, awakes the twin gods; she drives in a brilliant, well-adorned chariot drawn by ruddy steeds or kine. The Açvins follow Ushas in their car, and thus their relative time seems to have been between dawn and sunrise. The twin brothers have the particular function of coming to the rescue of people in distress, and are constantly praised for such deeds. The story most often referred to in the Rig-veda is that of the deliverance of Bhujyu, son of Tugra, who was abandoned in the midst of the ocean or in the water-cloud and who, tossed about in darkness, invoked the succor of the youthful heroes. They rescued him with animated, water-tight ships which traversed the air (a sort of hydro-aeroplane), or with an animated winged boat (compare Lana's flying ship, Plate V), or with three flying cars having a hundred feet and six horses. The twins are wedded to the sun-maiden or the daughter of the sun, and in the marriage rite they are invoked to conduct the bride home in their chariot.

The Maruts, gods of the winds, children of the storm-cloud, born from the laughter of lightning, speed in cars which gleam with lightning, drawn by spotted coursers; brilliant as fire, they carry spears on their shoulders, anklets on their feet, golden ornaments on their breasts, fiery lightnings in their hands, and golden helmets on their heads. They are also described as having yoked the winds as steeds to their pole; that is, their chariot is driven by the winds.

Sūrya, the sun god, the far-seeing spy of the whole world, who beholds all beings and the good and bad deeds of mortals, moves in a chariot drawn by one steed or by seven horses or mares. In various passages, however, he is conceived as a bird traversing space, is represented as flying, and is compared with a flying eagle. The god Agni, the personification of the sacrificial fire, which is the centre of the ritual poetry of the Veda, drives in a lightning chariot, described as golden and luminous, drawn by two or more horses impelled by the wind. He yokes them to summon the gods to the sacrifice, and then acts as their charioteer, bringing Indra from the sky, the Maruts from the air. Agni is also likened to or directly designated a bird, and in one passage is spoken of as the eagle of the sky.

Another Vedic god, Pūshan, who is closely connected with the sun, moves in golden ships sailing over the aerial ocean (the sky is conceived as an ocean: thus, also, the Açvins' chariot approaches from the celestial ocean), acting as messenger of Sūrya, the sun-god; making his abode in heaven, he moves onward, beholding the universe. He is praised as the best of charioteers or air-pilots, and drives with a pair of goats, presumably because the goat is a bold climber and appears fittest to clamber the dizzy heights of heaven. The sun likewise appears as a boat in which Varuna, the god of the encompassing sky, navigates the aerial sea. This primitive conception naturally arose from the experience of seeing the sun set in the ocean, and being animated with a personality, he required a ship to guide him out of the sea toward his path along the sky. On the one hand, therefore, the sun functions as a charioteer, and is symbolized by the horse and the wheel; on the other hand, as a boat and boatman. Hence, in India, the idea of an airship developed from that of a solar ship. Similar notions occur among other peoples. Re, the Egyptian sun god, is the owner of two barks, changing from one to another in the morning and evening.

Greek philosophers style the sun "boat-shaped," and Helios rides in a golden boat from sunset to sunrise. In songs of the Letts the setting sun must be rescued and taken in a boat to save his life; or the sun sadly bewails the sinking of the golden ship into the sea, and is consoled by the wish that God might build a new one.

In one of the Jātakas (No. 159) or stories of the Buddha's former births, a king of Benares owns a jewelled car in which he used to race through mid-air. Gunavarman (A.D. 367-431), a Buddhist monk from India, sailed from Ceylon to Java, where he was to convert one of the kings to Buddhism. A day prior to his landing in Java, the king's mother had a dream to the effect that a religious friar had embarked on a flying ship and would enter the kingdom; on the following morning Gunavarman indeed arrived in person. In the prelude to the fifth act of Bhavabhūti's drama Mālatīmādhava, a sorcerer's maid appears in "a chariot traversing the air."

In post-Vedic literature, the vehicle of the god Vishnu is Garuda, the chief and lord of birds, a celestial bird,—originally a solar bird. This purely mythological conception proved very fertile in stimulating imagination and, according to Indian stories, led to constructions of airships and attempts at flying.

The Panchatantra (I, 5), the most popular collection of Indian folk-lore, contains the story of the Weaver as Vishnu. A weaver became infatuated with the king's daughter; and his friend, a carpenter, made for him a wooden airship in the shape of a Garuda, the mythical bird and vehicle of the god Vishnu, which was set in motion by means of a switch or spring. Equipped with all paraphernalia of the god, the weaver mounted the machine; and when the carpenter had explained to him how to manipulate the switch, he hopped off and dropped in on the seventh story of the palace, where the princess had her apartment. Seeing him astride the Garuda in the splendor of Vishnu's regalia, she took him for the creator of the three worlds, and he married her instantly according to the rites of the Gandharva marriage (i. e. by mutual consent, without ceremony), and then continued his relations with the princess for some time until her guards suspected her and made a report to the king. He questioned her, but her explanation that she is the consort of a god gratified his vanity. Believing himself in alliance with Vishnu who would grant him the rule over the world, the king became overbearing toward his neighboring kings, who consequently made war on him. Through his daughter he implored the pseudo-Vishnu to come to his rescue. This one, in despair, appeared in the air above the battle-field, armed with bow and arrow and ready to die; but Vishnu himself, fearing that if the weaver disguised as Vishnu were killed his own authority among men might suffer, entered into the weaver's body and scattered the king's enemies. After the victory was obtained, the weaver descended from the sky; and recognized by the king, his ministers and the people, told the whole story, whereupon the king highly honored him, solemnly married him to his daughter, and rewarded him with a large estate. The most interesting point of this story is that the bird-plane is utilized for military purposes to defeat and rout an army. When we read that Abhayākara, a saint of the ninth century from Bengal, assumed the form of a Garuda to disperse an army of Turushkas (Turks), we must understand that he was mounted on a Garuda-plane which functioned as a war-plane.

A dirigible airship is described in the celebrated old collection of Indian stories known as "The Twenty-five Tales of a Vetāla," which is as well known in India as the Panchatantra and which was translated into Tibetan and Mongol; it is usually quoted under its Mongol title Siddhi Kür. The heroes of this tale are six young men,—the son of a rich man, a physician's son, a painter's son, a mathematician's son, a carpenter's son, and the son of a smith, who leave home in quest of adventure in a foreign land. The first of them won the hand of a beautiful woman of divine origin, but she was soon kidnapped by a powerful king who took her into his harem. The six youths conspired to rescue the stolen wife from her captivity, and the carpenter's son hit upon the scheme to construct a wooden bird, called Garuda, whose interior was equipped with an elaborate apparatus which allowed the machine to fly in various directions and to change its course at will: it was provided with three springs. When the spring in front was touched, the aeroplane flew upward; when the springs on the sides were tipped, it floated evenly along; when the spring beneath was pressed, it made its descent. The painter's son decorated the Garuda in various colors, so that it could not be distinguished from a real bird. The rich youth boarded the machine, pressed the spring, and crossed the air in the direction of the king's palace, where he soared above the roof. The king and his people were amazed, for they had never before seen such a gigantic bird. The king bade his consort to ascend the palace and offer food to the strange visitor. So she did, and the bird descended. The aviator opened the door of the machine, made himself known, seated his former wife inside, and hopped off with her, navigating his way back to his companions—in the same manner as we have all seen it in the movies with modern airships.

In the Sanskrit version of the same story, as embodied in Somadeva's Kathā Sarit Sāgara ("Ocean of Streams of Stories"), an excellent Brahman, Harisvāmin, has a beautiful daughter who wants to marry only a man possessed of heroism and knowledge, or magic power. The first suitor, thus informed, professed to command magic power. At the father's request to demonstrate it, he immediately constructed by his skill a chariot that would fly through the air, and in a moment he took Harisvāmin up in that magic chariot and showed him heaven and all the world, and he brought him back delighted to the camp of the king of the Deccan to whom he had been sent as an ambassador to negotiate a treaty. Then Harisvāmin promised his daughter to that man possessed of supernatural power. His son promised her to a man skilful in the use of missiles and hand-to-hand weapons, and his wife promised her to a man who professed to have supernatural knowledge. When the three bridegrooms appeared on the wedding day in Harisvāmin's house, it happened that the intended bride had disappeared in some inexplicable manner. The possessor of knowledge soon found out that an ogre (Rākshasa) had carried her off to his habitation in the Vindhya forest. The possessor of magic power prepared, as previously, a chariot that would fly through the air, provided with all kinds of weapons. Harisvāmin, the man of knowledge, and the brave man jumped into the airship with him. In a moment they were carried to the Rākshasa's dwelling-place. The giant was duly slain by the brave man, the Brahman's daughter was released, and they all returned home in the flying chariot.

A fundamental document referring to airships is found in Budhasvāmin's Brihat Kathā Çlokasamgraha (edited and translated into

French by F. Lacôte, 1908), a collection of stories written during the eleventh century. Vāsavadattā desired to mount an aerial chariot and thus to visit the entire earth. Vasantaka, the master of games, exclaimed laughingly, "The wives of the king's servants had just the same craving. I said to all, 'Suspend a swing with long poles, mount it, and you will ascend into the air. Your husbands do not know of any other way of satisfying your desire.' If she had a notion to travel through the air, she must be content with the same medium." All burst out laughing. "Stop joking, and come to the point," Rumanvat said. "What good is it to dream thereof?" Yaugandharāyaṇa spoke, "This is exclusively an affair of artisans." Rumanvat summoned the carpenters and enjoined them to make without delay a machine capable of traversing the air. After a long deliberation these said frightened and stammering, "We know only four kinds of machines, but as regards flying-machines, they are known only to the Yavanas (the Greeks), but we never had occasion to see them."

Farther on, in the same story, Viçvila mounts a mechanical cock and makes a trip through the air. At his return he speaks to the king's ambassadors thus: "It is not proper to reveal to any one, artisan or any other person, the secret of the aerial machines, which is difficult to obtain for whoever is not a Yavana (a Greek). It is the same matter as the secret of the manufacture of beds; if it leaks out, it would become common property, and the public would treat it with disdain, for fashion is the creation of the moment. To bring this respectable art into disfavor is a grave sin, so let us drop this matter." A month later Pukvasaka, a clever carpenter and craftsman, said to Viçvila, who was his son-in-law, "The king took me aside to-day and told me with a gentle smile that I must reveal to him this science of aerial machines. I replied that I did not reveal it to you, but that it was Greek artisans with whom you had curried favor. The king waxed angry and threatened my life. Save my life and my sons, therefore, by revealing to the king the secret of these machines, since it is his desire!" Viçvila consented, but during the night awakened his sleeping wife, Ratnāvalī, and said, "I have to inform you that I am returning to my country. Your father employs intrigues to banish me from this place: he wants to wrest from me the secret of the flying-machines which it is our duty to keep concealed, as a miser guards his treasure. But enough, it is my life or your father who is dear to you. In order to guard my secret, I shall go so far as to forsake you!" He and his wife mounted the machine in the form of a cock, and during the night made an ascent and escaped into the country whence he had come.... The commander-in-chief assembled all the artisans, gave them a flogging, and ordered them to construct a flying-machine. Meanwhile a stranger appeared and said, "I shall construct the machine for you; do not flog the artisans! Give me immediately the necessary appliances." These were at once furnished by Rumanvat. One of the artisans said to the stranger, "Ask the commander-in-chief for the number of passengers. Because they did not know how many passengers they could transport, kings saw their chariot sink, and more than one artisan, it is said, has therefore suffered their wrath." The other responded, "These must have been wretched village artisans! But it is of no avail to waste so many words. Wait a moment!" In the nick of time he produced a flying chariot in the form of a Garuda, adorned with Mandāra flowers, and said to the king, "Oh king, Vishnu of the kings, mount the Garuda and traverse the earth formerly traversed by Vishnu!" "Madam," the king said to the queen, "what do you tarry now? Mount this chariot and depart in accordance with your wish!" "My consort," she responded, "without you I do not even go into the garden; without you, how could I support myself in the vacuum of the celestial space?" The king reported these words to the craftsman, who exclaimed, "But this chariot can carry the entire city!" Thereupon the king took his seat in the chariot together with the personnel of the harem, his wives, his officials, and a section of each urban corporation. They gained the pure spaces of the firmament, and finally took the route of the winds. The chariot circumambulated the earth girt by the ocean, and then was directed toward the city of the Avantis, where the festival of the offering of the water was held. The machine was stopped, so that the king might enjoy the spectacle. After a brief stay the king departed for Kauçambi under the eyes of the multitude which admired the chariot, and acquitted himself of his obligations toward the immortals, the priests, the sacred fires, his parents, his servants and burghers. Then he commanded to do honors to the craftsman.

In a Sanskrit romance, the Harshacharita ("Deeds of King Harsha"), written by Bāna in the seventh century, mention is made of a king, Kākavarna. Being desirous of marvels, he was carried away, no one knows whither, on an artificial aerial car made by a Yavana (Greek) who had been taken prisoner. The term used in this passage means literally, "a mechanical vehicle (*yantrayāna*) which travels on the surface of the air;" that is, an airship.

From the preceding texts it follows that the Indians profess to have had two distinct types of flying-machines,—the Garuda airship of native manufacture constructed on the principle of bird-flight, and the Yavana airship ascribed to the Yavanas or Greeks, the manufacture of which was scrupulously guarded as a secret.

The first question to be raised is, Did the ancient Indians really navigate the air? Are their dirigibles realities or fiction, merely the upshot of a poetic imagination? To my way of thinking this point is irrelevant. The main point is: they had the idea; and their ideas about aeronautics were not worse or more defective than those entertained in Europe from the sixteenth to the first part of the nineteenth century.

The Indians saw two points clearly—that aircraft must operate on the principle of the flight of birds and that a mechanism is required to start the machine, to keep it in mid-air, and to make a descent. Whether they actually flew or not, whether they succeeded or failed, the stories cited (and another will be given below) are sufficient evidence of the fact that they devoted considerable thought to problems of the air and aeronautics and that as a sequel of a highly developed mechanical science efforts were made to construct aircraft of various types.

The second query that we may revolve in our mind is, Did the Greeks, as asserted by the Indians, really supply them with flying-machines? A direct response to this inquiry is not forthcoming from the Greek camp. The Greek mechanicians, in the ingenuity of their inventions perhaps the greatest of all times, are silent as to aircraft. Perhaps the information is lost; for myself I see nothing impossible in the assumption that the Greeks of the Alexandrian epoch should have made successful experiments in mechanical flights.

Greek mechanics and artisans enjoyed a high reputation in India, and marvellous inventions were ascribed to them. The Indians have numerous stories of wonderful automata set in motion by intricate machinery; for instance, movable figures of beautiful women who may even assume life, tempt men and cause a quarrel among them, wooden figures of men able to strut, sing and dance, artificial elephants moving by means of a mechanical apparatus, or artificial fishes which appear to swim under a floor of rock-crystal looking like water. It is noteworthy, again, that in some tales such mechanical marvels are attributed to the Greeks; thus, a painter from central India once travelled on business to the land of the Greeks (Yavana), and took up his abode in the house of a mechanician who made an artificial maiden to wait upon the painter. She washed his feet and then stood still; he called to her to draw near, but she made no reply. He seized her by the hand, and when he tried to embrace her, the figure collapsed and turned into a heap of chips. In another tale, an Indian carver in ivory travels to the land of the Greeks and settles in the house of a Greek painter. The great mechanicians of Alexandria were very proficient in the construction of mechanical toys and figures, and we still have Heron's work on the Automatic Theatre (Automatopoietika), written in the second century before our era. Heron was the founder of a school, surveyor, mechanician, and the greatest physicist of ancient times.

Maybe, because so many wonders of technique were created by the Greeks, the poets of India reasoned that aircraft also must have been due to their genius. We do not know, but what we do know at present is that in the records of ancient Indian lore are distinguished two types of flying apparatus—the native Garuda airship and the imported Yavana airship. Here remains a fascinating problem which more abundant documentary material that may come to light in the future will help us solve.

As regards winged flight, only one example is known to me from Indian literature. The Kathā Sarit Sāgara ("Ocean of Streams of Stories") contains the following good tale: "Once upon a time there was a young Brahman who one day beheld a prince of the Siddhas flying through the air. Wishing to rival him, he fastened to his sides wings of grass, and continually leaping up, he tried to learn the art of flying in the air. As he continued to make this useless attempt every day, he was at last seen by the prince while he was roaming through the air. The prince thought, 'I ought to take pity on this boy who shows spirit in struggling earnestly to attain an impossible object, for it is my duty to patronize such.' Thereupon, being pleased, he took the Brahman boy, by his magic power, upon his shoulder, and made him one of his followers." In Indian art, particularly in the sculpture of the Buddhists, winged beings in the act of flying are frequently represented, and such types like the Apsarases and Kinnaras have also been adopted by the Chinese. Plate X illustrates a fine Chinese marble sculpture from the Blackstone Chinese Collection of the Museum: two Apsarases or heavenly nymphs are flying down from Indra's celestial abode to guard the Buddha Amitābha.

I shall not dwell at length on the alleged power of flightacquired by magical practices or witchcraft, first taught in the Yoga system and from it transplanted into Buddhism. Among the marvellous abilities promised as a reward of Yoga practice there were understanding of the speech of animals, assuming any shapes, resuscitating the dead, descent into the inferno, fast locomotion, penetrating everywhere as air does, being poised cross-legged in the air, and traversing the air. What has been observed of "flying" among the modern Yogins proved to be walking or hopping close to the surface of the ground without seemingly touching it. A few examples from Buddhist literature may suffice.

In a Buddhistic story entitled The Magician's Pupil (Schiefner-Ralston, Tibetan Tales, p. 288) a man of the Chandāla caste (the lowest and most despised Pariah class) is versed in spells and magic lore, and obtains by means of spells from the Gandhamādana mountain fruits and flowers as are not in season, and these he presents to the king. A Brahman youth becomes his pupil, and when taught the art of magic, immediately makes a trial of his art on the spot and soars into the sky, reaching the fabulous mountain where he plucks fruits and flowers.

In one of the Jātakas (No. 186) is mentioned a gem endowed with magic power and capable of raising into the air whoever holds it in his mouth. By his own miraculous power the Buddha is able to rise in the air, to be poised in mid-air, and to travel through the air wherever it pleases him.

What is more interesting are two charming motifs of folk-lore presented by India to the world— the magic boots and the enchanted horse.

The Kathā Sarit Sāgara contains the following tale (in the translation of C. H. Tawney):—

"King Putraka, faithful to his promise, entered the impassable wilds of the Vindhya, disgusted with his relations. As he wandered about, he saw two heroes engaged heart and soul in a wrestling match, and he asked them who they were. They replied, 'We are the two sons of the Asura Maya, and his wealth belongs to us,—this vessel, this stick, and these shoes; it is for these that we are fighting and whichever of us proves the mightier is to take them.' When he heard this speech of theirs, Putraka said with a smile, 'That is a fine inheritance for a man!' Then they said, 'By putting on these shoes one gains the power of flying through the air; whatever is written with this staff turns out true; and whatever food a man wishes to have in the vessel is found there immediately.' When he heard this, Putraka said, 'What is the use of fighting? Make this agreement, that whoever proves the best man in running shall possess this wealth.' Those simpletons said, 'Agreed,' and set off to run, while the prince put on the shoes and flew up in the air, taking with him the staff and the vessel. Then he went a great distance in a short time and saw beneath him a beautiful city named Akarshikā and descended into it from the sky. Subsequently the king fell in love with the daughter of a king, and one night flew up through the air to the window of her room by the aid of his magic shoes. Later on he eloped with her by taking her in his arms and flying away through the air, finally descending from heaven near the banks of the Ganges."

This story is also extant in a Chinese translation (S. Julien, Avadanas, No. 34, and E. Chavannes, Cinq cent contes, Vol. II, p. 185), and has likewise migrated to Europe (Grimm, Märchen, No. 92, where the three wondrous objects are almost identical).

The legends of later Buddhist saints, as related in Tibetan records, frequently mention a "swift-foot apparatus" for rapid locomotion. In one case we are informed that such boots were made of the leaves of trees, the underlying idea apparently being that they should be as light in weight as possible. The owner of this footgear was the saint Vararuchi who had obtained it by virtue of magic spells; whenever he donned it, he was able to penetrate into the abodes of gods and serpent-demons (Nāgas) and to abstract many treasures with which he delighted the hearts of the poor. Once he had a quarrel with the king who suspected him of using evil spells against his life and who sent a messenger to do away with him. Vararuchi put on his magic boots and fled to the city Ujjayini. The king employed a woman to trick him out of his boots, and when unable to flee, he was overpowered by the royal henchman.

The thousand-league boots are well known to European folk-lore; they belong, for instance, to the equipment of the ogre in the tale of Petit Poucet.

In a Swedish story entitled "The Beautiful Palace East of the Sun and North of the Earth" a youth acquires boots by means of which he can travel a hundred miles at every step, and a cloak that renders him invisible in a very similar way.

A recipe for making magic boots is thus given in an Icelandic story: "A giant told a woman that Hermodr was in a certain desert island which he named to her; but could not get her thither unless she flayed the soles of her feet and made shoes for herself out of the skin; and these shoes, when made, would be of such a nature that they would take her through the air, or over the water, as she liked."

The enchanted horse of later Indian folk-lore is doubtless evolved from the solar horse of early Vedic mythology. From India this motif spread westward and was adopted into the Arabian Nights and many other collections of stories. In the Nights the flying horse, made of ebony, can perform in a single day a journey which under normal conditions would take a year; for the purpose of making the horse descend it is necessary to rub the switch on its left shoulder.

In an Armenian story of Persian origin, entitled "Solomon's Garden and Its Mysteries," Ġül, a servant of Solomon, possesses two wonderful steeds,—the horse of the wind and the horse of the clouds, which Solomon had bequeathed to him. The cloud-horse was not so fleet, but always followed in the track of the wind-horse.

In a collection of Jaina stories (Samyaktvakaumudī) it is narrated that Samudradatta was a groom in the service of a horse-dealer and received from him in compensation two horses which he was allowed to choose himself. He won the love of his master's daughter at whose advice he picked two ill-shaped horses: one of these was capable of running through the air, the other was able to go through water. On the air horse he returned home together with his wife.

In Jātaka No 196 the Bodhisatva comes into the world as a flying horse, white all over and with a beak like that of a crow, possessed of supernatural power, able to fly through the air. From the Himalaya he made a non-stop flight to Ceylon. There he passed over the ponds and tanks of the island and lived on wild-growing rice. Then he took back a number of ship-wrecked traders who had fallen into the hands of flesh-devouring ogres,—some climbing up on his back, some laying hold of his tail,—and conveyed all of them to their own country, and set down each in his own place.

The collection of stories known as "The Thirty-two Tales of the Lion-throne" or "Tales of King Vikramāditya" contains the account of an air-journey on an enchanted horse (No. 8), treated very much like the Garuda airships aforementioned, and is remarkable for the vividness of impressions received by the traveller in the air.

A carpenter once appeared before a king, leading a wooden horse richly caparisoned and in every respect looking like a live animal. The king did not think much of it, except that it was a clever model of a horse as any workman could accomplish, when the carpenter called his majesty's attention to the mechanical apparatus in the interior, which would allow him to reach any place he wished in a few moments. The king, who was interested in every uncommon thing and had never before seen a mechanical contrivance of such wonderful make, bade the carpenter to mount the horse. In an instant the man was seated on its back, and before any one had time to notice what he did in setting the machine in motion, both horse and rider had flown up and vanished. In a quarter of an hour he alighted on the ground, guiding his horse to the foot of the throne, and dismounted. When the king had seen the amazing speed at which the horse could fly through the air, he was seized by the ardent desire to possess the magical steed and paid the carpenter a large sum for it (two lakhs of rupees; that is, two hundred thousand rupees).

The following evening the king mounted the hippoplane, and turning the starting switch, took to the air and was out of sight in a moment. The rapid movement took the king by surprise: he felt dizzy and saw nothing around him but blue ether, wishing he had never made the ascent. For an hour he continued to rise higher and higher till the mountains below could not be distinguished from the plains, and in a moment all earthly landmarks passed out of sight. Then he thought it was time to descend, and imagined that all he had to do was to turn the same switch in the opposite direction, but to his horror he found that, turn as he might, he did not at all change his course. In his impatience to acquire the horse he had forgotten to inquire how to descend to earth. He set about to examine the horse's neck, till at last he discovered a tiny switch close to the right ear. This he turned and the next moment found himself dropping down toward the earth, somewhat more slowly than he had ascended. It was dark, and as he was unable to see, he was fain to allow the horse to take his own course. It so happened that the machine struck against the top of a tree, and the king, bruised and bleeding, fell to the ground, but escaped serious injury. He had landed in a dense jungle, where he discovered an enchanted princess with whom he fell in love.

It is interesting to note his experiences in the air, as he relates them to the princess: "In an instant I was soaring much faster than the speed of an arrow, and I felt I was approaching the sky so closely that I should soon hit my head against it. I could discern nothing beneath me, nothing around me save the invisible air, and for some time I was so confused that I did not know in what direction I was travelling. At last when it grew dark, I found a second switch near the horse's right ear, and on turning it, I began slowly to sink toward the earth. I was forced to trust to chance, content to abide by whatever my destiny had in store for me, and it was just midnight when finally I found myself landed safely on firm ground. I soon discovered that I was in the midst of a forest, and passed the remaining hours of the night among the branches of a tree. I thank the tree for having afforded me the means of discovering this palace, and still more, of discovering you."

Finally the king escaped with the princess, mounting the magic horse and seating the princess on a pillion behind him, and when she was firmly seated with her hands holding tightly to his belt, he touched the button, and the horse began to ascend heavenward like a rocket. They raced through the atmosphere like a flash of lightning, and the king, now an experienced air-pilot, guided his horse so skilfully that in a few hours the temples and towers of Ujjain appeared beneath the horse's feet. They alighted outside the city gate and walked to the royal palace.

FROM BABYLON AND PERSIA TO THE GREEKS AND THE ARABS

The earliest traditions of the Euphrates Valley carry us back to a mythical age in which rulers are pictured as deities or of divine descent. Among these is the legendary sovereign, Etana, a shepherd, who is the hero of various tales of which large fragments, though not all, have been recovered. This is not the place for a discussion of the entire myth. An eagle has a struggle with a serpent who badly tears his wings and feathers, and leaves him in a mountain-pit to die. The eagle appeals to Etana to release him from his prison, and as a reward promises to fly with Etana to the dwelling of the gods. Etana mounts on the back of the eagle, and together they fly upward. They reach the heaven of Anu and halt at the gate of the ecliptic. At this point there is a gap in the narrative, and when the thread is taken up, the eagle urges Etana to continue the journey in order to reach the place where Ishtar (the planet Venus) dwells. As in the case of the first flight, a distance of three *kasbu* or six hours is covered. Whether at this point the eagle's strength becomes exhausted or whether the goddess herself intervenes, the precipitous descent begins. The eagle drops through the space of three double hours and reaches the ground. The close of the story is wanting, but clearly the purpose of the flight has failed.

It seems that there is no other myth relating to a flight preserved in cuneiform literature, and G. Hüsing is probably right in evaluating the Etana myth, together with many others preserved in Babylonian records, as non-Babylonian and hailing from the Caucasus region. Be this as it may, the Etana myth is Aryan, not Semitic, and may also be derived from Iran.

A Babylonian seal cylinder, which is preserved in the Berlin Museum, represents the story of Etana, the flyer. He is shown being carried on an eagle's back, soaring between heaven and earth. The crescent of the moon is to his left, the sun to his right. A man standing on the ground looks up at the strange spectacle in amazement, and two dogs bark at the flying pair. On the left side of the seal impression appears a flock of sheep in a fold, guarded by a shepherd—obviously the herd belonging to Etana. The British Museum owns another seal illustrating the same subject: here Etana is seated on the eagle, who is bearing his burden aloft in the sight of an admiring and upward-gazing dog. See, further, note on p. 91.

An ancient Persian tradition is of especial interest, as it was transmitted to Europe at an early date and exerted no small influence on those occupied with dreams of aviation. This story forms a chapter of its own, and its fate will be traced down to recent times.

In the ancient semi-legendary history of Iran, Kavi Usan (in Persian: Kai Kawus) is the second king of the dynasty of the Kaianians. He was not a wise ruler, but was a rather imperfect character, easily led astray by passion. He ascended Mount Alburz, where he built seven palaces, one of gold, two of silver, two of steel, and two of rock-crystal. He then endeavored to restrain the demons of Mazandaran. One of these evil spirits retaliated by a ruse and sowed in his heart the seeds of discontent with his sovereignty on earth, so that he set his mind on aiming at the supremacy of the celestial abode. Yielding to the temptation of the Evil One, he seated himself on a throne which was supported and raised by four eagles, and as an incentive to fly up four pieces of flesh were fastened to the top of four spears planted on the sides of the throne. In this manner he sought to be transported into the empyrean; but the flight was of brief duration: the strange vehicle soon came down in a crash, and the grandees found the king unconscious in a forest.

In his great epic poem, the *Shahnameh* ("Book of Kings"), Firdausi (935-1025) describes this event as follows (in the translation of Warner):—

The Shah mused how to roam the air though wingless,
And often asked the wise, "How far is it
From earth to moon?" The astrologers replied.
He chose a futile and perverse device:
He bade men scale the aeries while the eagles
Were sleeping, take a number of the young,
And keep a bird or two in every home.
He had those eaglets fed a year and more
With fowl, kabab, and at some whiles with lamb.
When they were strong as lions and could each
Bear off a mountain-sheep, he made a throne
Of aloe from Komor (Khmer) with seats of gold.
He bound a lengthy spear at every corner,
Suspended a lamb's leg from every spear-head,
Brought four strong eagles, tied them to the throne,
And took his seat, a cup of wine before him.
The swift-winged eagles, ravenous for food,
Strove lustily to reach the flesh, and raising
The throne above earth's surface bore it cloudward.
Kawus, as I have heard, essayed the sky
To outsoar angels, but another tale
Is that he rose in this way to assail
The heaven itself with his artillery.
The legend hath its other versions too;
None but the All-wise wotteth which is true.
Long flew the eagles, but they stopped at last,

Like other slaves of greed. They sulked exhausted,
They dropped their sweating wings and brought the Shah,
His spears and throne down from the clouds to earth,
Alighting in a forest near Amul.
The world preserved him by a miracle,
But hid its secret purposes therein.
Instead of sitting on his throne in might
His business then was penitence and travail.
He tarried in the wood in shame and grief,
Imploring from Almighty God relief.

An illustrated Persian manuscript of the *Shahnameh*, dated 1587-88, which is preserved in the Metropolitan Museum of Art, New York, vividly depicts this aerial flight (Plate XI). The ambitious Shah clad in a pink robe, with feathered turban, is seated on a green mat spread on the bottom of a yellow hexagonal couch, holding a bow in his left hand and an arrow in his right, a fully laden quiver resting in front of him. He is ready for an attack from the air at any price against any enemy who might dare oppose his will. Four black eagles, on the wing, are harnessed to the sides of the throne, eagerly looking up and striving toward the flesh tied to four spears with fluttering red flags. The flyers are soaring in yellow and black clouds set off from the blue ether, leaving beneath them the highest mountain top from which a goat and another animal gaze at them in bewilderment.

This Iranian motif of an aerial conveyance lifted by starved eagles, like many other Oriental motifs, was adopted by the Greek Romance of Alexander the Great, which during the middle ages was translated into most European languages and thus became widely known. I quote from Dunlop's classical book "History of Fiction." Having reached the extremity of the world, having received homage from all nations who inhabit its surface, and being assured that there remained nothing more to conquer, Alexander formed the inconsiderate project of becoming sovereign of the air and deep. By the conjurations of the eastern professors of magic, whom ne consulted, he was furnished with a glass cage of enormous dimensions, yoked with eight griffins well matched. Having seated himself in this conveyance, he posted through the empire of the air, accompanied by magicians, who understood the language of birds, and asked of the most intelligent natives the proper questions concerning their laws, manners, and customs, while Alexander received their voluntary submissions. So far Dunlop. The common version of the story is that the birds of prey were first starved for three days and then put to a carriage, while a horse-liver was stuck on a spit in front of them. Greedy for the flesh, the birds drew the vehicle and in it Alexander up into the air until a bird with human face met him and bade him return to earth. Dunlop has justly remarked, "This aerial journey, like most of the fictions concerning Alexander, is of eastern origin. An old Arabian writer, in a book called Malem, informs us that Nimrod being frustrated in his attempt to build the tower of Babel, insisted on being carried through the air in a cage borne by four monstrous birds." Mediaeval miniatures illustrating this air voyage show Alexander with full regalia seated in a palanquin impelled by sixteen gryphons.

Francis Godwin (1562-1633), bishop of Hereford, wrote a romance entitled " The Man in the Moone, or a Discourse of a Voyage Thither by Domingo Gonsales the Speedy Messenger," first published in 1638 after his death. In this story Gonsales, on account of sickness during a voyage, is abandoned on the then uninhabited island St. Helena, and passes his time by training a number of wild swans to obey his call and gradually to carry small burdens while flying. He conceived the idea of harnessing several birds together and devised a mechanism whereby the difficulty of distributing the weight equally at the start of the flight might be overcome. With a team of seven birds Gonsales experimented on a lamb, and by increasing their number to about twenty-five, he was himself carried aloft to his great satisfaction. "For I hold it far more honor," he says, "to have been the first flying man than to be another Neptune that first adventured to sail upon the sea." On his return to Spain Gonsales was saved from shipwreck by his birds, who subsequently flew with him to the moon—a journey which lasted eleven days. He finally learned that the birds he had trained were not really denizens of St. Helena, but of the moon. "The Man in the Moone" had a considerable influence on literature. Swift is said to have derived from it the idea of the flying island in Gulliver's Travels (1727). Several features of Godwin's romance were borrowed by Capt. Samuel Brunt in his "A Voyage to Cacklogallinia," which recounts adventures among a nation of bird-men and a voyage to the moon. The frontispiece to this book shows the voyager conveyed through the air on a palanquin supported by four large birds—the same conception as found in the Shahnameh (Plate XII).

Hodgson adds the following interesting comment to Godwin's romance: "Godwin's name is now seldom remembered save by scholars.. , but his name deserves to be kept in remembrance in aeronautical history. For though flight had been an aspiration and an object of achievement long before Godwin's time, the idea that 'the first flying man' would be greatly deserving of honor, finds its earliest expression in the sentiments above quoted. Moreover, that ingenious pioneer of flight, Domingo Gonsales, though an imaginary creation, is inspired with an admirable spirit of enthusiasm for aerial adventure of a kind that has since inspired a countless succession of real pioneers in aeronautical endeavor. To suggest that Godwin's book created that spirit would be to press the point unduly—the motif of a first success has ever been a strong one, and one which usually predicates the impulse of enthusiasm."

In April, 1786 M. Uncles announced that he was constructing a balloon to be drawn by "four harnessed eagles, perfectly tame, and capable of flying in every direction at their master's will." In May he disclosed that nothing prevented an ascent save the unsettled weather, the birds being "well-practiced." The trial was deferred, however, until August when the eagles were ready on the ground at Ranelagh, the inflation proved a fiasco, and the balloon did not rise from even the ground.

In July, 1835 Thomas Simmons Mackintosh proposed a scheme by which balloons may be conducted in moderate weather with safety by having a sufficient number of larger birds, such as hawks—eagles would do better if they could be tamed, but perhaps strong pigeons would do very well,—and let them be harnessed to the balloon to draw it along. In a "sketch of an aerial ship" Mackintosh shows his balloon formed like the hull of a ship with an additional frame-work keel on either side of which are "harnessed" eight eagles, immediately controlled or "driven" by the aeronauts seated in a small car between the two keels. A colored reproduction of this airship is contained in Hodgson's book (Fig. 113).

J. Kaiserer published in 1801 at Vienna a pamphlet on his invention to direct an air balloon through eagles ("Ueber meine Erfindung einen Luftballon durch Adler zu regieren"). A plate depicts the inventor in his balloon driving (as it were) a pair of harnessed eagles. The same idea was revived in France in 1845, and as it has been demonstrated, this idea has its root in an ancient Persian tradition transmitted to Europe through the medium of the Romance of Alexander the Great.

Daedalus ("Cunning Worker") was an ingenious craftsman of an inventive turn of mind; he is the representative of the mechanical arts of the later Minoan age. While in the service of Minus, king of Crete, he built the labyrinth for the confinement of the Minotaur, but incurred the king's wrath; and to escape imprisonment, he fashioned a pair of artificial wings coated with wax for himself and his son Icarus. Thus they fled and flew westward across the sea. The father enjoined his son not to fly too low lest the wings dip in the brine and the wax which held them together be softened, nor too high, lest the heat of the sun melt the wax. Icarus disregarded the paternal admonition, came too near the sun in his lofty flight, the wax which fastened the wings to his shoulders melted, and he fell headlong into the sea which is still named for him the Icarian Sea. The more cautious Daedalus landed safely on Sicily. Of all flying stories of classical antiquity it is this one which has left a lasting impression on future generations and fired the ambition of many imitators; and it is on this point, its moral effect, that the importance of the story rests. Daedalus was an historical personage, a many-sided artisan who surely made some attempts to fly. Like many others of his type he was not understood or even was misunderstood by his contemporaries, and his story has been handed down in the form of poetic romance and exaggerated legend.

And what, after all, is the difference whether the Daedalus story is true or not? It is not the gray, cold, naked objective truth that counts in the history of mankind and will advance the cause of civilization, but it is the flight of human imagination, the impulses and visions of a genius, very often his errors and miscalculations, which have stimulated inventions and progress. Ever since Daedalus' alleged or real flight men in Europe have tried and died until finally success was insured.

Daedalus' adventure finds an echo in the Germanic saga of Wayland the Smith (Anglo-Saxon Wêland, Old Scandinavian Völundr), the artificer of marvelous weapons extolled in Icelandic, English, French, and German poetry. King Nidung endeavored to keep him in his service by cutting the sinews of his feet and thus laming him forever. Wayland forged a feather robe and, flying up to the highest tower of the royal castle, revealed his purpose to the king, and flew off to his home on Seeland. Wayland is represented on an Anglo-Saxon box of walrus bone of the eighth century, covered with Runic inscriptions; in this carving, his brother, Egil, is engaged in capturing birds from whose skins the clever smith will prepare his feather-shirt.

To mention all the winged gods of Greece and their flights through space would mean to pass in review a substantial portion of Greek mythology which is a subject of common knowledge. Suffice it to refer to Hermes or Mercury, the messenger and herald of the gods, of supernatural swiftness, often represented with winged shoes and cap; and to Perseus, who received from the nymphs a pair of winged sandals, a pouch, and the cap of Hades which rendered the wearer invisible.

On the Greek stage theatrical machines (*mechanē*, *geranos*) were used to convey the illusion of persons descending from the air or being lifted upward; for instance, in Aeschylus' *Fettered Prometheus*, where the choir descends on a winged chariot and where the god Okeanos arrives on a fantastic conveyance.

Archytas was a Greek who lived at Tarentum in southern Italy (about 428-347 B.C.). He was a philosopher of the Pythagorean school, mathematician, statesman, and general. Numerous fragments and titles of works are ascribed to him, but the authenticity of some is doubted. He attained great skill as a practical mechanician, and his flying dove of wood was one of the wonders of antiquity. He lost his life drowning on a voyage in the Adriatic. What his flying dove was is not clear from the few succinct and unsatisfactory accounts we have. It is described as having consisted of a wooden figure balanced by a weight that was suspended from a pulley; it is said to have soared in the air and to have been set in motion by a current of air "hidden and enclosed" in its interior, or by compressed air escaping from a valve. Some scholars incline to the opinion that it was an anticipation of the hot-air balloon; others think that it was an aerostat or glider, as it is said it could fly, but not rise again after falling. It may also have been on the order of Lu Pan's wooden kite (p. 23), but assuredly it was not a paper kite, as sometimes assumed.

Reference has been made to the Indian traditions of Greek airships (p. 51), but thus far no confirmation of such flying-machines has been found in Greek sources.

It will not be amiss to cast a glance at the writings of Lucian, that delightful satirist and divine liar of the second century of our era, as his imageries of air voyages have inspired such eminent authors as Rabelais, Cyrano de Bergerac, and Swift. In his "Icaromenippus or the Journey above the Clouds" Lucian introduces the flyer Menippus as a persiflage of Daedalus, who goes one better than his predecessor by refraining from the use of wax. He took an eagle and a vulture of the largest kind, clipped their wings off together with the shoulders, and fastened to himself the eagle's right wing and the vulture's left wing by means of strong leather straps, to the ends of which two handles were attached as a grip for his hands. Thus he essayed to fly, first timidly, leaping with a movement of his hands and, as geese do, keeping close to the ground on tiptoe and flapping his wings. Seeing that he succeeded, he attempted a bolder stunt, ascended the citadel, plunged downward, and flew to the theatre without a mishap. After several minor trials and exercises he scaled the Olympus, and carrying a supply of victuals as light as possible, started his flight skyward, crossed the clouds, and reached the moon.

In another work known as *The True History,* Lucian relates how, prompted by curiosity, he sailed from the pillars of Hercules and launched into the western ocean. A whirlwind carried him with his mariners toward a resplendent island which turned out to be the moon. There they were met by a curious class of creatures who styled themselves Hippogypes ("Horse-vultures"),—men riding on huge vultures which they rode like horses. These vultures were of enormous size, almost all of them provided with three heads; each of their feathers was longer and thicker than the mast of a large transport-vessel. The Hippogypes had the duty of encircling the island and conducting any stranger they encountered to the court of the king. This was Endymion, king of the moon, who at that time was engaged in a war with Phaeton, king of the sun. Lucian and his crew were graciously received by his lunar majesty, who requested their cooperation in the ensuing campaign, and as an inducement offered to furnish each with one of his royal vultures and the equipment pertaining to it.

The importance of Lucian's work rests on the fact that it gave the impetus in France to a class of fiction known as "voyages imaginaires" in which are recounted imaginary excursions to the planets or moon, like Cyrano de Bergerac's adventurous journey to the lunar world. Following Godwin's example (p. 61), Cyrano de Bergerac (1620-55) wrote the "Histoire comique de la lune et du soleil," relating aerial journeys to the lunar and solar worlds, wherein flight is achieved by such chimerical contrivances as the ascensive power of the dew when contained in glass balls and subjected to the sun's rays (an idea probably borrowed from Francesco Lana, see p. 21), or the use of a "very light machine of iron" drawn upward through the atmosphere by the attractive power of the loadstone.

The Arabs, the heirs of Greek philosophy and science, were clever mechanicians, and independently made considerable progress in mechanical devices. They were, as Washington Irving characterizes them, a quick-witted, sagacious, proud-spirited, and poetical people, and were imbued with Oriental science and literature. Wherever they established a seat of power, it became the rallying-place for the learned and ingenious.

About the year 875 of our era an Arabic mechanician of Spain, Abu'l Qāsim Abbās Ibn Firnās, called the Sage of Spain, devised a contrivance to make his body rise into the air; he made a pair of wings, clothed himself with feathers, and flew quite a distance through the air, but, as he had not taken into consideration what would happen during his descent, he fell and injured his buttocks. He was ignorant of the fact, the Arabic chronicler adds, that a bird falls only on his rump, and had forgotten to make a tail for himself. This man was the first who manufactured glass in Spain and who constructed clepsydras. In his house he made a model of the heavens in which he showed the stars, clouds, lightning and thunder. It is therefore credible that a man of his mechanical ability was led to make attempts at flying.

The story of a flying architect is handed down by Ibn al-Faqih, an Arabic geographer, who lived in the tenth century. He erected in Hamadan, Persia, a huge tower for King Shapur I, son of the founder of the Sasanian dynasty of Persia. The jealous king decided to leave the master-builder on the top of the tower, as he did not want any one else to profit by his genius. The architect consented, but asked one favor of the king; he was permitted to erect a wooden hut on the tower to protect his corpse from the attack of vultures. The king granted the request and ordered to supply him with as much timber as he needed. Then the architect was abandoned to his fate. He took up his tools, made a pair of wings from the wood left with him, and fastened them to his body. Driven by the wind he rose into the air and landed unscathed at a safe place, where he kept in hiding. This tradition exhibits a striking affinity with the Daedalus story. The same Arabic author, in describing the scenes represented in the chamber of Perwiz near Behistun, mentions the figure of Fattūs, a celebrated Arabic architect, outfitted with the wings of a bird,—presumably an emblem of architecture and sculpture.

Under the emperor Manuel Comnenus a Saracen tried to show his skill in flying before a large audience at Constantinople. An eye-witness relates the story as follows: "It was on the occasion of the festivities held in honor of a Sultan of the Seljuks, who had come on a visit. The Saracen clambered a tower of the hippodrome where the horse-races were held, and announced that he would fly across the race-course. There he stood on the tower, clad in a very long and wide garment of white color braced with rods of willow-wood laid over a frame-work. The cloth was loosely draped over this frame, and he intended to fly like a ship with its sail, hoping that the wind would catch in the folds of his garment. All eyes were intently fixed upon him, and the onlookers, enjoying the spectacle, kept on shouting, 'Fly, fly!' and 'How long will you put us off, Saracen, and estimate the wind from the tower?' The emperor sent a messenger over to detain him from the adventure. The Sultan, who was among the spectators, was filled with fear and hope, and worried about his compatriot. He, however, remained undisturbed, frequently examined the wind, and put the audience off. He often raised his arms, used them like wings, and lowered them to catch the wind. When the wind appeared to him favorable, he soared like a bird and seemed to fly in the air."

Oliver (also Eilmer or Elmer) of Malmesbury, an English astrologer and mechanician, who lived early in the eleventh century, is said to have fitted wings to his hands and feet and to have attempted to fly off from a tower with the help of the wind. He fell and broke his legs, and attributed his failure to the lack of a tail. Milton, in his "History of Britain" (1670), thus relates the story of the attempted flight: "He in his youth strangely aspiring, had made and fitted wings to his hands and feet; with these on the top of a tower, spread out to gather air, he flew more than a furlong; but the wind being too high, came fluttering down, to the maiming of all his limbs; yet so conceited of his art, that he attributed the cause of his fall to the want of a tail, as birds have, which he forgot to make to his hinder parts." Hodgson regards this story as legendary. Be this as it may, it bears such a striking resemblance to the Arabic accounts of flying aforementioned that a connection between the two must inevitably be assumed: either Oliver made his attempt in imitation of the Arab of whose experiment he had heard or read, or the story itself is patterned after the Arabic model.

Giovanni Battista Danti, a mathematician of Perugia, is said to have attempted about 1490 winged flights over the lake of Trasimeno in Umbria.

A similar adventure is ascribed to John Damian, abbot of Tungland, an Italian by birth, who came to Edinburgh from France in 1501 and became the favorite of King James IV, residing at the Scottish court in the capacity of physician to the king's household. In the autumn of 1507 when an embassy had been sent to France,

Damian averred he could overtake it by flying and to arrive in France before the ambassadors. He made from bird-feathers a pair of wings which he fastened on to himself, and hopped off from the top of Stirling Castle, but shortly fell to the ground and broke his legs. This failure he attributed to the fact that some hen feathers were contained in his wings and showed a natural affinity to return to the barnyard instead of maintaining flight skyward. John Lesley, bishop of Ross, who records this story in his "History of Scotland" (1578), winds up with the remark that in this adventure Damian was endeavoring to outdo King Bladud (p. 14). At any rate Damian was not so wrong from the standpoint of his time in laying the blame for his misfortune on the chicken feathers. During the middle ages it was a wide-spread superstition that one could not sleep well on a feather pillow, nor could one easily die on it. Bird-feathers were believed to retain the soul, hence the pillow had to be pulled away from under a moribund. In Ireland the belief prevailed that when a dying man suffered great agony, it was due to the presence of chicken-feathers in his bed, and his friends would sometimes lift him up and place him upon the floor to relieve him. In Norway it was a rule not to have chicken-feathers in one's pillow, for the chickens have a certain feather known as "restless feather" on which no one can sleep or die.

Roger Bacon (1214-94), the Franciscan monk, one of the few great scholars of the middle ages, merits a place in the prehistory (I say advisedly prehistory, not history) of aviation, as he points to the possibility of a flying-machine. In his "Epistola de secretis operibus," written about 1250, he affirms in the chapter "Of Admirable Artificial Instruments," "Likewise flying-machines con be made in such a way that a man is seated in the midst of the machine, revolving some sort of device by means of which wings artificially composed may beat the air after the fashion of a flying bird." The Latin original is as follows: "Item possunt fieri instrumenta volandi, ut homo sedeat in medio instrumenti revolvens aliquod ingenium, per quod alae artificialiter compositae aerem verberent ad modum avis volantis." Discussing other mechanical devices, Bacon continues that "all these were made in ancient times and have also been made in our times, as it is certain, with the sole exception of the flying-machine which I have not seen, nor do I know any one who has seen it, but I know an expert who has thought out the way to make one" (Haec autem facta sunt antiquitus, et nostris temporibus facta sunt, ut certum est, nisi sit instrumentum volandi, quod non vidi, nec hominem qui vidisset cognovi; sed sapientem, qui hoc artificium excogitavit explere, cognosco).

Roger Bacon has been greatly overestimated in modern times, until Professor Lynn Thorndike in his "History of Magic and Experimental Science during the First Thirteen Centuries of Our Era" (1923) has successfully refuted the exaggerated and distorted estimate of his importance and uniqueness and has presented the man and his work in a critical and just attitude. In fact Bacon was as superstitious and credulous as the majority of his contemporaries. In the chapter just cited he speaks, for instance, "of machines that can be made for walking in the seas and rivers, even to the bottom without danger; for Alexander the Great employed such that he might see the secrets of the deep." The story of Alexander diving into the sea in a sort of submarine is, of course, not historical, but appears only among the fictions of the Alexander Romance, which Bacon evidently swallowed as historical truth. What his real notions of flying were appears from the following passage inserted in the midst of his discussion of experimental science (characterized by Thorndike as "an instance of his gullibility"): "It is certain that Ethiopian sages have come into Italy, Spain, France, England, and those Christian lands where there are good flying dragons; and by an occult art that they possess, excite the dragons, and drive them at top speed through the air, in order to soften the rigidity and toughness of their flesh, just as boars, bears, and bulls are hunted with dogs and beaten with many blows before they are killed for eating. And when they have tamed the dragons in this way, they have an art of preparing their flesh...which they employ against the accidents of age and prolong life and inspire the intellect beyond all estimation. For no education which man can give will bestow such wisdom as does the eating of their flesh, as we have learned without deceit or doubt from men of proven trustworthiness." This much the Chinese knew centuries before our era. The preceding quotation shows that Bacon's mind was steeped in Oriental lore, and there is no doubt that his notions of flying are traceable to this source. In particular, the legend of men who tame flying dragons by their incantations and magic appears among the thirteenth century additions to the famous letter of Prester John in which the marvels of India and adjacent territories are recorded, and this must be the source of Bacon's version.

We know that Bacon to some extent was under the influence of Arabic science. His mathematical ideas are based on Latin translations of Arabic works, particularly through the medium of Witelo, a Polish scholar, his contemporary, who studied the writings of Alhazen (Ibn al-Haitham, 965-1038) and Avicenna (Ibn Sina, 980-1037).

It is therefore an exaggeration to say with Hodgson that "the first dawn of a rational idea of flight and of a belief in the possibility of achieving it is revealed in the writings of Roger Bacon," or with Brown that "with prophetic vision he saw the wonders that the future might hold." After quoting the above passage with reference to a flying-machine, Brown continues, "This single observation can hardly justify us in regarding Roger Bacon as a student of aeronautics, and the thought behind it was alien to the thought of the time. Nevertheless, it was a portent that the mental attitude of the middle ages would not last for ever." On the contrary, the thought was not at all novel or alien to his time, but was merely the echo of an ancient idea that we have traced in China and India as well as among the Persians and Arabs. Bacon is very far from being the herald of a new era and opening the historical period of air navigation; his place is at the end of the line of its prehistoric age.

The modern history of aviation begins with Leonardo da Vinci (1452-1519), who was a true pioneer of science by studying the flight of birds and left several sketches of aeroplanes in his manuscripts which were hidden in obscurity for nearly three hundred years until their existence was revealed in 1797. This subject, however, as well as the modern development of aircraft is beyond the scope of this inquiry.

THE AIR MAIL OF ANCIENT TIMES

To one who look'd from upper air
O'er all the enchanted regions there,
How beauteous must have been the glow,
The life, the sparkling from below!
Fair gardens, shining streams, with ranks
Of golden melons on their banks,
More golden where the sun-light falls;—
Gay lizards, glittering on the walls
Of ruin'd shrines, busy and bright
As they were all alive with light;—
And, yet more splendid, numerous flocks
Of pigeons, settling on the rocks,
With their rich restless wings, that gleam
Variously in the crimson beam
Of the warm west—as if inlaid
With brilliants from the mine, or made
Of tearless rainbows, such as span
The unclouded skies of Peristan!

Thomas Moore, *Paradise and The Peri*

Air-mail service was first established in the United States in the year 1918 when the New York-Washington mail route (218 miles) was inaugurated on May 15. A year later the Cleveland-Chicago route (325 miles) was opened. The New York-Cleveland service (430 miles) followed on July 1, 1919. On August 16, 1920, the Chicago-St. Louis service (300 miles) was inaugurated, and on September 8 of the same year New York was connected by air mail with San Francisco (2,651 miles).

While our air mail is one of the epoch-making innovations and achievements of modern times, there was also a "prehistoric" air mail which is no less admirable—carried on the wings of pigeons. This prodigious institution we also owe to the Orient. I propose to survey it from China and India to Persia and the Near East and to show how it was transmitted from there to Europe.

The first Chinese who has gone down in history as having made use of carrier pigeons is Chang Kiu-ling (A.D. 673-740), who flourished as a statesman and poet under the emperor Ming Huang of the T'ang dynasty. In his youth he was in the habit of corresponding with his relatives by means of a flock of carrier pigeons which he trained in large numbers and which he called his "flying slaves." *Fei nu* ("flying slaves") is still a designation of a carrier pigeon. The messages were attached to the feet of the birds, and they were taught how to deliver them.

It is singular that the government organs of China never saw this opportunity and never availed themselves of pigeons for conveying important messages, as it was done by the kings of India and by the Mohammedan rulers in the Near East. The employment of carrier pigeons remained restricted to private correspondence, chiefly for commercial purposes. They were of great service to merchants in conveying intelligence to the producing districts, or bringing news of the arrivals of cargoes and the ruling prices of the markets. In the old days of the Manchu empire merchants of Hongkong used pigeons in sending news to their business partners at Canton of the arrival of the English, French, or American mails. In Canton they are termed *ch'ün shü kop* ("letter-transmitting pigeons").

Up to the time of the introduction of telephones in Peking, carrier pigeons (called *sung sin,* "letter-carriers") were used to send quotations of money exchange rates from the banks located in the Chinese City to those in the Manchu City.

The Chinese say that carrier pigeons are difficult to train and that it takes two or three years before they can be employed for long distance flights, which quite agrees with our own experiences. It takes about three years to determine the qualifications of a good homing pigeon for a five-hundred-mile flight.

While the Chinese have never bred carrier pigeons on a large scale or intensively, they have added to the art of pigeon-training an attractive means of amusement: in the same manner as they were the first who communed with the air by means of kites, they also were the first who created "music on the air." This was accomplished by means of whistles extremely light in weight, attached to the pigeon's tail-feathers. These whistles consist of two, three, or five reed tubes of graded length in the shape of a Pandean pipe, varnished yellow, brown, or black; or of a small gourd into which reed pipes are inserted. A collection of these whistles, some engraved with the names of the makers, is on view in a case illustrating the musical instruments of China in the West Gallery of the Museum; there also a mounted pigeon outfitted with the whistle and photographs of live pigeons thus equipped and taken in Peking may be seen. When a flock of pigeons circles the air, the wind strikes the apertures of the instruments which are set vibrating, and produce a not unpleasing, open-air concert whose charms are heightened by the fact that the whistles used in a flock are tuned differently. The Chinese explain that the sounds of the whistles are intended to keep the flocks together and to protect the birds from onslaughts of hawks and other birds of prey. This rationalistic interpretation, however, is not convincing. It is not known and at least doubtful whether such music makes an impression on either pigeon or hawk, and would it really prevent the famished princes and pirates of the air from making a swoop at their quarry? Even supposed this might happen once in a while, we must consider that this music constantly fills the atmosphere year by year, and the unrelenting foes of the pigeon will gradually become accustomed to it and treat it with disdain or disregard. It seems more plausible that this quaint custom has no rational origin, but that it rather is the outcome of purely emotional and artistic tendencies. Psychologically, the pigeon whistles move along the same line as the musical bows attached to kites. It is not the pigeon that profits from the aerial music, but the human ear that feasts on the wind-blown tunes and derives esthetic enjoyment from them.

The pigeons which fly about with whistles attached to them are termed "mid-sky beauties" (*pan t'ien kiao jen*).

In India the use of carrier pigeons goes back to a great antiquity, and may with certainty be assumed as having been in full swing in the beginning of our era. The Arthaçāstra, an ancient handbook of polity and state wisdom written in Sanskrit by Kautilya, a minister of state, gives us the specific information that the kings of India received news about the movements of hostile troops by air mail, through domesticated pigeons which brought them stamped and sealed letters.

In Indian stories various kinds of birds appear as harbingers of messages. A white wild goose, for instance, who had been with a prince all his life carries to him a letter from his parents into a remote kingdom, and returns there with a response from him (in the legend of Kalyānamkara and Pāpamkara). Aryadeva received an invitation to come to Nalanda by a letter attached to the neck of a crow (in Tāranātha's History of Buddhism in India). Parrots frequently appear in the role of winged messengers.

Linschoten, who travelled in India in the seventeenth century, mentions the fact that he met in India a Venitian who had brought carrier pigeons along to try them out and naturalize them. John Fryer, who travelled in the East from 1672 to 1681, notes in his description of Surat carrier pigeons with blubbered noses and of a brown color to carry letters. The fact that Darwin received carriers from Madras would seem to point to their use in southern India.

As regards Persia, an interesting bit of evidence is preserved by Twan Ch'eng-shi, author of the *Yu yang tsa tsu* (ninth century), to the effect that on the sea-going vessels of the Persians many pigeons were kept, capable of flying several thousand *li* (Chinese miles); these were released and at a single flight returned to their homes, bearing as it were the tidings that everything on board was well.

Ch'ang Te, a Chinese traveller, was sent in 1259 by the Mongol emperor, Mangu, as envoy to his brother Hulagu, king of Persia. He kept a diary of his journey which was edited in 1263 by Liu Yu. Speaking of the postal service of Persia in his time he mentions a special kind of swift camel trained for the service of couriers, as well as pigeons which transmit news to a distance of a thousand *li* (Chinese miles) in one day. In mediaeval times Persian authors repeatedly refer to the conveyance of letters by pigeon-mail in western Asia, even in time of war. In 1262 when the Mongols besieged the city of Mosul, they caught a tired pigeon which was perching for rest on one of their catapults and which carried a message for the beleaguered. The letter was intercepted, and was found to contain news of the approach of an army for the relief of the city. This enabled the Mongols in time to throw an army against the onmarching enemy.

The pigeon appears in love-songs of the Baluchi, an Iranian tribe inhabiting Afganistan. One of these love-messages begins, "Oh dove! Oh pigeon, among the birds be thou a messenger of my state to my love. Travel over the long distance, I beg of thee, blue bird, fly from the cliff where thou dwellest night, from the rugged rocks of the fowls of the air, go to my beloved's home and perch on the right side of her bed." In another love-song it is said, "Oh pigeon, peahen among the birds, be a messenger of my state to my true-love, to that modest fair one." (M. L. Dames, Popular Poetry of the Baloches.)

Pigeons were used by the ancients for sending love messages (Anacreon IX, 15; Martialis VIII, 32), news of a victory in the Olympic games, or letters into a besieged city. The earliest Greek allusion to a carrier pigeon is found in one of the fragments of Pherecrates, a writer of comedies, who lived in the fifth century before our era. Greek seafarers are said to have carried on their ships pigeons for the purpose of sending home tidings of their welfare.

Aelianus, who lived in the second century of our era, tells this story, "When Taurosthenes won the laurels in the Olympic games, intelligence of his victory was conveyed to his father at Aegina on the same day by means of a pigeon whom he took away from her young ones who were still unfeathered. He attached a purple piece of cloth to the bird and released her; she sped away to her young ones and in a day returned from Pisa to Aegina."

Pisa was an ancient town in the territory of Elis in the western part of the Peloponnesus, not far from Olympia, where the celebrated athletic games and contests were held. The distance from Pisa to Aegina amounts to about twenty-three and a half geographical miles. It will be noticed that in this case not a letter, but merely a pre-arranged token of victory was attached to the flying messenger; purple was a symbol of victory.

Pliny (X, 110) relates that pigeons have acted as messengers in important affairs (internuntiae in magnis rebus fuere) and cites as example that during the siege of Mutina, Decimus Brutus, who was beleaguered in that city by Antonius from December, 44, till April, 43 B.C., sent into the camp of the consuls (Hirtius and Pansa) dispatches fastened to pigeon's feet (epistulas adnexas earum pedibus). A somewhat different version of this event is contained in the work of Frontinus (Strategemata III. 13, 8) of the first century of our era: the Consul Hirtius attached letters to the neck of pigeons by means of strong hair; previously he had starved the pigeons in a dark room; thereupon he released them near Mutina, where they settled on the roofs of the houses, and were caught by Brutus, who in this manner was duly informed of the events. Caesar is said to have been advised of a revolt in Gaul by pigeon-post just in time that he could lead his legion down the Alps to suppress the rebellion.

However, what is known about the use of carrier pigeons among Greeks and Romans is restricted to isolated instances. We must not generalize that it was a customary practice, for there are no records of carrier pigeons having been kept and trained for such purpose in large numbers, nor was there anything like a regular pigeon-mail. The curious fact remains that carrier pigeons were not transmitted from Italy to northern Europe in the wake of Roman civilization. The North-European nations first made the acquaintance of carrier pigeons in the Orient during the Crusades, and from that time onward they appeared inEurope, inclusive of Italy, as a novel affair. Therefore it is reasonable to conclude that among the ancients the whole business was of no great significance and that it was extinct in the days of the declining Roman Empire. The consensus of opinion is that the Greeks derived the institution from the Near East, and we have to wend our way back again to the Orient to learn more about its history.

Mesopotamia appears to be the home of the domesticated pigeon, and the domestication of the bird was accomplished as early as pre-Semitic times by the Sumerians. In Sumerian documents the pigeon is referred to as a domestic bird. Among the Semites pigeons were closely connected with religious practices. They are sacred to the goddess Ishtar (Astarte), the mother goddess or great goddess personifying the productive powers of the earth, life, generation, and death.

Lucian, in his treatise on The Syrian Goddess, informs us with reference to Syria, "Of birds the dove seems to be the most holy to them, nor do they think it right to harm these birds, and if any one have harmed them unknowingly, they are unholy for that day; so when the pigeons dwell with the men, they enter their rooms and commonly feed on the ground."

It is unknown, however, when and where pigeons were first trained for conveying messages. Nothing to this effect has as yet come to light in the cuneiform literatures or on Egyptian monuments; both in Egypt and Mesopotamia the practice was unknown. At the outset it is improbable that it might have been developed in the Euphrates valley, where clay tablets were the common writing-material, which on account of their weight could not have been attached to pigeons; in later times, of course, parchment and papyrus were also used in Mesopotamia.

The dove which Noah sent forth from the ark three times has frequently been classified among carrier pigeons, but this notion is erroneous. Noah's dove represents an entirely distinct class: it is not sent out with a message, but belongs to the category of land-spying birds, such as navigators of ancient times used to keep on board their ships and which were released by them when in quest of land if they had lost their bearings, on the supposition that the birds would fly in the direction of land; these birds, of course, never returned to their ships. In the Pāli Bāveru Jātaka, which echoes ancient commercial relations of India with Bāveru or Babiru, i.e. Babylon, the Indian seafarers are assisted by a crow which serves for the purpose of directing their way in the four quarters. The crow has a well-developed sense of locality, and in all ancient systems of divinations crow or raven auguries are correlated with the cardinal points. According to Pliny, the mariners of Taprobane (Ceylon) did not take recourse to the observation of stars for the purpose of navigation, but carried birds out to sea, which they sent off from time to time, and then followed the course of the birds who flew in the direction of land. When the people of Thera emigrated to Libya, ravens accompanied them ahead of the ships to guide their way. In the ninth century when the Vikings sailed from Norway, they kept on board birds who were set free from time to time amid sea, and with their aid they succeeded in discovering Iceland. Land expeditions also were accompanied by land-spying birds, and tribes on the path of migration would settle in a territory where birds carried along by them would descend. The Celts, as Justinus informs us, were skilled beyond other peoples in the science of augury, and the Gauls who invaded Illyricum were guided by the flight of birds. The legendary emperor Jimmu of ancient Japan when engaged in a war expedition marched under the guidance of a gold-colored raven.

It is asserted by many authors that the ancient Hebrews were acquainted with carrier pigeons, but there is no direct evidence to this effect in the Old or New Testament.

In the present state of our knowledge we can only assert with safety that the highest development in the use of pigeon messengers was reached in the empire of the Caliphs and under the Mohammedan dynasties of Egypt when the whole business was organized and systematized on a scientific basis, while, of course, isolated cases occurred many centuries earlier. The Arabs, on their part, were only to a small extent original or inventive, but exceedingly clever in absorbing and digesting the ideas and cultures of subject nations, and thus created an imperialistic civilization as a result of their far-flung conquests. Indo-Iranian peoples may very well have given the first impetus to the training of carrier pigeons.

Damiri (1341-1405), in his Book of Animals (*Hayāt al-hayawān*), writes in regard to the pigeon, "It may be mentioned as a part of its nature that it seeks and finds out its nest even if it be set free at a distance of a thousand leagues; it carries news and brings it from a very distant place in a very short time. There are some pigeons which can fly three thousand leagues in a day. It may sometimes happen that it is caught, and may be thus away from its native place for ten years or more, but it still retains its intelligence and power of memory, and is desirous of returning to its native place, so that when it finds an opportunity, it flies back to it."

Damiri likewise informs us that the Caliph Hārūn al-Rashid (786-809) was very fond of pigeons and sporting with them. The Arabian Nights (No. 698) introduce to us the father of Dalila who was postmaster and guard of the carrier pigeons at the court of this illustrious Caliph at a monthly salary of a thousand dinars; he used to train the pigeons so that they conveyed letters and messages, and to the Caliph each of these birds was dearer at a time of distress than any of his sons. After her husband's death, Dalila and her daughter took care of the forty pigeons, and she would daily visit the state council to find out whether the Caliph had a message to transmit by pigeon-mail.

In another story of the Arabian Nights (No. 96), Afridūn, king of Constantinople, is advised of the movements of the Mohammedan army in Asia Minor by means of a letter sent "on the wings of a bird" and brought to him by the Guardian of the Pigeons.

According to Masudi (tenth century), news of the victory over a rebel army was conveyed to the Caliph Motasim (838-847) by pigeon-post. In 1171 the Sultan Nūr-ed-dīn established a regular air mail in Syria, actuated by the desire to obtain as quickly as possible intelligence of everything that happened in all of his provinces. For this purpose he ordered pigeons to be maintained in all castles and fortresses of his empire, also had special towers erected for breeding and postal purposes, and devoted the greatest care to the training of the birds. After his death this mail service declined until in the year 1179 it was re-established by the Caliph Ahmed Naser-lidin-allah, who had a veritable passion for pigeons and bestowed a special name on each bird. In sending a letter by pigeon-mail he was in the habit of designating in the letter the exact name of the feathered messenger, thus: "This bird, son of . . ." or "this bird, mother of. . . ." In this manner he conducted a voluminous correspondence with the remotest parts of his dominion. The air mail developed into a general institution in his time, and although many engaged in the business of raising pigeons whose number was enormous, their prices reached amazing figures: a well-trained pair sold at a price up to a thousand gold pieces. Baghdad was the central station of the air mail until it was conquered by the Mongols in 1258.

The price of a pigeon of the first quality amounted to seven hundred dinars (gold coins), and the egg of such a bird sold as high as twenty dinars. Genealogies of renowned pigeons were kept on special registers.

One of the most curious incidents in the history of the pigeon-mail, as reported by Makrizi, refers to the rapid transmission by air of a consignment of cherries. The Caliph Aziz (975-996) of the Fatimid dynasty, distinguished by his tolerance and his love for science, had a great desire for a dish of cherries of Balbek. The Vezir, Yakub Ben-Kilis, caused six hundred pigeons to be dispatched from Balbek to Cairo, each of which carried attached to either leg a small silk bag containing a cherry. This is the first example of parcel post by air mail recorded in history.

In his "History of Egypt in the Middle Ages" Stanley Lane-Poole writes, "The most famous and energetic of all the Bahri Mamluks, Beybars (1266-77), established a well-organized system of posts, connecting every part of his wide dominions with the capital. Relays of horses were in readiness and answered reports from all parts of the realm. Besides the ordinary mail, there was also a pigeon-post, which was no less carefully managed. The pigeons were kept in cots in the citadel and at the various stages, which were farther apart than those of the horses; the bird was trained to stop at the first post-cot where its letter would be attached to the wing of another pigeon for the next stage. The royal pigeons had a distinguishing mark, and when one of these arrived at the citadel with a dispatch, none was permitted to detach the parchment save the Sultan himself; and so stringent were the rules, that were he dining or sleeping or in the bath, he would nevertheless at once be informed of the arrival, and would immediately proceed to disencumber the bird of its message." Beybars connected Damascus and Cairo by a postal service of four days, and used to play polo in both cities within the same week. Pigeons contributed to the complete defeat of the Mongols after the decisive battle of Hims (Emesa in Syria) in 1281, when they were beaten back by Kalāūn, who harassed their retreat and sent orders by pigeons to his governors at the Euphrates to bar the fords to the fleeing enemy.

The letters, which were written on a fine tissue paper with specifications of place, day, and hour, were fastened beneath the wings, at a later time to the tail-feather.

The caretakers brought the incoming birds directly to the Sultan who alone had the right to take the letters off. The pigeons were therefore called by the Arabs "angels of the kings." An Arabic scholar says, "The carrier pigeons are arrows which reach their goal despite the resistance offered them by the clouds. There is no mistake in styling them the prophets among the birds, because like the prophets they are dispatched with scriptures." An Arabic poet has this line: "In the marvellous speed of their flight they rush ahead of the winds; swiftly like a moment they bear under their wings in rapid flight tidings of what happens in places distant a month's journey."

Another Arabic author writes, "The pigeons who forward messages are a wonder of divine almightiness worthy of being admired and praised by us. In faithfully executing their commissions they confirm the proverb which calls them birds of auspicious foreboding. Indeed they often surpass the itinerant messengers: the clouds are their bridles, the air is the course they race through, the wings are their equipment, the winds are their escorts. They fear on their flights neither brigands of the desert nor the perils threatening from accidents on the roads."

In 1323 Symon Semeon, an Irish Franciscan, and his companion, Hugo Illuminator, were on a pilgrimage to the Holy Land and stopped at Alexandria. An entry in Symon's diary reads as follows: "The admiral [of Alexandria], on learning of the affair [the arrival of the two pilgrims], immediately dispatched a message to the Sultan at Cairo by means of a carrier pigeon. These pigeons were trained in the Sultan's Castle at Cairo and sent in cages to the governors of the various maritime cities, who whenever they wish to make something known to the Sultan dispatch one with a letter tied under its tail, which never stops until it has reached the castle from which it was brought originally; and so the Sultan and his governors are informed daily of what is going on in the country and of the necessary measures to be taken."

A German pilgrim, von Bodmann, paid a visit to the Holy Land in 1376-77, and when his ship neared Alexandria, she was boarded by two officials from the city who drew up two lists of the vessel's cargo. These, he relates, were tied to the wings of two pigeons who were dispatched to the court of King Soldan at Babylon, a distance of two hundred miles.

In the second half of the fifteenth century the governmental pigeon-post expired in consequence of political troubles, but as a means of private communication it has survived in the Orient long after and even until the present time, especially in commercial correspondence when transactions had to be made quickly or when perishable merchandize like drugs and perfumes were at stake. Travelling merchants also availed themselves of the air mail to advise their families of their safe arrival at a place.

In 1599 Thomas Dallam, the organ-builder, during his voyage from London to Constantinople, made the following observation on the use of carrier pigeons: "The firste of June thare was letters convayede varrie straingly from Alippo to Scandaroune, the which is thre score and twelve myles distance. After I hade bene thare a litle whyle, I persaved that it was an ordinarie thinge. For, as we weare sittinge in our marchantes house talkinge, and pidgons weare a feedinge in the house before us, thare came a whyte cote pidgon flyinge in, and lyghte on the grounde amongeste his fellowes, the which, when one of the marchantes saw, he sayd: Welcom, Honoste Tom, and, takinge him upe, thare was tied with a thred under his wynge, a letter, the bignes of a twelve penc., and it was Dated but four houres before. After that I saw the lyke done, and always in 4 houres."

Linschoten, the great Dutch traveller of the seventeenth century, describes the pigeon-mail in the Turkish empire extending from Bassora and Babylonia to Aleppo and Constantinople, and writes that the letters were fastened to a ring placed around the bird's leg.

Pietro della Valle (Viaggi in Turchia, Persia e India, Vol. I, p. 284) wrote in a letter dispatched from Ispahan in 1619, "From the Province of Babylon whither I addressed a letter I am awaiting some pigeons which convey letters from one place to another and which Tasso styles 'flying carrier' (*portator volante*). They have thus been used in Asia from the earliest times down to the present."

The Jesuit father Philippe Avril (about 1670) relates how pigeon messages were sent from Alexandria to Aleppo. "No sooner had we got ashore," he writes, "but we had the pleasure to see dispatched away before us one of the messengers which they make use of in those parts to carry such intelligence as they would have speedily made known. For the doing of which, their most usual way is this. A merchant of Aleppo, who desires to have the most early information of what merchandizes are come from France or any other parts, takes particular care by an express to send away a pigeon that has young ones, much about the time that the ships are expected at Alexandretta, where he has his correspondent; who as soon as any vessel comes to an anchor, goes and informs himself of what goods the vessel has brought most proper for his turn; of which when he has given a full account in his letter, he fastens the paper about the neck of the winged courier, and carrying her to the top of a little mountain, gives her her liberty, never fearing her going astray. The pigeon which we saw let go, after she had soared a good height to discover, doubtless, the place from whence she had been taken some few days before, and pushed forward by that instinct, which is common to all birds that have young ones, took her flight toward Aleppo, and arrived there in less than three hours, tho that city be very near thirty leagues from the place from whence she was sent. However, they do not make use of any sort of pigeons to carry their dispatches, in regard that all pigeons are not alike proper for that service. For there is a particular sort of these birds, which are easily trained up to this exercise, and which as occasion serves, are of extraordinary use, especially for the swift management of business, and where speed of intelligence is required, as in the factories of the Levant, far remote one from the other. This was the only piece of curiosity which we could observe during our stay in this same first port of the East."

During the middle ages, the European nations became acquainted with the pigeon air mail when the cross and the crescent clashed during the crusades. In the history of the enterprises against the infidels there are several stories on record which depict the wonder and amazement of the Christian soldiers at this novel experience. In A.D. 1098 the commander of the Turkish castle Hasar disobeyed his liege lord, Rodvan of Aleppo, who declared war on him. The Turkish chief was unable to resist when one of his Emirs counselled him as follows: "Recently when Christian pilgrims marched against Edessa, I captured the wife of a knight, Fulcher (also called Foulqe) of Bouillon, and married her on account of her beauty. She is acquainted with our perilous situation and advises us to seek assistance from the Duke of Lorraine, the most powerful of the victorious Franks." The aversion toward an alliance with Christians was suppressed by the apprehension of graver consequences, and a Syrian was dispatched to the Duke with a ready proposal. Succor was promised by the latter, and the son of the Turkish chief retained by him as hostage. Meanwhile Rodvan beleaguered the fortress Hasar with an army of forty thousand, and the Franks were at a loss as to how to send the tidings of the pact into the fortress. To their amazement the Turkish envoys brought pigeons forward and tied papers to the under side of their wings. The birds were released, and the Franks assured that the good news would reach the fortress and encourage the Emir in his resistance till the arrival of the relief army.

Another episode is related thus: In A.D. 1099 when the Christian army advanced from Akkon to Caesarea, a wounded pigeon, who had a narrow escape from the claws of a hawk, dropped lifeless in the camp of the Christians. The bishop of Apt picked the bird up and found under its wings a letter addressed by the Emir of Ptolemais to the Emir of Caesarea, reading as follows: "The cursed rabble of Christians has just traversed my territory, and is passing on to yours. All chiefs of Musulman towns should be informed of their onward march and take measures to crush our foes." This letter was read aloud in the council of princes and before the entire army. Surprise and joy seized the Crusaders who did not doubt that God protected their enterprise, since he sent them the birds of heaven to reveal the secrets of the infidels.

This incident has inspired Torquato Tasso (1544-95), the great Italian poet of the Renaissance, to a poetic composition, which is inserted in his *La Gerusalemme Liberata* (Jerusalem Delivered, XVIII, 49-53). I give a literal prose rendering of my own:—

"While the camp prepares for assault and the city for defence, a pigeon is seen towering high along aerial paths over the host of the Franks. Agitating her swift pinions, she sails the clear air with outstretched wings, and the strange messenger (*la messaggiera peregrina*) is just about to alight from the high clouds into the city.

"A falcon (I do not know whence) swoops downward, armed with curved beak and large claws, and obstructs her path between the camp and city-wall. She does not wait for the tyrant's claws, but he pounces upon her and chases her to the main tent. He seems to reach her now, and holds his foot over her tender head when she takes refuge in the lap of the pious Godefroy of Bouillon.

"Godefroy takes her up and protects her, then, while looking at her, notes a strange thing suspended from her neck and fastened with a thread,—a letter concealed under a wing. He opens it and unfolds it, well comprehending the terse message it contains. 'To the Lord of Judea,' the epistle read, 'the Captain of Egypt sends greetings.

"Despond not, my lord, resist and hold out for four or five days, and I will come to liberate these walls, and you will soon see your foe vanquished.' This was the secret conveyed in pagan script and confided to the winged courier, as the Levante employed such messengers at that time.

"The prince released the pigeon who, since she had revealed her master's secrets and fancied that she had betrayed him, did not dare to return as an unlucky harbinger."

The poet had evidently read about carrier pigeons in the documents of the crusades, and was profoundly impressed by this ingenious device of postal service. His detailed description, as well as his observation that this was customary in the Levante, seem to hint at the fact that letter-carrying pigeons were still unknown in the Italy of his time,—the sixteenth century.

Lodovico Ariosto (1474-1533), in his *Orlando Furioso* (XV, 90), also refers to the pigeon-post: The giant Orrilo was slain by the duke Astolfo on the lower Nile, and this event was air-mailed by the Castellan of Damiette to Cairo. This is the custom there, the Italian poet adds, and in a few hours the news was broadcast to the whole of Egypt that the bandit had met his death.

Shakespeare alludes to pigeons as letter-carriers in *Titus Andronicus* (IV, 3), where upon the entry of a clown with two pigeons Titus exclaims,—

> News, news from heaven! Marcus, the post is come.
> Sirrah, what tidings? Have you any letters?

Another interesting reference, though not to carrier pigeons, occurs in *Venus and Adonis,* where Venus rides in a chariot drawn by doves:—

> Thus weary of the world, away she hies,
> And yokes her silver doves; by whose swift aid
> Their mistress, mounted, through the empty skies
> In her light chariot quickly is convey'd;
> Holding their course to Paphos, where their queen
> Means to immure herself and not be seen.

The Crusaders brought carrier pigeons along from the Orient. Mediaeval knights used them in sending communications from one castle to another; the convents also availed themselves of pigeon messengers.

A study of the various breeds of carrier pigeons has led Darwin to the conviction that nearly all the chief domestic races existed before the year 1600 and that the names for them applied in different parts of Europe and in India to the several kinds of carriers all point to Persia or the surrounding countries as the source of this race. Certain it is that the common European breeds of pigeon were not fit for air-mail purposes, but that all varieties used in Europe for messenger service are of Oriental origin and in the last line are traceable to the bagdotte which under the name of carrier was bred to perfection in England. The Baghdad pigeons won renown everywhere, and were known simply as a Baghdad, or Babylonian pigeon. Thus Thomas Moore, in *The Fire-Worshippers*, has the line:—

> As a young bird of Babylon,
> Let loose to tell of victory won—

The great Rabelais (1483-1553), in his Gargantua and Pantagruel (IV, 3), makes Pantagruel correspond with his father Gargantua by means of a pigeon called "Gogal [the Hebrew word for a pigeon], the heavenly messenger." Whenever the son was well or successful, he tied a white ribbon to the bird's foot; in case something untoward should happen to him, they had agreed on a black ribbon. Rabelais describes in detail this manner of communication, the bird's desire to return to her young ones as swiftly as possible and the rapidity of her flight, which seems to hint at the fact that this was a novel feature in his time.

The first employment of pigeons for military purposes in Europe took place during the war of liberation of the Netherlands in the sixteenth century. During the siege of Harlem by the Spaniards in 1573, the garrison received several advices by pigeon-mail, announcing the approach of a relief army under the command of the Prince of Orania, and therefore persevered in its resistance. In commemoration of this event the Prince caused these pigeons to be cared for until their end, and after their death they were stuffed and preserved in the town-hall of Leiden.

The breeding of carrier pigeons was given special attention in Belgium as early as the beginning of the eighteenth century. From Belgium the experience thus gained was transmitted to France and Germany. In Belgium, Holland, and France the fondness of carrier pigeons developed into a sport.

In the beginning of the nineteenth century the pigeon-mail took a new development, chiefly for commercial purposes. The story goes that Rothschild of London had his agents join Napoleon's army and received from them first-hand war news by air mail. He was advised of the emperor's defeat at Belle-Alliance three days earlier than the British Government, and correspondingly arranged his financial speculations. In the organization of the modern press and news agencies pigeons also rendered useful services. Reuter, who subsequently founded Reuter's Bureau in London, started his career by founding a pigeon-post from Aix-la-Chapelle to Brussels, and the Gazette of Cologne at first maintained such an aerial news service. In England also, a newspaper reporter equipped with a small pigeon-cage was formerly not a rare sight at public meetings from which he sent his reports immediately to his paper by a pigeon messenger. The press availed itself of pigeons especially for the purpose of reporting yacht races, and some yachts were actually fitted with lofts. I am informed by Japanese friends that pigeons were likewise employed by newspapers in Japan.

The French were the first who ingeniously used carrier pigeons for military purposes. During the siege of Paris in 1870 (till January 28, 1871) several hundred pigeons were placed at the disposal of the military service by a Carrier Pigeon Club. The sole advices that arrived at Paris from the outside world at that time were conveyed by the wings of pigeons. A hundred and fifteen thousand official dispatches and about a million private messages are said to have reached their destination in this manner. The dispatches were reproduced on both sides of small films by means of microphotography; eighteen such films weighed a half gram, and contained from twelve to sixteen large folio pages of news on an area of about a hundred square centimetres. The contents of a complete number of the Times could be accommodated in this space. About three thousand dispatches could be copied on each film. The films were rolled and placed in a quill which was sealed and fastened to a tail-feather of a pigeon by means of a fine wire fortified by a silk thread. For the purpose of deciphering the incoming dispatches they were projected, considerably enlarged, on a screen, so that they could be easily read and copied. The price of these air dispatches was half a franc (ten cents) each word. Money orders also were sent out to the extent of three hundred francs each. The average income from every flight of a carrier pigeon amounted to 35,000 francs ($7,000).

During the World War, as is still within the memory of every one, pigeons were extensively utilized and achieved brilliant records of flight under great difficulties. A case of supreme endurance was noted on October 21, 1918, when a carrier pigeon was released with an important message at Grand Pré at 2:35 p.m. during intense machine-gun and artillery fire. This bird delivered its message to the loft at Rampont, a distance of 24.84 miles in twenty-five minutes. One of its legs had been shot off, and its breast was injured by a machine-gun bullet, but even under these conditions the bird did not fail to reach its destination. For more information on the valorous deeds of pigeons in our army the *National Geographic Magazine* (January, 1926, pp. 86-91) may be consulted. The same article contains twelve beautiful colored plates representing various breeds of pigeons.

In warfare the service of pigeons will always remain indispensable. Telephone and wireless communication are often interrupted in the zone of advance, or may be put out of commission. Scouts and couriers may be delayed or intercepted, optical signals obscured by rain, smoke, or dust, and aerial observation hampered by unfavorable weather conditions. Pigeons are not disturbed by bombardments, fog, smoke, or dust, and will work regularly under almost any conditions. In 1919 an area of Texas was wrecked by a storm, and a United States Army relief-train was dispatched to Corpus Christi. Pigeons carried on this train were released and braved storm and rain, bringing the first news of conditions in the stricken area. Even for two days after a radio had been set up and put in operation, the pigeons were the only means of conveying news from that district, as atmospheric conditions crippled radio communication.

Pigeons are still bred and kept in large numbers for messenger service and racing. They are useful for transmitting messages whereever communication by telegraph or telephone is not available. In the beginning of this century there was still a real pigeon-mail between New Zealand and Great Barrier, a solitary and inhospitable isle about ninety km distant from Auckland, whose colonists are engaged in mining operations. A land-owner of this isle, Fricker by name, hit upon the idea to establish a permanent daily pigeon-mail with Auckland, as the mail-steamer ran but once a week. The letters had to be written on a special form, and the postage from Great Barrier to Auckland was twelve cents, in the opposite direction twenty-five cents. The Dutch Government established a pigeon-post system in Java and Sumatra early in the nineteenth century, the birds being obtained from Baghdad.

At a trial flight conducted from Compiegne in France to Antwerp a swallow previously distinguished by a special mark was released simultaneously with several carrier pigeons. The bird immediately took up its flight in the direction of Antwerp whence it had been taken, while the pigeons, as they always do, first fluttered around to find their bearings. The swallow made the way from Compiegne to Antwerp (255 km) in sixty-eight minutes, which means that in one minute it covered 3 ¾ km, whereas the first pigeon arrived only after three hours. The swallow therefore was about three times faster; it would be the swiftest winged messenger, but unfortunately it cannot be trained like a pigeon.

Amazing records of speed and endurance have been achieved by pigeons. In good weather young birds will fly about three hundred miles in from seven to nine hours, and flights of six hundred miles in one day have been accomplished by older birds. This is the maximum of a day's flight; in fact, only a very small percentage of the birds will make five hundred miles in one day. During favorable weather some pigeons will fly five hundred miles without stopping to eat or drink. The distance from Dover to London (76 ½ miles by rail, 70 miles by air-line) was once covered by a carrier pigeon beating by twenty minutes the express train which ran at a speed of sixty miles an hour.

NOTES

In regard to Shun as a flyer and user of a parachute compare E. Chavannes, Les Mémoires historiques de Se-ma Ts'ien, Vol. I, p. 74; and J. Legge, Chinese Classics, Vol. III, Prolegomena, p. 114.

The *K'ai yüan t'ien pao i shi* (ch. A, p. 9) relates that the magician Ye Fa-shan, who lived under the T'ang dynasty, had an iron mirror which reflected objects like water; whenever a person was ill and looked into this mirror, his interior organs became completely visible, and revealed any obstructions that might be there; then he was treated by means of drugs until he was completely cured. Cf. above, p. 11.

The same work also contains the first notice of carrier pigeons (p. 5) alluded to above on p. 71: "In his youth Chang Kiu-ling kept in his house swarms of pigeons. When he had to correspond with his relatives, he tied the letter to a pigeon's foot. The bird, relying on the localities to which it had been trained, flew off and delivered the letter. Chang Kiu-ling styled them 'flying slaves.' His contemporaries were all filled with admiration."

The same work (ch. B, p. 24b) contains a curious story of a swallow transmitting a letter: "At Ch'ang-an there was a man of the people, Kwo Hing-sien by name, who had a daughter called Shao Lan. She was married to a big merchant, Jen Tsung, who pursued his trade in Siang (Hu-nan). For several years he was absent from home, and no news from him had reached his family. One day Shao Lan was in the living-room of her house and observed a couple of swallows playing on the ridge-pole of the roof. She heaved a long sigh and addressed the swallows, 'I have heard that you swallows come from the east of the sea and return there and in your constant migrations must pass through Siang. My husband left home several years ago, and has not returned. There is no tidings as to whether he is dead or alive, and I have no means of knowing whether he exists or not. I trust to you to deliver a letter to my husband.' When she had finished her speech, she burst into tears. The swallows fluttered around, uttering sounds as though responding to her request. Again, Lan spoke to them, 'If you wish to be loyal to me, descend into my lap!' The swallows thereupon flew on her lap, and with many sighs Lan recited the following stanza: 'My husband has gone far away beyond the lakes; I am almost in despair, mingling bloody tears with this message. Confidently I trust to the swallow's wings to transmit this letter to my unfeeling husband.' Thereupon Lan committed this brief message to writing and tied it to the foot of one of the swallows. These emitted a sound and flew off. Jen Tsung then happened to be at King-chou and suddenly noticed a swallow flying above him. He was astounded when he saw the bird who alighted on his shoulder. He observed that a tiny letter was attached to the bird's foot; he released it and read his wife's message in verse. He was deeply moved and shed tears. The swallow rose into the air and flew off. The following year Jen Tsung returned home and showed Shao Lan the verses which she had written. Subsequently the scholar Chang Yüe (a well-known poet, A.D. 667-730) recorded this story to have it preserved as a curiosity of literature."

Chao, an emperor of the Han dynasty, while hunting in a park, shot a wild goose and found a piece of cloth attached to one of its feet. It contained a message to the effect that Su Wu and his companions were in a certain marsh in the country of the Hiung-nu. Messengers were at once dispatched to the Hiung-nu to demand the release of the prisoners who had been believed to be dead (Pétillon, Allusions, p. 505; Giles, Biogr. Dict., p. 685).

The first who made the passage from the *Ti wang shi ki* known was G. Schlegel (Chinesische Bräuche und Spiele in Europa, p. 32, Breslau, 1869). Schegel, in the same manner as I, takes Ki-kung-shi (wrongly written by him Ki-kwang-shi) as the name of an individual, but draws an erroneous conclusion from this text when he observes the "the air-balloon invented in Europe in 1872 was assuredly known to the ancient Chinese." The Chinese "flying chariot" has nothing to do with a balloon which is based on the principle of a bag filled with heated air or hydrogen gas; such a contrivance was not known to the Chinese at any time, notwithstanding what has been written to the contrary. Professor Giles, in the article quoted below, justly remarks, "No credence whatever should be given to the absurd story of the French missionary, Father Besson, who is said to have written in 1694, stating that a balloon had ascended from Peking at the coronation of Fo Kien in 1306. No emperor was crowned in 1306, and no such emperor is known to Chinese history as Fo Kien."

H. A. Giles, Traces of Aviation in Ancient China (in his Adversaria Sinica, Vol. I, No. 8, 1910, pp. 229-236), cites all texts relative to the Ki-kung flying chariot, save the one from the *Ti wàng shi ki*. It is noteworthy that the *Ts'e Yüan* (under "flying chariot") quotes only the latter, which apparently is the most important, but omits the *Po wu chi, Shu i ki,* and *Kin lou tse.*

As in European folk-lore, so in China also rocks of peculiar shape, bells, statues, swords, and other objects are credited with a magic power of flight. "A rock which arrived flying" (*fei lai shi*) is shown on the sacred Mount T'ai in Shan-tung. Flying swords are mentioned in the romance of the Three Kingdoms (Brewitt-Taylor, San Kuo, Vol. II, p. 311). "Flying scissors" (*fei lai tsien*) of cast iron are figured and described by L. Gaillard (Croix et Swastika en Chine, 1893, p. 217).

Good information on Korean and Japanese kites will be found in the interesting book of Stewart Culin, Korean Games with Notes on the Corresponding Games of China and Japan, pp. 9-21 (Philadelphia, 1895); see also W. Müller, Der Papierdrachen in Japan (Stuttgart, 1914), who deals well with the construction of Japanese kites.

Those interested in the subject of kite-fishing may consult H. Balfour, Kite-fishing, in Essays and Studies Presented to William Ridgeway (1913), pp. 583-608, and H. Plischke, Der Fischdrachen, published by Museum für Völkerkunde, Leipzig, No. 6, 1922. In this monograph the distribution of kite-fishing and the use of kites for fishing in Indonesia, Melanesia, and Micronesia are set forth in detail. The author also regards China as the home of the kite whence it spread to Indonesia and the South Sea Islands on the one hand and to Europe on the other hand. His statement (p. 36) that the earliest Chinese references to the kite belong to the second and fifth centuries B.C., however, is erroneous; he has been misled by De Groot (Religious System of China, Vol. III, p. 665), who misinterprets the wooden bird mentioned by Mo Ti as a kite.

In the Panchākyānaka, a Jaina recension of the Panchatantra, we also find the story of the Weaver as Vishnu (translated by J. Hertel, Indische Märchen, 1921, p. 92); in this version, the Garuda airship is set in motion by a push of the elbows.

The story of the Bodhisatva as a divine horse rescuing merchants from flesh-devouring ogres by carrying them from Ceylon to India, traversing the clouds and passing the sea to the other side, is contained in the Valahassa Jātaka (Jātaka No. 196) and Hüan Tsang's account (S. Beal, Buddhist Records of the Western World, Vol. II, p. 242).

The influences of Greece on India are set forth in a good summary by Count Goblet d'Alviella in his book Ce que l'Inde doit à la Grèce: des influences classiques dans la civilisation de l'Inde (2nd ed., Paris, 1926). While art, medicine, mathematics, and astronomy are duly considered, mechanics and references to airships are passed over with silence.

Sylvain Lévi, however, in his treatise Quid de Graecis veterum Indorum monumenta tradiderint (Paris, 1890, p. 24), has thus referred to the Yavana airship: "Memorandus tandem ille Yavana qui machinam per aera volantem construxerat, ut Candīs principem tolleret."

In regard to the myth of Etana see G. Hüsing, Zum Etana Mythos, *Archiv für Religionswissenschaft*, 1903, p. 149, and Die iranische Ueberlieferung (1909), pp. 39, 100; M. Jastrow, Another Fragment of the Etana Myth, *Journal American Oriental Society*, 1910, pp. 101-129; B. Meissner, Babylonien und Assyrien, Vol. II (1925), pp. 189-191. The British Museum seal representing Etana's bold flight is figured and described by P. S. P. Handcock, Mesopotamian Archaeology (1912), pp. 297-298. W. H. Ward, The Seal Cylinders of Western Asia, pp. 142-148 (Washington, Carnegie Institution, 1910), describes five seals with this subject.

The story of Kai Kawus is also recorded in the Bundahishn (translated by E. W. West in Sacred Books of the East, Vol. XXXVII, pp. 220-223) and in the Arabic History of the Kings of Persia by Al-Tha'alibi, translated by H. Zotenberg (Histoire des rois des Perses, 1900, p. 167). The Arabic chronicler gives the case a more theological flavor by making Kai Kawus construct the tower of Babylon whence he takes his skyward flight. After his fall he demands milk and water from the people who have come to his rescue, and that locality was therefore called Siraf ("Milk and Water"). In the same work (p. 13) is found the story of King Jemshed constructing a chariot of teak and ivory which is transported by demons through the air and in which he flies from Donbāwand to Babylon in a single day.

Hodgson errs in tracing Godwin's bird-airship (p. 61) to Lucian, who in fact has nothing of the kind. All that Lucian offers in regard to air voyages is given above (p. 64), and these are effected by means of wings, not of birds. Feldhaus is mistaken in permitting the Babylonian tradition of the flying Etana to migrate into Persia without even knowing the story of the Shahnameh. Etana, however, accomplishes flight merely by mounting a bird, while the Persian king Kai Kawus flies comfortably seated in a vehicle drawn by four eagles who supply the motor. The two traditions are entirely distinct and not interrelated.

The old yarn of Simon the Magician as having attempted, at the time of Nero, a flight which ended in failure, is still warmed up in many books, recently again by C. L. M. Brown (p. 7). Suetonius (Nero, 12) reports nothing about a flight, still less lisps a word about a flight of Simon. He writes merely that at a performance of the story of Icarus in the theatre an actor (a petaurista or petauristarius) had a fatal accident and collapsed on a spot near the emperor whom he covered with his blood; the question is of a stage disaster, not of a flight. Only mediaeval legend connects Simon with a flight achieved with the devil's assistance. Arnobius, writing about the year 300 of our era, says that the people of Rome saw the chariot of Simon Magus and his four fiery horses blown away by the mouth of Peter and vanish at the name of Christ. Cyril of Jerusalem (315-386) speaks of Simon's being borne in the air in the chariot of demons, and is not surprised that the combined prayers of Peter and Paul brought him down. Finally in the *Didascalia Apostolorum,* an apocryphal work extant in Syriac and Latin, Peter finds Simon at Rome drawing many away from the church as well as seducing the gentiles by his "magic operation and virtues." Peter then states that one day he saw Simon flying through the air, but by virtue of his prayer Simon fell and broke the arch of his foot. In another, Greek version of the legend Simon announced his flight in the theatre. While all eyes were turned on him, Peter prayed against him. Meanwhile Simon mounted aloft into mid-air, borne up, Peter says, by demons, and telling the people that he was ascending to heaven, whence he would return bringing them good tidings. The people applauded him as a god, but Peter stretched forth his hands to heaven, supplicating God through Jesus to dash down the corrupter and curtail the power of the demons. He asked, however, that Simon might not be killed by his fall, but merely bruised. Thereupon Simon fell with a great commotion and bruised his bottom and the soles of his feet (compare L. Thorndike, History of Magic and Experimental Science, Vol. I, p. 422). All this is freely invented legend for a dogmatic purpose and has nothing to do with a real attempt at flight.

In regard to the letter of Prester John see the critical discussion of L. Thorndike, History of Magic and Experimental Science, Vol. II, pp. 240-245, a book that is to be highly recommended for its thorough, judicious, and critical scholarship.

The chapter "The Air Mail of Ancient Times" is the most comprehensive historical study of carrier pigeons thus far written. An interesting article on Chinese lore of pigeons is by T. Watters, Chinese Notions about Pigeons and Doves, in *Journal China Branch Royal Asiatic Society*, Vol. IV, 1868, pp. 225-242. No reference to carrier pigeons is made in this article, although the name of Chang Kiu-ling is mentioned. Those interested in Chinese pigeon whistles may consult my article on the subject in *The Scientific American*, 1908, p. 394, where also the process of making the whistles is described with illustrations of examples and of the tools used in making them. In regard to pigeon breeds in general see W. B. Tegetmeier, Pigeons: Their Structure, Varieties, Habits, and Management, London, 1868 (with colored plates).

The Oriental origin of Greek carrier pigeons is upheld by H. Diels, Antike Technik (1914), pp. 68-69; see also my review of this book in *American Anthropologist*, 1917, pp. 71-75. There is an interesting article by F. Kluge, Die Heimat der Brieftaube, reprinted in his Bunte Blätter (Freiburg, 1908), pp. 145-154. The author of this article quotes chiefly from early German pilgrimages to Palestine to prove the Oriental origin of the pigeon-mail. He justly emphasizes the point that the ancients employed pigeons as messengers only incidentally and occasionally. L. Rauwolf's Beschreibung der Raiss inn die Morgenländer (1583), p. 215, may be added to his German sources. Compare also Gaudefroy-Demombynes, La Syrie à l'époque des Mamelouks d'après les auteurs arabes (1923), pp. 250-254.

In his charming story "Legend of Prince Ahmed Al Kamel or, The Pilgrim of Love" inserted in his *The Alhambra*, Washington Irving has skilfully combined the Oriental motives of talking birds, knowledge of birds' speech on the part of the prince, the courier pigeon carrying love letters (that "trustiest of messengers"), the enchanted horse, and the flying carpet of Solomon on which the lovers elope. Thomas Moore, in *The Veiled Prophet of Khorassan,* alludes to Solomon's silken rug in the lines—

> Waved, like the wings of the white birds that fan
> The flying throne of star-taught Soliman—

and comments that when Solomon travelled, he had a carpet of green silk on which his throne was placed, being of a prodigious length and breadth, and sufficient for all his forces to stand upon, the men placing themselves on his right hand and the spirits on his left; and that when all were in order, the wind, at his command, took up the carpet, and transported it with all that were upon it, wherever he pleased; the army of birds at the same time flying over their heads, and forming a sort of canopy to shade them from the sun. The same motif is frequently referred to in the mediaeval Midrash literature.

BIBLIOGRAPHICAL REFERENCES

BELL, ALEXANDER GRAHAM.—The Tetrahedral Principle in Kite Structure. National Geographic Magazine, Vol. XIV, 1903, pp. 219-251.

Aerial Locomotion. National Geographic Magazine, 1907, pp. 1-34.

Dr. Bell's Man-Lifting Kite. National Geographic Magazine, 1908, pp. 35-52.

BROWN, C. L. M.—The Conquest of the Air, an Historical Survey. London, Oxford University Press, 1927.

CHANUTE, O.— Progress in Flying Machines. New York, 1899.

FELDHAUS, F. M.—Ruhmesblätter der Technik. Leipzig, 1910.

Leonardo der Techniker und Erfinder. Jena, 1913.

HART, IVOR B.—The Mechanical Investigations of Leonardo da Vinci. Chicago, Open Court Publishing Co., 1925. 240p.

Chap. VII: Leonardo da Vinci as a Pioneer of Aviation.

HODGSON, J. E.—The History of Aeronautics in Great Britain from the Earliest Times to the Latter Half of the Nineteenth Century. 150 illustrations. Oxford University Press, London, Humphrey Milford, 1924. 436p.

LANA, FRANCESCO (Bresciano).—Prodromo overo saggio di alcune inventioni nuove premesso all'arte maestra. Brescia, 1670. 252p. 20 plates.

Copy in Library of Armour Institute of Technology, Chicago.

Chap. VI: Fabricare una nave, che camini sostentata sopra l'aria a remi e a vele; quale si dimostra poter riuscire nella prattica ("To manufacture a ship which travels supported above the air by means of oars and sails: it is demonstrated that this is feasible in practice").

WILKINS, JOHN.—Mathematicall Magick or, The "Wonders that may be performed by Mechanicall Geometry." In Two Books. Concerning Mechanicall Powers, Motions. Being one of the most easie, pleasant, usefull, (and yet most neglected) part of Mathematicks. Not before treated of in this language. By J. W. M. A. London, printed by M. F. for Sa. Gellibrand at the brasen Serpent in Pauls Church-yard, 1648. 269p.

Copy in Library of Armour Institute of Technology, Chicago.

Book II is entitled "Daedalus, or Mechanicall Motions."

INDEX

Aelianus, 74.
Aerial top, Chinese, 42.
Air bombardments, 12-13, 22, 47.
Airship, Chinese conception of, 17; Indian conception of, 46
Alchemy, in China, 26, 28; in Asia and Europe, 30.
Alexander the Great, romance of, 60, 62, 69.
Alexandria, carrier pigeons at, 80, 81.
Anti-aircraft practice, first example of, 35.
Arabian Nights, 77, 78.
Arabs, attempts at flying by, 65-67; organizers of pigeon-posts, 77-80.
Archytas, flying dove of, 23, 37, 64.
Ariosto, 83.
Aziz, Egyptian Caliph, first recipient of parcels by air mail, 78.
Avril, P., 81.

Bacon, Roger, 68-70.
Baden-Powell, man-lifting kite of, 42.
Baghdad, centre of pigeon-mail, 78; pigeons of, 84, 87.
Bate, J., description of kite by, 38.
Baveru Jātaka, 76.
Belgium, carrier pigeons in, 85.
Bell, A. G., 13, 42.
Beybars, air mail organized by, 79.
Bird-men, Chinese, 15.
Birds, used by navigators to spy land, 76.
Bladud, aerial flight of, 14.
Boots, magic, 53-54.
Bow, musical, attached to kites, 33, 37.
Breathing, art of, 29.
Brihat Kathā Çlokasamgraha, 48.

Caesar, 75.
Cairo, carrier pigeons at, 79, 80.
Carrier pigeons, history of, 71-87.
Cayley, Sir George, 42.
Centipede kite, 32.
Chang Kiu-ling, first Chinese who kept and trained carrier pigeons, 71, 88.
Chanute, O., 31, 32.
Cock, Indian airship in the form of, 49.
Crane, vehicle of flyers, 26, 27.
Crusades, carrier pigeons used during, 75, 82-84.
Cyrano de Bergerac, 64, 65.

Daedalus, 9, 62-63.
Dallam, T., 80.
Damian, John, 67-68.
Danti, G. B., 67.
Darwin, C., 84.
Darwin, E., 13.
Dirigible airships, in Indian tradition, 47.
Dragons, as vehicles of aerial flights, 18, 69.

Eagles, carrying an airship, 59, 62.
Elixir, promoting flight, 26, 30.
Enchanted horse, 55.
England, carrier pigeons in, 85; kite used in, 38.
Etana, 58, 91.

Fire-crackers, in connection with kites, 33.
Flutes, connected with kites, 33.
Flying elixir, 30.
Flying horse, 55-56.
Flying shoes, 28.
Flying Taoist saint, on painting, 28.
France, use of carrier pigeons in, 85-86.
Franklin, Benjamin, experiments of with a kite, 39.
Frontinus, 75.

Gandharva marriage, 47.
Garuda airship, 6, 47, 50, 51.
Giles, H. A., on Chinese airship, 20, 89.
Glanvill, J., 13.
Godwin, F., bird airship of, 61, 65, 91.
Goose, as messenger, 73, 89.
Greeks, carrier pigeons among, 74-75; flying among, 62-65; regarded in India as the inventors of a type of airship, 49-52.
Gunavarman, 46.

Harshacharita, 50.
Harun al-Rashid, pigeon-post of, 77.
Hawthorne, Nathaniel, 30.
Hodgson, J. E., author of History of Aeronautics in Great Britian, 8, 34, 43, 61, 67, 70, 91.
Holland, carrier pigeons in, 85.
Huang Ti, 18, 19.

India, carrier pigeons in, 73; conception of airships in, 46-52, 55-57; kites in, 37.
Irving, Washington, 65, 93.

Jātakas, 46, 53, 55.
Jivaka, 11.

Kai Kawus, flying Persian king, 59, 91.
K'ai yüan t'ien pao i shi, three texts translated from, 88.
Kathā Sarit Sāgara, 48, 52, 53.
Ki-kung, maker of flying chariot, 19.

Kibaga, flying warrior of Uganda, 12.
Kircher, A., acquainted with kites, 37, 41.
Kite, contest, 32; for catching fish, 36; in England, 38; in India, 37; in Italy, 37; in Polynesia, 36; in Siam, 37; paper, 31, 35; ridden by men, 40; wooden, 23, 25.
Kites, history of, 31-43.
K'ü Yüan, Chinese poet, air-journey of, 17.
Kung-shu Tse, 23, 24.

Lana, Francesco, airship of, 2, 21, 22, 65; flying birds made by, 25; on kites, 37.
Lanterns, attached to kites, 30.
Lei Kung, Chinese god of thunder, 16.
Leonardo da Vinci, aeroplanes of, 70; parachute of, 15.
Li Sao, first description of an air-journey in, 17.
Li Ye, inventor of musical kite, 33.
Linschoten, 73, 81.
Logan, J., 13.
Lu Pan, 23, 24.
Lucian, 64, 76, 91.

Mackintosh, bird airship of, 62.
Magic boots, 28, 53-54.
Mao Mong, 18.
Masudi, 78.
Mercury, in alchemy, 30.
Mesopotamia, flying in, 58; home of domesticated pigeon, 75.
Mirror, showing interior organs of body, 11, 88.
Mo Ti, 23.
Mongols, 74, 79.
Moore, Thomas, 71, 84, 93.
"Music on the air," produced by kites, 33; produced by pigeon whistles, 72.

Noah, dove of, 76.
Nur-ed-din, air mail organized by, 78.

Oliver of Malmesbury, 67.

Panchatantra, 46.
Parachute, first used by Shun, 14; of Leonardo da Vinci, 15.
Parcel post by air mail, first example in tenth century, 78.
Passenger airship, in India, 50.
Pei Ti, 18.
Persia, carrier pigeons in, 73-74; flying architect of, 66; tradition of airship drawn by eagles in, 59-60.
Pigeon whistles, 72, 93.
Pliny, 75, 76.
Po Kü-i, flying shoes of, 28.
Po wu chi, 19.
Pocock, G., kite-chariot of, 21, 41.

Rabelais, 64, 84.
Regiomontanus, 25.
Reuter, pigeon-post of, 85.
Roentgen rays, idea of anticipated in China and India, 11-12, 88.
Romans, carrier pigeons among, 75.
Rukh, the giant bird, first air-bombardier, 12.

Schlegel, G., 89.
Shahnameh, 59, 60.
Shakespeare, 84.
Shan hai king, 19.
Shun, Chinese emperor, first flyer recorded in history, 14, 88.
Si Wang Mu, flying on crane's back, 27.
Siddhi Kür, 47.
Signalling, by means of kites, 34, 35.
Simon the Magician, 91-92.
Solar ship, conception of in India, Egypt, and Greece, 46.
Spinning Damsel, 16.
Stanley, story of flying Uganda warrior recorded by, 12.
Strutt, J., 38.
Sulphur, in alchemy, 30.
Sun Pin, flying shoes of, 28.
Swallow, Chinese story of transmitting a letter, 88; speed of compared with carrier pigeon, 87.
Swift, 61, 64.
Symon Semeon, 80.
Syria, pigeon-post in, 78.

Tasso, 83.
Thorndike, L., 69, 92.
Thousand-league boots, 28, 54.
Ti wang shi ki, 19, 89.
Tiger, vehicle of flyers, 27.
Ts'in Shi, mirror of, 11.
Turks, kite-flying among, 37.

Uganda warrior, flyer and air-bombardier, 12.
Uncles, M., eagle airship of, 62.

Wayland the Smith, 63.
Weaver as Vishnu, story of, 46, 90.
Wells, H. G., on Daedalus story, 9.
Wilkins, J., 9, 22.
Wilson, A., scientific experiments of with kites, 39.
Wind-driven chariot, in China, 19; in India, 45.
World War, carrier pigeons in, 86.

Yavana airship, 49, 50, 51.
Ye Fa-shan, magician, 88.
Yoga practice, 53.
Yogins, 30.
Yu yang tsa tsu, 73.

WINGED DEITY ATTENDED BY BIRD-MEN (p. 17)
Stone Bas-relief of Han Period, A.D. 147. Shan-tung, China

AERIAL CONTEST OF DRAGON-CHARIOT AND DRAGON-RIDERS (p. 18)

Stone Bas-relief of Han Period, A.D. 147. Shan-tung, China

AERIAL CONTEST OF DRAGON-CHARIOT AND DRAGON-RIDERS
Continuation of the Panel shown in Plate II

KI-KUNG'S FLYING CHARIOT (p. 20)
Chinese Woodcut from T'u shu tsi ch'eng

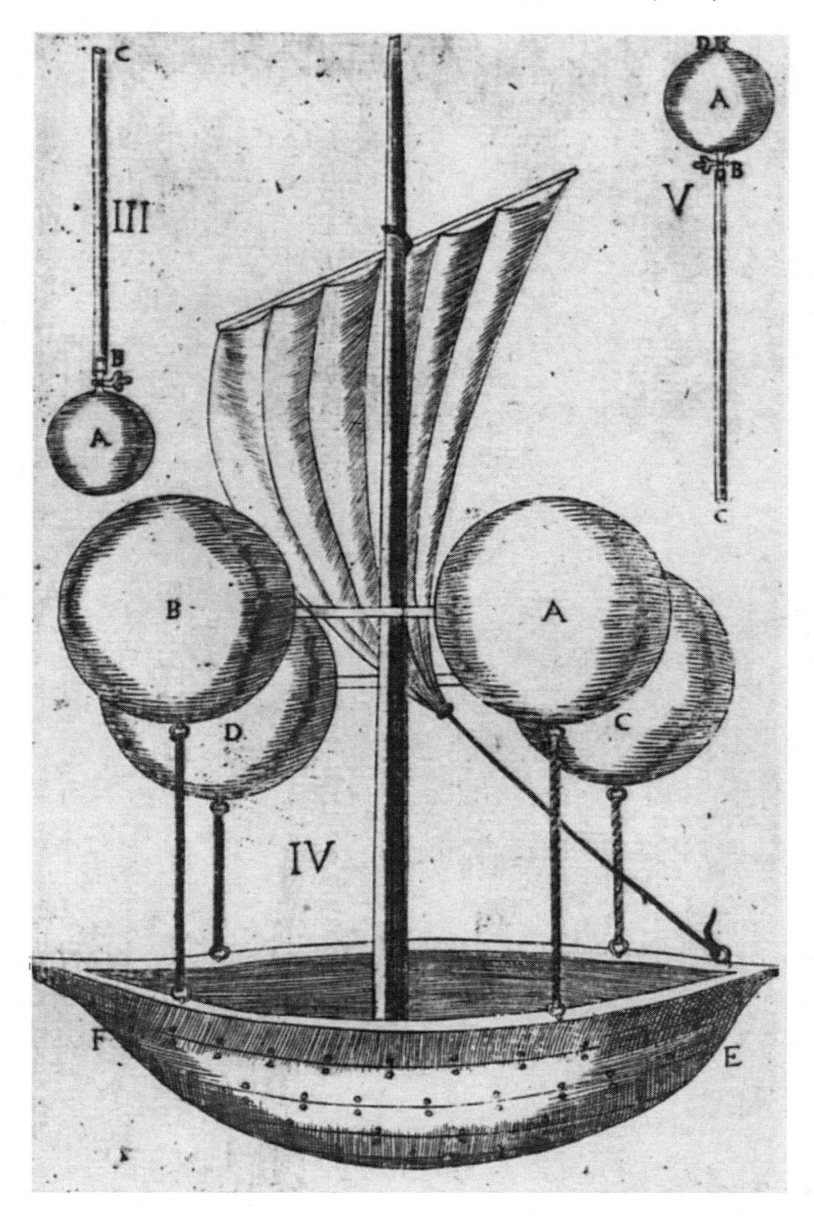

FRANCESCO LANA'S FLYING BOAT (p. 21)

From Lana's Prodromo, 1670

FLYING TAOIST SAINT (p. 28)
Chinese Landscape in Ink from General Munthe Collection
now in Los Angeles Museum

THE GODDESS SI WANG MU FLYING ASTRIDE A CRANE (p. 27)
Scene from an Embroidered Chinese Screen of the K'ang-hi Period (1662-1722) in Blackstone Collection of Field Museum

BOYS FLYING A KITE (p. 36)

Scene from a Chinese Painted Roll on Silk by Su Han-ch'en of the Twelfth Century (Sung Period)
in Collections of Field Museum

EARLIEST ENGLISH ILLUSTRATION OF A KITE (p. 38)
From John Bates' The Mysteries of Nature and Art, 1634

TWO APSARASES OR HEAVENLY NYMPHS FLYING DOWNWARD AND SURROUNDING THE BUDDHA AMITABHA (p. 52)
Marble Sculpture with Votive Inscription Yielding Date A.D. 677
Blackstone Chinese Collection of Field Museum

KAI KAWUS FLIGHT TO HEAVEN (p. 60)

From a Persian Illustrated Manuscript of the Shahnameh, Dated 1587-88

Courtesy of Metropolitan Museum of Art, New York

THE AERIAL VOYAGE OF DOMINGO GONSALES (p. 61)
From F. Godwin's Man in the Moone, 1638

185

电视的历史背景

THE SCIENTIFIC MONTHLY

EDITED BY J. McKEEN CATTELL

VOLUME XXVII
JULY TO DECEMBER

NEW YORK
THE SCIENCE PRESS
1928

It is hardly necessary to expatiate on these data. They are self-explanatory, and any attempt to adorn them would be presumptuous. I can not resist suggesting, however, that in view of this evidence of early biological wisdom we should look carefully to our laurels in order to be sure that they are not indurated with the dust of the ages.

THE PREHISTORY OF TELEVISION

By Dr. BERTHOLD LAUFER

CURATOR OF ANTHROPOLOGY, FIELD MUSEUM OF NATURAL HISTORY, CHICAGO

In his drama "Back to Methuselah" George Bernard Shaw depicts the following scene as a reality in the year 2170: the head of the British government is holding a conference with his cabinet ministers, who are several hundred miles distant, in this manner—that he operates a switchboard near his desk and, by pressing a certain key, makes appear on a silver screen a life-size picture of the person to whom he desires to talk and whose voice is simultaneously transmitted.

In April, 1927, the Bell Telephone Laboratories gave the first practical demonstration of transmitting over electric wires the pictures and voices of moving persons, voice and image being perfectly synchronized. A two-way telephonic communication was maintained between Washington and New York. Secretary Hoover in Washington opened the demonstration, and the illuminated image of himself cast over the wire on a screen in New York synchronized perfectly with his voice that was heard over the telephone at the same time. One telephone line was used for transmitting the voice, another for transmitting the television current and a third for synchronizing the electric driving-motors at each end of the lines. It has since been demonstrated that both the voice and the "picture" currents can be sent over the same wire or, in the case of radio transmission, over the same wave-length.

While television at present is a fact, it has been the dream of mankind for hundreds and thousands of years. In my monograph, "The Prehistory of Aviation," recently published by Field Museum, Chicago, I have emphasized the fact that human imagination has been of paramount importance and proved most fertile in the development of mechanics and inventions and that the trend of man's mind toward the romantic and adventurous has resulted in the conquest of the air. The essential point is that many fundamental contrivances have not been reasoned out logically through progressive scientific thought, but that man's mind conceived them through visionary reveries as an accomplished fact and then proceeded to work toward this imaginary goal. Thus television also has its prehistory in the domain of oriental folk-lore, a brief outline of which is given on the following pages.

In Firdausi's great epic poem, the Shāhnāmeh ("Book of Kings"), figures a cup which mirrors the world and distant persons. It is the property of King Kai Khosrau and appears in the love-story of Bizhan and Manizha. The king, while holding a feast, receives a petition for succor from the people of Irmān, whose country is being ravaged by wild boars, and sends Bizhan and Gurgin to scour the country of them. Through the machinations of Gurgin, who envies him, Bizhan falls in love with Afrasiyab's daughter, Manizha, who carries Bizhan off to Turan and hides him in her palace. He is discovered and imprisoned in a pit with Manizha as his attendant. In the meantime Gurgin has returned to Iran, where his lame story rouses suspicion. By means of the divining-cup Kai Khosrau ascertains Bizhan's situation and dispatches the hero Rustam to deliver him. The king says to Giv, Bizhan's afflicted father (in Warner's translation):

Then will I
Call for the cup that mirroreth the world,
And stand before God's presence. In that cup
I shall behold the seven climes of earth,
Both field and fell and all the provinces,
Will offer reverence to mine ancestors,
My chosen, gracious lords, and thou shalt know
Where thy son is. The cup will show me all.

Then the poet narrates how Kai Khosrau saw Bizhan in the cup that shows the world:

When jocund New Year's Day arrived, Giv yearned
For consultation with that glorious cup,
And came, bent double on his son's account
But hopeful, to Khosrau who, seeing him
With shrunken cheeks and sorely stricken heart,
Went and arrayed himself in Ruman garb
To seek God's presence. Then before the Maker
He cried and oft-times blessed the Shining One,
Imploring of the Succorer succor, strength,
And justice on pernicious Ahriman,
And then returning to his throne, assumed
The Kaian crown, took up the cup, and gazed.
He saw the seven climes reflected there,
And every act and presage of high heaven,
Their fashion, cast, and scope, made manifest.
From Aries to Pisces he beheld
All mirrored in it—Saturn, Jupiter,
Mars, Leo, Sol and Luna, Mercury,
And Venus. In that cup the wizard-king
Was wont to see futurity. He scanned
The seven climes for traces of Bizhan,
And, when he reached the Kargasars, beheld him
By God's decree fast fettered in the pit,
And praying in his misery for death,
With one, the daughter of a royal race,
Attending him. The Shah, with smiles that lighted
The daïs, turned his face to Giv and said,—
"Bizhan is yet alive; be of good cheer!"

In one of the stories of the Arabian Nights (No. 271) the three sons of a Sultan of India—Prince Husain, Prince Ali and Prince Ahmed—undertake a year's journey into a distant part of the world to find some unusual treasure for their royal father, who had promised the hand of a princess to him who would bring back the rarest jewel. Prince Ali traveled to Shiras, capital of Iran, and while rambling in the bazar of the city, one day met a man who carried in his hand an ivory tube about a yard long and offered it for sale at the price of thirty thousand sequins. The prince thought him to be a fool, as he demanded so enormous a sum for such a wretched thing, but was soon informed that this broker was wiser and more sensible than all others of his profession. He examined the ivory telescope, which was equipped with a piece of glass at either end and which if placed in front of the eye brought anything close to it, even though it may have been many hundreds of miles away. Moreover, this tube had the miraculous power of showing any object or any person its owner desired to see. Prince Ali wished to see his father whom he had left in India, and no sooner did he hold the ivory tube close to his eye than he espied him hale and hearty seated on his throne and giving judgment to the people of his land. Then he demanded to behold his beloved princess, and immediately he caught sight of her as she was leisurely reclining on a couch, chatting and laughing and attended by a flock of maids.

In Grimm's tale, "The Four Skilful Brothers," the situation is very similar to that of the preceding story. Four brothers go out into the world to earn their living and to learn a craft. One becomes an expert thief. The second meets a man who asks him what he wishes to learn in the world. "I do not know it yet," he replies. "Come along with me and become a star-gazer; there is nothing better than that, nothing will be hidden to one." He consented and became so clever a star-gazer that his master, when the boy had finished his apprenticeship and would leave him, presented him with a telescope and said, "This will enable you to see what occurs on earth and in heaven, and nothing will remain concealed to you." He meets with his three brothers, who have also acquired an extraordinary art, and soon there is an opportunity for them to put their knowledge to the test. The king's daughter was kidnapped by a dragon, and the king made it known that he who would bring her back should receive her as his consort. The four brothers decide to deliver her from the dragon. "I shall soon know where she is," said the star-gazer, looked through his telescope and announced, "I behold her, she is seated far away on a rock in the sea and beside her the dragon who guards her." Then he went to see the king and requested a ship for himself and his brothers, and crossed with them the sea till they arrived at the rock. They return with the king's daughter, and naturally engaged in a quarrel as to which should have her as his wife. The star-gazer said, "If I had not espied her, all your arts would have been futile; therefore she is mine." As all their claims were of equal merits, the king decided that no one should get her, but assigned to each a half kingdom as his reward.

Lucian, in his "True History" (I, 26), relates that in the palace of Endymion, king of the moon, he saw a large mirror placed above a well of mediocre depth. In descending into this well, it was possible to hear whatever was talked on earth; and in lifting one's eyes toward the mirror, one saw all towns and all nations as though one were in their midst. "I saw there my parents and my country," he adds, "I do not know whether they saw me too. I do not venture to affirm it, but he who declines to believe me might go to the moon, and will then convince himself that I am not an impostor."

Zosimus, an alchemist who lived in Egypt during the third and fourth centuries A.D., discusses the electron, an alloy of gold and silver, and mentions a magic mirror which Alexander the Great had made of it and which subsequently was exhibited in the Temple of the Seven Gates (corresponding to the seven heavens) above all spheres. In this mirror one beheld one's own future and destiny until one's death. This was a divine mirror symbolizing God (compare Epistle of James, I, 23–24; I Corinthians XIII, 12; II Corinthians III, 18).

In 331 B.C. Alexander the Great founded the city named for him—Alexandria. About 280 B.C. the famed Pharus was constructed there by Sostratus of Cnidus—the earliest lighthouse known in history. It was about three hundred feet high, a three-storied structure; the lower story was square, the middle one octagonal, the upper one, which contained the light, circular and surmounted by a colossal statue of Poseidon (Neptune). The Pharus of Alexandria became widely known in the Islamic world, but Mohammedan authors erroneously attribute its foundation to the great world-conqueror himself, designating it Menarat Iskanderiah ("Lighthouse of Alexander"). They describe it as one of the marvels of the world. On the top of this lighthouse, they say, Alexander placed a magic mirror in which could be sighted all incoming ships, the country Rum (the Byzantine empire), the islands of the sea and whatever was done by the inhabitants. By virtue of this mirror, as long as it existed, the city of Alexandria was said to preserve its grandeur and power. The Persians called this lighthouse Mirror of Alexander (Aineh Iskenderi), believing that the fortunes of Alexandria depended on it, as it was a talisman constructed under the influence of a certain constellation. It is said to have broken to pieces shortly before the city was conquered by the Arabs in A.D. 641.

Rabbi Benjamin of Tudela, who traveled in the Orient between the years 1159 and 1173, mentions the high tower of Alexandria surmounted by a glass mirror by means of which the approach of a ship or a hostile fleet could be noticed even when it was a twenty days' voyage off. The city was therefore prepared for the reception of a hostile ship from whatever direction she approached. Once, however, when Greece was still subject to the Alexandrians, Benjamin continues, a Greek vessel cast anchor in the port of Alexandria. The captain, a Greek, Theodorus by name, instructed in all sciences, brought to the Egyptian king valuable presents of gold and silver, silk and purple. His ship was at anchor opposite the lighthouse. Every day the captain invited the guard of the lighthouse with his servants on board his ship until they were on friendly terms. One day he treated them to an opulent feast and filled them with wine till they were intoxicated and fell into a deep slumber. The captain then ordered his crew to smash the mirror, and set sail the same night. From that time onward Christian ships, small prowlers as well as large vessels, entered the port, and snatched away two large islands, Crete and Cyprus, which are still under Christian rule. Egypt was henceforward unable to resist the Greek power.

Leo Africanus, in his "History of Africa," writes that the mirror of Alexandria was of "steel glass" by the hidden virtue of which passing ships, while the glass was uncovered, should immediately be set on fire; but when the glass was broken by the Mohammedans, its secret virtue vanished.

The fame of Alexandria's Pharus and television mirror even spread to the Far East. Chao Ju-Kwa, who was stationed as inspector of maritime trade at the port of Ts'üan-chou in Fu-kien and collected there from the lips of foreign traders much interesting information on the countries of the Indian Ocean, which he published in his "Chu fan chi," written in A.D. 1225, gives a brief notice of Alexandria and its lighthouse. "On the summit of it," he writes, "there was a wondrous large mirror. In the event of a surprise attack by foreign warships they would be detected beforehand by this mirror, and the troops on guard duty were ready to meet the situation. In recent years Alexandria was visited by a foreigner who asked for work in the guardhouse of the tower and who was employed as a janitor. He was not suspected for years, when suddenly he seized an opportunity of abstracting the mirror and flung it into the sea, whereupon he disappeared." A late Chinese cyclopedia (*San ts'ai t'u hui*) has disfigured this tradition considerably by stating that Tsu-ko-ni (Alexander) erected in Egypt a temple on the top of which there was a mirror which when pirates of other countries made a raid reflected them and thus announced their arrival.

In the famous letter of Prester John (71) purporting to have been written by him to the Byzantine emperor, Manuel (1143–80), is described a marvelous mirror which is reached by ascending a hundred and twenty-five steps over an elaborate structure of pillars. In this mirror all plots and machinations and everything that was done in the adjacent and subjected provinces either on behalf of Prester John or against him would be clearly revealed and recognized; day and night it was guarded by twelve thousand armed men that it might not be broken by an accident. In the work of Johannes Witte de Hese (1389) this mirror of Prester John is also mentioned: three precious stones are deposited in it; one of these directs and sharpens the vision; another, the senses; the third, experience; three very capable doctors have been elected to examine the mirror and see in it everything that is done in the world.

A "History of the World" published in Neo-Greek by Dorotheos, metropolitan of Malvasia, at Venice in 1763, alludes to a magic mirror in the imperial palace of Constantinople made by the emperor Leo the Philosopher: whatever existed or happened or was intended to be done in the world could be most clearly visualized in this mirror. The Emperor Michael, who was given to a voluptuous life, was informed one day by a messenger that he had beheld in the mirror the war preparations of the Turks against Constantinople. Michael, who just pampered at a banquet, did not like to be disturbed and ordered a servant to smash the mirror to atoms.

According to an Arabic tradition, Saurid was the wealthiest king on earth and had a mirror made from various alloys, wherein he could scan whatever occurred in the seven zones, whether good or bad, and what land was irrigated or not. This mirror was placed in the city of Amsus on the top of a green marble column. In the city of Sa on the bank of the Nile stood a pillar of white marble and upon it a mirror in which King Sa, for whom the city was named, was able to discern whatever happened in the seven zones.

Spencer, in his "Fairy Queen," has Merlin make a magic mirror in which a girl beholds the image of her swain. Walter Scott, in his "Lay of the Last Minstrel," relates that Cornelius Agrippa showed the Count of Surrey, during his sojourn in Italy, his sweetheart Geraldine in a mirror as she was reclining on a couch and reading her lover's poems by the light of a wax candle.

Nathaniel Hawthorne, in his story "Dr. Heidegger's Experiment," refers to a looking-glass hung in the doctor's room, "presenting its high and dusty plate within a tarnished gilt frame. Among many wonderful stories related of this mirror, it was fabled that the spirits of all the doctor's deceased patients dwelt within its verge, and would stare him in the face whenever he looked thitherward."

A SUBSTITUTE FOR ARSENIC

By S. MARCOVITCH

TENNESSEE AGRICULTURAL EXPERIMENT STATION

DON MARQUIS in his book "Archie and Mehitabel" puts the following words in the mouth of Archie the cockroach:

i am going to start
a revolution
i saw a kitchen
worker killing
water bugs with poison
hunting pretty
little roaches
down to death
it set my blood to
boiling
i thought of all
the massacres and slaughter
of persecuted insects
at the hands of cruel humans
and i cried
aloud to heaven
and i knelt
on all six legs
and vowed a vow
of vengence
i shall organize the insects
i shall drill them
i shall lead them
i shall fling a billion
times a billion billion

186

马伯乐《古代中国》书评

THE AMERICAN HISTORICAL REVIEW

VOLUME XXXIII

OCTOBER 1927 TO JULY 1928

NEW YORK

THE MACMILLAN COMPANY

LONDON: MACMILLAN AND CO., LTD

1928

Le Travail dans la Préhistoire. Par G. Renard, Professeur au Collège de France. [Histoire Universelle de Travail.] (Paris, Alcan, 1927, pp. 278, 30 fr.) This is one of a series of twelve volumes on the *Histoire Universelle du Travail*, to be published under the direction of Professor Renard; eight volumes by as many authors have already appeared. After an introduction on methods to be employed in prehistory the author takes up in turn: food; two great inventions, fire and language; the first industries; shelter; clothing and defensive arms; man and animals (domestication); the beginnings of agriculture; the first means of transport; commerce and war; origin of art; origins of science; the first human societies—clan and totemism; and the finale of prehistory when writing passed from the ideographic to the phonetic stage.

In both style and content this work is worthy of high praise. The author justly stresses two of the greatest prehistoric achievements—the invention of fire and of language. In this class he might well have placed the wheel. The cord which binds together all the chapters is labor, both physical and mental, without which no people could ever have made prehistory what it is—the logical background for history.

Each chapter is followed by a short bibliography in which standard works, rather than the most recent, play a dominant rôle. This matters little, however, for the average reader's sense of satisfaction with the author's text will be so deep that he may fail to appreciate the need of further enlightenment on the special subjects under discussion.

George Grant MacCurdy.

La Chine Antique. Par Henri Maspero, Professeur au Collège de France. [Histoire du Monde, ed. E. Cavaignac, tome IV.] (Paris, Boccard, 1927, pp. xvi, 624, 40 fr.) This is the first critical and sensible history of ancient China ever written on which the author and the public alike merit sincere congratulations. The former histories of China were content to give a digest of the fabulous and nebulous traditions of ancient records whether these were authentic or not. Hirth's *Ancient History of China,* for instance, is not a history of China, but merely presents a summary more or less uncritical of what from the viewpoint of Chinese scholars or from our own angle may be regarded as historical in the contradictory chaos of native myth and makeshift. Maspero discards the native tradition completely and concentrates his efforts on reconstructing the social organization, the religious life, the philosophical systems and literature of ancient times. He has succeeded admirably in restoring a truthful picture of ancient Chinese society, as far as it can be done with the deficient state of our sources and the limited material at our disposal. The only hope for increasing our knowledge of ancient China rests on the spade—intelligent and systematic excavations in which merely a rudimentary beginning has been made. In his concluding chapter Maspero demonstrates that in the fourth and third centuries B. C. foreign influences, chiefly coming from Iran and India, gradually penetrated into

China and tended to establish the first scientific notions of geography, astronomy, astrology, and geometry; also, it may be added, the first progress, in engineering and mechanics.

The book should be translated into English and made the fundamental text-book in all university courses on the history of China.

B. LAUFER.

L'Économie Antique. Par J. Toutain, Directeur d'Études à l'École des Hautes Études, à la Sorbonne. [L'Évolution de l'Humanité, dirigée par Henri Berr, vol. XX.] (Paris, Renaissance du Livre, 1927, pp. xxvi, 439, 30 fr.) This survey of economic activities from Homeric times to the fall of Rome, while not intended for the specialist, is a well-proportioned and conservative summary of the best work that has been done in this field. On the Greek side the rather abundant evidence has been so well worked over during the last twenty years that Toutain could produce a consistent story; the Roman part—where the data are less satisfactory—reveals the hesitations of one who has had to avoid minute details and is averse to setting down hypotheses. Only in the chapters that deal with Gaul and Africa—provinces which Toutain knows exceedingly well—will the historian of Rome find new materials. After sketching the "domestic economy" of Homeric days, the author dwells especially on the growth of commerce in the Greek world, which he attributes directly to Hellenic love of adventure and liberty. The economic effects of Alexander's penetration far into Asia are well described. The extension of international trade as a consequence of the Pax Romana is perhaps overstated though the author is at least to be thanked for not finding paternalistic policy at every turn. When he attributes the decay of the economic system to governmental interference necessitated by the barbarian invasions he is conservative rather than persuasive. The whole argument is perhaps overschematized and, as is often the case in summaries of this kind, smooth phrases here and there take the place of disturbingly inconsistent records. However the book fully deserves a place in Henri Berr's very respectable series of surveys.

TENNEY FRANK.

Byzance et Croisades, Pages Médiévales. Par Gustave Schlumberger, Membre de l'Institut. (Paris, Geuthner, 1927, pp. 365, 60 fr.) In this volume M. Schlumberger has reprinted seven articles, sometimes with considerable additions, originally published between 1902 and 1918, mainly in the *Revue des Deux Mondes*. The subjects range from the Palace Revolution at Constantinople in 1042 to the visit of Emperor Emanuel V. to Paris at the close of the fourteenth century. The essays are not usually the result of original research; the author says of one, "I have followed Röhricht step by step". In these articles intended for the general public M. Schlumberger has not been so careful as in his works of erudition; for example, he quotes William of Tyre for an event which

187

唤起科学家兴趣的海龟化石

SCIENTIFIC AMERICAN

May 1929 35¢ a Copy

THE GUNS AGAINST THE AIRPLANES

THE NEW OUTLOOK IN PHYSICS

HAVE PLANTS A HEART BEAT?

with a high-pressure lubrication fitting, dirt-proof and permanently installed, in less than two minutes.

A reverse feather-edged bushing is driven into the open oil hole. These bushings are made in one sixty-fourth inch sizes from one eighth to onehalf inch. Special bushings are made for countersunk holes and for thin housings.

Into the top of the bushing a nipple with dirt-proof ball check valve is installed. An ordinary hammer and a set of three special drive tools form the complete equipment for installing this system in open oil holes.

The Alemite plant lubrication system provides for handling the lubricant from its original barrel to the bearing without exposure even to daylight. This system precludes possibilities for waste, unnecessary labor, or failure of the lubricant to reach the desired spot between bearing surfaces.

Disappearing Closet-Bed

DWELLERS in apartment houses are familiar with beds that fold up and tuck away into closets. Whether or not they have used them, they have seen these beds and no doubt noted that when the bed is opened out and made ready for the night, the door to its closet cannot be closed.

The mechanism of a recently invented disappearing bed is so constructed as to allow the door to be closed at any time—that is, when the bed is in the closet or when it is down. The door is swung on a vertical rod as shown in one of the accompanying illustrations in such a manner that, although when the bed is ready for use it covers half the door space, the door itself easily slides shut.

The new invention is manufactured by Marshall Stearns Company, of San Francisco.

Hearing the Eye See

PROFESSOR E. L. CHAFFEE, of Harvard University, has for some time been conducting experiments endeavoring to learn how the eye functions. Instead of attacking the problem psychologically as many have done, he has been following the methods of some early experimenters, namely Einthoven and Jolly, using a direct method of measuring the electrical changes produced in the eye when light shines on the retina. He has considerably improved and developed the methods used by these early experimenters and has been carrying on these experiments since about 1921.

The eye of the animal is taken out and cut in two, thereby exposing the sensitive retina surface. Electric terminals, consisting of fine threads, are connected to this retina and to a vacuum tube amplifier which, in turn, operates a sensitive galvanometer and recording apparatus. When light shines on the retina, the galvanometer moves in a rather complicated manner showing that the light falling upon the retina sets up a complicated effect in the retina and the connecting nerves. He has been studying these complicated electrical changes in an endeavor to unravel the intricacies of the processes of vision.

Although he and his associates have learned a great deal concerning the manner in which light sensations are transformed into nerve impulses and hence carried to the brain, their experiments have given so far very little aid to curing of troubles of the eye. One never knows the outcome of investigations of this sort and although it may be hoped that their results will be of benefit in diagnosing eye troubles, that is not the main purpose of the experiment.

The electrical changes produced in the eye consist in part of very rapid electrical vibrations and these vibrations, after passing through an additional amplifier, can be made to give out sound in a telephone. This leads to the interesting experiment of being able to hear the eye see. This experiment is perhaps more sensational than useful.

Professor Chaffee has been aided in this work first by Mrs. E. L. Chaffee and, during the last year and a half, by Miss Evelyn Sutcliffe.

Professor E. L. Chaffee, who has been studying the eye, with the apparatus used in his experiments. With this apparatus he has heard the eye see

Turtle Fossil Arouses Interest of Scientists

ONE fossil turtle, about a foot long, recently acquired by Field Museum of Natural History, has aroused intense interest on the part of distinguished scientists in three departments of that institution specializing in research along widely different branches of science, it was learned recently from Stephen C. Simms, director of the museum. The fossil turtle which is receiving so much attention is a gift to the museum from Mrs. Chauncey B. Borland, prominent Chicago society woman.

The new disappearing closet-bed. The door of the closet in which it disappears closes after bed is down

The disappearing bed folded away. In this photograph may be seen the vertical rod on which the door swings

The department of anthropology is concerned because on the back of the turtle, which was found in China, are six mysterious ancient Chinese inscriptions believed to have been carved on the shell almost 4000 years ago. Dr. Berthold Laufer, curator of anthropology and noted Orientalist, is engaged in deciphering the archaic script. The turtle was regarded by the Chinese as a sacred animal with the magical power of accurately predicting the future, according to Dr. Laufer.

The department of zoology is deeply interested because study of the fossil indicates that it is an entirely unknown species of the genus Testudo. There are no zoological records in existence of any other turtle of its identical species, it was determined after an investigation by Karl P. Schmidt, assistant curator in charge of reptiles and amphibians. Mr. Schmidt made a thorough examination of the specimen before his recent departure as scientific leader of the Crane Pacific Expedition of Field Museum. Paleontologists in the department of geology are likewise interested because their study of the turtle indicates that it lived in the Miocene epoch some 19,000,000 years ago. Like the zoologists, paleontologists have no previous record of a fossil of a turtle of its species, according to Professor Elmer S. Riggs, associate curator of paleontology.

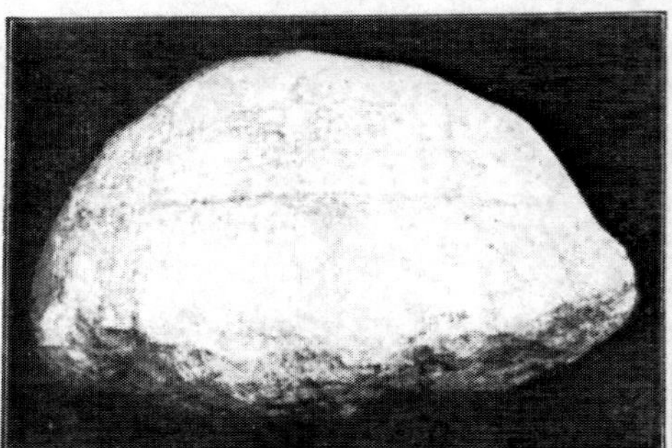

Looking like a simple piece of stone with a peculiarly marked, dome-shaped top, this large fossil turtle aroused the intense interest of three departments of a great museum. It is covered with Chinese inscriptions dating back, it is believed, 4000 years. Study of it indicates that it is of an entirely unknown species

Courtesy Field Museum of Natural History.

Apparently, millions of years after the turtle's death, its fossil turned up to attract the attention of some Chinese mystic, who carved the inscriptions on its back, says Dr. Laufer. The turtle was found in Shensi province, China, and was brought to this country by C. F. Yau. The inscriptions, delicately traced, are in the earliest stage of Chinese script, identical with that on the famous oracle bones of Ho-nan. At a very remote period the shell of a turtle was one of the chief elements in the art of divination in China, declares Dr. Laufer. For this purpose, shells were scorched over a fire, and the cracks occurring yielded a picture believed to foreshadow future events. The oldest examples of Chinese writing are preserved on tortoise shell fragments which contain questions addressed to soothsayers and the answers given by them, Dr. Laufer states.

"Divination was one of the dominating influences in ancient Chinese life," says Dr. Laufer. "The diviner was not a priest, but his position was comparable to that of a lawyer in our society. In the same way that the modern business man consults his lawyer on important questions, the Chinese did not make a step without asking the advice of a diviner. The oracles and sometimes the questions were carved on bones, turtle shells, and other such objects, many examples being extant."

The turtle will be placed on exhibition in the near future, after the inscriptions have been deciphered.

Aquatic Animal Remembers for Sixteen Generations

THAT "the elephant never forgets" is a favorite tradition of the circus lot and the zoo. But the big pachyderm's record has been badly scratched by a lowly animal that lives in the water, whose great-grandchildren of the 22nd generation remember a complex instinct possessed by their ancestor but which their more immediate forebears have never had an opportunity of exercising.

At the meeting of the American Society of Zoologists in New York recently Prof. W. A. Kepner presented the results of experiments which he and J. W. Nuttycombe, now of the University of Tennessee, performed on a tiny, almost microscopic, animal known as "microstomum." This creature is one of the many in the lower realm of nature that is armed with stinging cells in its body wall, partly for protection against its enemies and partly to assist in the capture of its prey. In the fierce economy of the lesser world of the waters, this animal feeds on a similar but smaller form, the hydra, which is also armed with stinging cells. It apparently does not like the hydra very much, for it will not eat it except when its supply of stinging-cells is low. Then it swallows hydra readily enough, and appropriates the ready-made cells.

The two experimenters grew a strain of microstomum for 22 generations, without ever giving them an opportunity to meet and feed on hydras. Yet at the end of that time the animals went through the same performance their ancestors had been used to: swallowed the hydra, turned over the stinging cells to certain wandering cells of their own bodies, and eventually set them in order among their own stinging cells. In another case, a 16th generation microstomum had its upper and lower ends cut off. The middle third of the body regenerated new ends, and the "revised" animal went through the long-disused performance as though it had been used to it all its life.—*Science Service.*

Royal Photographic Society's Annual Exhibit

THE 74th annual exhibition of the Royal Photographic Society of Great Britain will be held this year from September 14 to October 14.

The secretary wishes a strong American respresentation, since active support from representative scientific workers in this country contributes largely to the success of the exhibition.

Exhibits intended for the scientific section may be sent to the Eastman Kodak Company, Rochester, New York, in care of Mr. Alexander Murray, and should reach him not later than July 15.

Eighty-four Elements in Single Collection

SHORTLY after publication of an article in our February issue entitled "Best Collection of Rare Gases," Mr. Edward S. Candidus of the Brooklyn Edison Company, Inc., wrote us relative to a collection of chemical elements he has made. His comments are so interesting that we quote below a part of his letter:

A new collapsible power reel of aluminum or steel, recently announced by the Highway Trailer Company, Edgerton, Wisconsin. It collapses to permit placing or removing a coil of wire or rope. The winch takes power from the motor of the truck. It is used for pulling in aerial or underground cables

188

巴托尔德《蒙古入侵时期的中亚》书评

THE AMERICAN HISTORICAL REVIEW

VOLUME XXXIV
OCTOBER 1928 TO JULY 1929

NEW YORK
THE MACMILLAN COMPANY
LONDON : MACMILLAN AND CO., LTD.
1929

Reprinted with permission of the Original Publishers by
KRAUS REPRINT CORPORATION
New York
1968

tents of this volume. Inasmuch as Norway, so far as we know, has always been occupied by the same North Germanic people, the linguistic problems involved in the study of local nomenclature in that country are not so varied or difficult as they are in England, for example, where one has to deal with the contributions of successive races. But if the research is less baffling, it is certainly not less extensive; for the author estimates that Norway has at least five million place-names of respectable age. Most of these have, of course, little interest for the student; but there is a considerable remainder, the study of which has led to a series of significant conclusions.

Professor Olsen deals almost exclusively with farm names. His interest lies in the question whether these can be made to contribute in any way to our understanding of social or religious history. He concludes that many of these names, especially such as are simple in structure, have a history that goes far back into antiquity, perhaps in cases to the beginning of the Christian era. Professor Olsen sees an early settlement of large farms, each occupied by a single family of the patriarchal type. In the course of time it became necessary to divide these estates, or to form new farms with outlying fields as the nuclei; frequently it also happened that younger members of the family, or even freedmen, were sent forth to clear and develop holdings at some distance from the parent home. These new farms usually received names of a compound character, with such endings as *heimr* (home), *land, setr* (place of settlement), *ruð* (clearing), and the like, as the more significant part of the compound. Thus it seems that these later names can give information as to how successive groups of farms originated and how the settlement spread from the lower valleys to the higher ground.

Professor Olsen devotes a lecture to such farm-names as indicate the location of heathen sanctuaries, names ending in *hof* or *hǫrgr*. The work closes with a discussion of a few names which seem to cast light into certain obscure corners of Northern mythology.

L. M. L.

Turkestan down to the Mongol Invasion. By W. Barthold. Second edition, translated and revised by the author with the assistance of H. A. R. Gibb, M.A. [E. J. W. Gibb Memorial Series, new series, V.] (London, Luzac and Company, 1928, pp. xx, 514, 25 s.) The work of W. Barthold first appeared in Russian in 1900 and thus remained known only to a few specialists. We are grateful for the English translation which has been thoroughly revised and amplified by the author himself who endeavored to bring his work up to date. It is a most erudite and painstaking piece of work which aims at presenting a digest of all Arabic, Persian, and Turkish sources relative to the historical geography and political history of Russian Turkestan from the end of the seventh century down to the death of Chingiz-Khan in 1227. Chinese sources, as far as accessible in translations, have also been utilized. The introduction contains a discussion of the sources bearing on the Pre-Mongol period, the Mongol invasion, and European works of reference. The book is divided into four chapters: geographical survey of Transoxania, which is very detailed, Central Asia down to the twelfth century, the Qara-Khitays and the Shahs of Khwārazm, and Chingiz-Khan and the Mongols; the last-named, especially the characteristic of Chingiz, being the best portion of the book. It is concluded by a chronological summary of events, bibliography, and index which might be more complete, and is accompanied by a good map.

It is not surprising that in a work of this compass, despite all care, many slips occur. Thus the word *bakhshi* used throughout Central Asia is not derived from the Sanskrit *bhikshu,* as asserted on pages 51 and 388. This etymology was disproved by me in *T'oung Pao,* 1914 (p. 411), and 1918 (pp. 485–487), I have given the correct derivation from Chinese *po-shi* (ancient form *bak-shi*).

The name of the Chinese pilgrim is not Hiuen Tsiang, as it is spelled on page 70, but Hüan Tsang. The Kin can not be called a Manchu dynasty (p. 381); Kin is the Chinese dynastic name for the Jurchi, a Tungusian tribe akin in language to the Manchu, but not identical with them. Can it truly be said that nomadic life and intellectual culture are two incompatible things (p. 461)? Despite its eminently geographical and historical character the Oriental sources translated in the work contain numerous data of culture-historical interest and references to commerce and products, but most of these are unfortunately not registered in the index.

B. LAUFER.

The Achievement of the Middle Ages. By W. E. Brown. (London, Sands and Company, 1928, pp. 240, 5 s.) By the Middle Ages the author refers to the period 1100 to 1500 A.D., and the achievement he notes is threefold. "The men of these generations achieved and maintained . . . a reign of law in the relations between man and man. Secondly, they developed their towns, i.e., their industries and their commerce, in a way which is unique in history and which was intimately related to the contemporary improvement in ordered liberty. Thirdly, these generations achieved a high culture, which did not decay but was developing into a yet more splendid form at the end of the period", to be checked by the wars of religion. A discussion of this threefold development is prefaced by an essay on "their tradition", *i.e.,* the background of the development from 1100 onward.

The author's point of view is shown by these statements. "The third quarter of the eleventh century saw the resurgence of the Catholic Church as the great European law giver", and "it remains to be shown how the Church, as a free juridical society, informed the institutions of the Middle Ages with the governance of law". There are no foot-notes and no references; "there is no pretence at a complete history, but only

189

李济《中国民族的形成》书评

Rebellion, our best account of that movement, and Steiger, *China and the Occident.* This is, however, not a particularly important defect. It is to be hoped that the book will have a wide use.

K. S. LATOURETTE.

The Formation of the Chinese People: an Anthropological Inquiry. By Chi Li, Ph.D., Lecturer in Anthropology in the Tsing Hua Research Institute. (Cambridge, Harvard University Press, 1928, pp. 283, $5.00.) Dr. Li's work is the most competent contribution hitherto made to the physical anthropology of the Chinese and is distinguished for the originality of approach to a difficult problem. It is equally notable as the first production of a Chinese scholar who received his training in the department of anthropology of Harvard University, under Professors Dixon and Hooton. The book does not interest the anthropologist exclusively, but will appeal (and to a higher degree) to the historian; for the author's investigation is a remarkable contribution to the history of race-mixtures on the soil of China. He destroys the old prejudice that the Chinese have been unchanging for several thousands of years and that the modern Chinese represent a physical homogeneity.

On the contrary, the physical formation of the Chinese is a most complex problem. The term "China" indicates merely a political unit, while the most diverse races and entirely distinct ethnical and cultural groups have been welded together into what we are pleased to call simply the "Chinese". The author introduces, after W. G. Sumner, a new terminology by speaking of the Chinese proper as the "We-group" and of the non-Chinese or barbarians as the "You-group"—to denote the psychological phenomenon involved in the perception of "meum" and "tuum". Personally, I see no advantage in the use of these affected terms; the same idea can be expressed by "Chinese" and "non-Chinese" just as efficiently. Dr. Li first describes the physical traits of the living Chinese based on the data of his predecessors and his own observations and measurements of Chinese students in the eastern universities of the United States and Chinese laborers from Kwang-tung province in Boston. The results are lucidly presented in sixteen tables and four maps. The author's most interesting investigation is contained in chapter IV., in which he endeavors to trace the origin and movements of ten clans, based on their family names. Of the history of Chinese families we know but little, and this contribution to family history will be gratefully appreciated. It should be extended, of course, to several hundred families. One of the interesting results is that many Chinese clans of northern China intermarried from early times with people of Turkish, Tungus, and Mongol descent. The centres and migrations of the ten selected families are charted on 59 skeleton maps on which the provinces are outlined but unfortunately not named. Another peculiar practice is to quote historical periods by letters, while nowhere a table identifying periods A, B, C, etc., is given.

In chapter VI., in which the aboriginal or so-called barbarian tribes of southern China are passed under review, the author is less fortunate and commits many errors, as he is not very familiar with this field. I regret that he did not utilize my article "Totemic Traces among the Indo-Chinese" (*Jour. Am. Folklore,* XXX. 415–426), where a better translation and interpretation of the P'an-hu myth is given than his on page 243.

The author's supposition that there are Negritoes in south China is most improbable, and is not warranted by the evidence produced by him (p. 259). Hala-wusu (read Khara-usu) is not Tibetan, as stated, but Mongol; in Mongol (Khara-usu) means "black water", while the corresponding Tibetan term is Nag ch'u. This name of a river has nothing to do, as alleged, with a dark-skinned population. The influence of tribes of the Mon-Khmer group on Chinese, if any, is negligible; nor is there any evidence for the bold assertion that the ancient Yüeh in southeastern China were Shan-speaking people or bear any relation to Shan by which he obviously means the Tai group.

B. LAUFER.

The Soul of China. By Richard Wilhelm, translated by John Holroyd Reese and Arthur Waley. (New York, Harcourt, Brace, and Company, 1928, pp. 382, $3.75.) Keen insight, warm sympathy, and charm of style raise this attempt to interpret China, the baffling, above most of its contemporary studies. The author, a German missionary and teacher who went out to the infant Kiaochow colony soon after its acquisition, spent twenty-five years in China, in Shantung province, and in Peking. His book contains a survey of political developments in the first quarter of the century, with descriptive sketches of persons and places and observations on religion and philosophy. The comments of this missionary who christened no converts, this representative of European culture who learned to esteem the culture of the East, this lover of the past who saw the old China crumble about his head, are full of interest and suggestion. In preparing the English text no attempt seems to have been made to adopt the usual English equivalents for Chinese place and proper names, while at least three different spellings of Szechwan province may be observed.

The Luna Papers: Documents relating to the Expedition of Don Tristán de Luna y Arellano for the Conquest of La Florida in 1559–1561. Translated and edited with an historical introduction by Herbert Ingram Priestley, Ph.D., Professor in the University of California. [Publications of the Florida State Historical Society, no. 8.] Volume II. (De Land, Fla., the Society, 1928, pp. xv, 383.) This second volume, added to that which was reviewed in the preceding volume of this journal (XXXIII. 920), completes the record of Don Tristán de Luna's unfortunate Floridian expedition of 1559–1561. The texts are presented with great care,

190

美国植物的迁移

THE SCIENTIFIC MONTHLY

EDITED BY J. McKEEN CATTELL

VOLUME XXVIII
JANUARY TO JUNE

NEW YORK
THE SCIENCE PRESS
1929

THE AMERICAN PLANT MIGRATION

By Dr. BERTHOLD LAUFER

CURATOR, DEPARTMENT OF ANTHROPOLOGY, FIELD MUSEUM, CHICAGO

ALPHONSE DE CANDOLLE'S famous book, "Origin of Cultivated Plants," ranks among those monumental productions of science which, although written a generation ago and on numerous points out of date, will always remain young and act as a stimulus to fresh research for generations to come. Botany and agriculture, as well as history and ethnography, have advanced so rapidly since his day that many are now qualified to add new contributions to the problems discussed by him, but no one at the present time would be able to recast his book so as to make it conform to modern progress. An alliance of many scholars—botanists of all shades and grades, agriculturists, horticulturists, archeologists, orientalists, historians, etc.—would be required to cope adequately with a task of such magnitude. Should such a plan ever be considered, it is hoped that a fundamental deficiency in de Candolle's book will be adjusted. One may peruse this great encyclopedic work many times, our memory may retain a number of facts pertaining to this or that plant, as our personal inclinations may lead us; yet no lasting impression will remain on our minds as to the significance of plant cultivations in the development of mankind.

The reason is obvious: the subject is not presented in correlation with human culture. The division of the book is purely mechanical: there are five main sections dealing with plants cultivated for their subterranean parts, for their stems or leaves, for their flowers, for their fruits and for their seeds. Under these botanical categories, the most heterogeneous plants are arranged and succeed one another, regardless of time and space, with no bond linking them together. We are thus confronted with a collection of historical essays or sketches in which subjects historically interrelated are scattered and widely separated, without any attempt at correlation, coordination or pragmatic history-writing. A wild plant is a given fact of nature; the cultivation of a plant presupposes human interference and care, and must therefore be regarded as life and movement in the long career of man. Cultivated plants are an essential element in the history of human economy and civilization, and their study must be grasped in the sense of a cultural movement. In the same manner as many other cultural ideas and inventions, plants have migrated at all periods, and continue to migrate and expand under our own eyes. The great plant-migrations mark the lines in the march of civilization even more clearly than other departments of human endeavor, because agriculture and all concomitant features represent the most stable and unchangeable factor of our economic life. In order to be significant and fruitful, the history of cultivated plants must be conceived of as the history of plant-migrations.

At the present time almost everything is almost everywhere, and we are anxious to learn whether it was there always, or, if not, how and when it has taken its place. Looking at our country, we recognize without difficulty four strata of plant-cultivations: (1) those peculiar to the aborigines of America, like maize, several varieties of bean, pumpkins, squashes, sunflower, *Nicotiana rustica*, etc., subsequently adopted by the white settlers, who also succeeded in cultivating wild species of North America, as, for instance, the grape; (2) plants introduced from England in colonial times, like wheat, barley, rye, oats, buckwheat, apple, pear, etc.; (3) American plants introduced from the West Indies in the seventeenth and eighteenth centuries; (4) numerous plants brought over from China and Japan from the eighteenth century onward to the present day.

In countries like India and China, conditions naturally are more complex, the stratification is deeper. In China we find from the earliest times a few cultivations typically Chinese, as, *e.g.*, the soy-bean, peach and apricot; others in common with the Sino-Tibetans or Indo-Chinese, the congeners of the great stock to which they belong, like oats, hemp and many species of pulse and Allium (leek and onion); others, like wheat and barley in common with western Asia, and rice in common with southeastern Asia. From the latter part of the second century B.C. the introduction of exotic plants sets in, inaugurated by alfalfa and grape-vine brought back from Fergana by the first Chinese explorer, General Chang K'ien, and followed by a long retinue of other Iranian and west-Asiatic plants, this great Iranian plant-migration, described in detail in my monograph, "Sino-Iranica," continuing for fourteen centuries. Simultaneously the Chinese absorbed what is now southern China and advanced to Tonking and Indo-China, adopting all the useful plants they encountered in that subtropical area and amalgamating the type of southern garden-culture with their old northern plow-ox culture. The last phase of this development is signalized by the introduction in the sixteenth and seventeenth centuries of a large number of species of American origin, which added a considerable plus to the already naturalized quota, with the striking result that almost all useful plants of the universe are now embraced by Chinese husbandry.

In determining the steps in the great American plant-migration, the records of the Chinese are of fundamental importance, as no other nation has preserved fuller and more exact accounts of agricultural events and innovations. It can hardly be fortuitous that all plants without exception, which are justly regarded as American, are placed on record by the Chinese as having been introduced into their country in the latter half of the sixteenth or beginning of the seventeenth century. Korean and Japanese traditions move along the same line; and wherever documents are at our disposal, as in Siam, India, Persia and Europe, the same result is echoed from every quarter. Let us suppose for a moment that there were no botany in America, that we were ignorant of the achievements in agriculture of the American Indian and that we did not even possess the European herbals of the sixteenth century which successively describe the novel plants as introduced from the New World—even this being the case, merely on the basis of our experience with the situation in Asia, we might justly claim that, from a historical view-point, all plants like maize, several species of Phaseolus, potato, batata, manihot, tobacco, pineapple, guava, papaya, Anonas, Capsicum, peanut, agava, sunflower, cashew, arnotto, cacao, tomato, prickly pear and many more, must have originated in and hailed from America. All these cultivations and their products were a revelation to the peoples of the Old World and entirely unknown there prior to 1492.

This conclusion is further linked with my conviction of the independence of aboriginal American agriculture. None of the Old World cultivated plants is found in pre-Columbian America; on the other hand, none of the American cultivated plants occurs in Europe, Asia or Africa prior to the age of discovery. There never was a direct prehistoric interchange of plants between China and Mexico, or between Oceania and Peru; there is not the slightest foundation in fact for such like speculations. The ingenuity and accomplishments of the American Indian in agricultural products are deserving of our highest admiration. The mere fact that the plants cultivated by the Indians had really been brought into a perfect state of cultivation as early as pre-Columbian times permitted the white man's colonization and made the propagation of these cultivations over the Old World possible; what the white man applied to them were solely improved methods of culture. From numerous biological data, as revealed by the differentiation and variation of corresponding cultivated and wild forms, it may be inferred that Indian agriculture must be not centuries, but thousands of years old.

Of all plant-movements, the American plant-migration, although the most recent, is the most extensive, the most prominent, the most universal and the most momentous in the world's history. It therefore merits profound study in every detail. It has encompassed the globe in its entirety, made its influence felt everywhere, changed the surface of the earth and brought mankind together into closer bonds. For the student of Old World agriculture it is essential to have a clear conception of these intrusions, if he is eager to know what plants originally belonged to a certain culture area.

The history of American plants is also more instructive and fascinating than that of any other plant-movement, for each American plant has its distinct and individual type of history: the migration was not a single event that might be told in a few pages, or a series of operations of a uniform character following a definite scheme, but it was a long, romantic chapter composed of an infinite variety of good plots and stories, set off from a picturesque background in a wide perspective. Nearly all the great nations of Europe assume their rôles—Spaniards, Portuguese, Italians, French, English and Hollanders. The scene is laid over a vast area stretching from the Atlantic coast of Canada, New England, Virginia, down to Florida, the Gulf of Mexico, the West Indies and Brazil, as well as along the Pacific coast from the shores of Mexico and Panama down to Peru and Chile. For the first time in history the oceans developed into a great artery of plant-communication, and plants crossed both the Atlantic and Pacific, directly to Europe and Africa eastward and to Asia and the Oceanic Islands westward. It was a world-wide crusade unparalleled in the microcosm of previous ages. The transatlantic and the transpacific migrations were almost coeval, and we face the curious spectacle that in India, Central Asia and Siberia, several American plants, notably tobacco, suddenly clashed in a sort of head-on-collision, by encircling the globe eastward and westward simultaneously.

The gifts of the New World were all of a democratic character and made a world-wide appeal; tobacco conquered all peoples of the globe without distinction, and I know of only a single tribe that does not practice smoking—the poor islanders of Botel Tobago. Tobacco is more universally consumed than any other narcotic, has profoundly influenced the economy of most nations and signally affected social customs and promoted sociability. In a spirit of gratitude, Chinese and Japanese have bestowed on tobacco the name "herb of amiability" (*ai-king ts'ao*), as they explain, "on account of the affectionate feelings entertained toward one another by all classes of mankind since its use has become general." Tobacco has proved the greatest peace-maker of mankind and contributed in a higher degree than all pacifist movements to the tranquillity, comfort and happiness of an overwhelming majority.

Maize, peanut, batata and potato have added a considerable share to the means of human sustenance, and consequently to the wealth of nations and the increase of population, notably in those countries where the ordinary cereals are of difficult or costly cultivation. Batata and potato, in particular, are invaluable as famine crops, and in times of drought and scarcity have saved the lives of millions of people. In many parts of western, central and southeastern Asia there are numerous destitute peoples cooped up in barren mountain-fastnesses, much like the tribes in the highlands of Peru, now subsisting exclusively on maize and potatoes; and we wonder how they ever managed to live prior to the discovery of America.

The subject of American cultivated plants is so vast that on this occasion I can merely give a brief abstract of some of my results. An extensive manuscript is almost ready and will be published at some later date. One of my results, I hope, may be of particular interest. It was hitherto unknown, at least to those who have widely written on the subject, when and how the potato reached North America. De Candolle merely dilates on speculations to the effect that some inhabitants of Virginia, perhaps English colonists, received tubers from Spanish or other travelers, traders or adventurers, during the ninety years which had elapsed since the discovery of America. Roze, Wittmack, Brushfield, Safford and others, to whom we owe monographs of the potato, are equally vague. It almost grieved me that we should be ignorant of the facts accompanying the introduction of the potato into our country, while almost all other nations have preserved a record of this event; hence I set out to delve in the early history of Virginia, but with no success. Finally, after several years' quest, I chanced to peruse the old "History of the Bermudas," and was at last rewarded by finding the desired information.

In 1613 the good ship *Elizabeth* brought potatoes from England to the Bermudas. The "Historye of the Bermudaes," ascribed to Captain John Smith (1580–1631), by others to Nathaniel Butler, governor of Bermuda from 1619 to 1622, reports this event as follows (p. 30): "In her wer first brought into thes partes certaine potatoe rootes sent from England, the which being planted and flourishinge very well, wer by negligence almost lost; at last, by a lucky hand, again revived from two cast awaye rootes; they have since encreased into infinite store, and serve at the present for a maine reliefe to the inhabitants." It was from the Bermudas that the potato was further transmitted to Virginia. On December 2, 1621, Captain Nathaniel Butler, governor of the Bermudas, sent from "St. Georges, in the Sommer Ilands," to the governor of Virginia (Francis Wyatt) two large cedar chests, "wherin were fitted all such kindes and sortes of the country plants and fruicts, as Virginia at that time and until then had not, as figgs, pomegranates, oranges, lemons, plantanes, sugar canes, potatoe, and cassada rootes, papes [papaya], red-pepper, the pritle peare [prickly pear], and the like" (*ibid.*, p. 277). In the following year, a Virginian "barcke took from the Bermudas twenty thousand waight of potatoes at the least" (*ibid.*, p. 285). All this is on record in the "History of the Bermudas." The fact that potatoes were actually planted in Virginia at the very moment of the first introduction is confirmed in letters sent from Virginia in

1621 and published by Purchas (vol. 19, p. 151): it is intimated there that "in December last they had planted and cultivated in Virginia potatoes and sundry other Indian fruits and plants not formerly seen in Virginia, which at the time of their said letters began to prosper very well."

The potato, accordingly, entered this country, not as surmised by de Candolle, through an alleged band of Spanish adventurers, but in a perfectly respectable manner—from England by way of Bermuda. It is a prank of fortune, of course, that the potato, originally a denizen of Chile and Peru, appears as a naturalized Englishman in the United States. This result is bound to modify to a large extent the entire early history of the potato as it has hitherto been conceived.

The potato had arrived in England about 1586 or a little later. For a long time the belief was entertained by botanists, even by H. Phillips and de Candolle, that the openauk, described among the wild roots of Virginia by Thomas Hariot in his "Brief and True Report of the New Found Land of Virginia" (London, 1588), was to represent our potato, that Hariot had brought his openauk-potato to England, and that from this source it was received by John Gerard, the first English botanist who raised the potato and described and figured it under the name "potato of Virginia." This speculation is erroneous: openauk is not the potato; Hariot does not claim that he ever took tubers of potatoes to England; indeed, he does not at all speak of potatoes, nor does Gerard mention Hariot's name or the openauk in connection with the newly introduced potato. Hariot described openauk as "a wild root found in moist and marsh grounds growing many together one by another in ropes." The potato certainly does not grow in swampy soil, and never occurred spontaneously in the United States or Mexico. Openauk, in fact, represents an entirely distinct plant, *Apios tuberosa*, popularly styled groundnut or Indian potato, a common article of food among the Indians. I am informed by Dr. Frank G. Speck, of the University of Pennsylvania, that in Penobscot there is a word *ponak* still applied to the groundnut. Be this as it may, from the data of the "History of the Bermudas" it is perfectly clear that "the potato was one of the plants which at that time and until then [that is, 1621] Virginia had not," so that it could not have been known to Hariot; and the fact remains that not a single account, report or letter from Virginia up to that date makes any mention of the potato, while after this date it is most frequently mentioned.

The question as to whom is due the honor of having first brought the potato to England is still deadlocked: the documents fail us. The principal testimony around which the history of the plant in England pivots is Gerard's who merely states, "It groweth naturally in America [this, in the language of the period, means South America], where it was first discovered, as reports Clusius, since which time I have received roots hereof from Virginia, otherwise called Norembega, which grow and prosper in my garden as in their own native country." Gerard's lengthy description and illustration of the potato has been harshly criticized, especially in England, but rather unjustly. In my forthcoming book, where the history of the potato in England is treated at great length, I have committed myself to a complete defense of Gerard's account, which is one of fundamental historical importance. Our imagination, of course, likes to attach the introduction of so useful a plant to the name of an individual, especially to one of historical fame. Sir Francis Drake and Sir Walter Raleigh have been considered in this connection, and it is possible that one or the other may have had a share in the introduction, though documentary evidence is wanting; Raleigh is credited by popular tradition also with the introduction into Ireland.

The potato had to struggle for over a century in Europe before it met with general recognition; a rapid advance in its propagation was made in England only during the eighteenth century, and in France and Germany still later.

The civilized nations of Asia, while they adopted numerous plants from America, have not yet cast their vote in favor of the potato, but treat it rather indifferently or even disdainfully. This attitude is not the outcome of prejudice or an instance of inherited conservatism, as is often insinuated, but has its cause in the system of nutrition which prevails among those peoples and in which there is no place or vital necessity for potatoes. To be sure, the potato is grown almost everywhere in Asia, and is a favorite dish of all poor mountain-tribes, even in China; but nowhere has it deeply affected agricultural economy, nor does it offer a continuous stream of logical development. Not being national, its history is purely local and split into a series of incoherent efforts of a sporadic and isolated character.

On the other hand, the batata or sweet potato was never duly appreciated in Europe, but proved a prize-winner in the Far East. The admiration and enthusiasm, nay, ecstasy with which the batata was received in China, Luchu and Japan has no parallel in the annals of plant-introductions, and its history in those countries is a little romance hardly excelled by any other useful plant. As it has never been recorded from Chinese sources, a condensed digest of the story is herewith presented. In 1593, the province of Fu-kien in southern China, presumably owing to the ravages of a typhoon, was stricken by a famine. The governor of the province, Kin Hio-tseng, dispatched a commission to Luzon in the Philippines with instructions to search for food-plants which might relieve the pitiable plight of his people. Luzon then was thickly settled with Fukienese, who advised their countrymen to take the sweet potato along. The Chinese chronicle has it that the men beyond the sea, that is, the Spaniards, had strictly prohibited the exportation of the species, so that the Chinese were compelled to have recourse to subterfuge.[1] They wrapped cordage around the tubercular roots of batatas, till they had the appearance of ship-cables, and pretended to load their ships with ropes. Thus they safely reached Fu-kien in 1594 and taught their compatriots the cultivation of the novel plant, which was greeted with unbounded joy and which stemmed the tide of famine. In a short time special agricultural treatises and poetical compositions in honor of the batata were produced, and like a prairie-fire its cultivation spread over all parts of the country. The economic value and the highly nutritive properties of the newcomer were at once recognized. It is unmistakably described and figured and carefully distinguished from the many native species of Dioscorea, with which the batata is so frequently confounded: the Chinese still call it the "foreign Dioscorea" or the "Dioscorea of Governor Kin," at whose initiative it was introduced, and state advisedly that it was previously unknown in their country. In 1786 an im-

[1] I have not been able to trace a confirmation of this charge in contemporaneous Spanish sources; while the regulation somewhat savors of Spanish policy, no other example of such rigid exclusiveness is known to me, and I am rather disposed to suspect some exaggeration on the part of the Chinese, prompted by the desire to aggrandize the perilous nature of their venture.

perial order was issued to encourage the cultivation of the batata as a means of preventing famine.

From Fu-kien it was transplanted on the one hand to Formosa about fifteen years later, and on the other hand to the Luchu Islands as early as 1605. At that time the Luchans still formed a kingdom of their own, though recognizing the sovereignty of the Chinese emperor. Nugun, superintendent of the Chinese settlement in Napa, the chief town of the archipelago, presented a native village chief, Masatsune, with cuttings of the plant; he eagerly studied its mode of cultivation and promoted it in his country. In front of Nugun's tomb a memorial pillar has been erected, and he is canonized under the name Mmu-ushume, that is, Ancestor of the Tuber. On Luchu, where famines are frequently caused by typhoons, the plant has served as a real life-saving device; and as early as the seventeenth century it became thoroughly nationalized and next to rice the most important article of food; it is still the daily bread of the islanders.

A Japanese farmer, Maeda Riuemon, a native of the province of Satsuma, made the acquaintance of the batata while paying a visit to Luchu in the latter part of the seventeenth century. On his return home he introduced its cultivation into Satsuma, and from there it was disseminated over the northern provinces of Japan. Riuemon's tomb is known as Kara-imo-den ("Temple of the Sweet Potato"), where every spring and autumn the soul of this simple farmer receives offerings from his grateful countrymen. The earliest Japanese treatise on agriculture, the Nogyo-zensho, written by Miyazaki Yasusada in 1696, gives a very full account of the nature and method of cultivation of the batata, and was followed by two substantial monographs in 1716 and 1734. During four years of scarcity, in 1832, 1844, 1872 and 1896, the people of Japan were saved solely by sweet potatoes. The nomenclature follows suit with the historical facts: in Luchu it is known as the Chinese Dioscorea, in Satsuma as the Luchu Dioscorea, and in the rest of Japan as the Satsuma Dioscorea.

De Candolle and all previous investigators, who blindly followed him, were misled by a superficial statement of Bretschneider, who according to de Candolle "has proved that the species is for the first time described in a Chinese book of the second or third century of our era." This, however, is a fallacy and refers to a species of *Dioscorea,* not to *Convolvulus batatas.* This bulwark of an alleged Asiatic origin of the batata, heralded in numerous books, has now fallen. In de Candolle's justice it must be added, however, that he clearly visualized "powerful arguments in favor of an American origin, and that the latter appeared to him much stronger." But there was no prehistoric interchange of the plant between the New and Old Worlds or *vice versa,* as further intimated by him; the plant is decidedly one of American origin and migrated into the Old World only after the discovery. In fact, there is not a single document from which a pre-Columbian existence in Asia or Africa of the batata might be reasonably deduced. In India it makes its appearance toward the end of the sixteenth century as an introduction due to the Portuguese, and is known in all Dravidian languages under the Portuguese-American term *batata.* On the Moluccas it is called *castilian,* on Java and Bali *catela,* based on Castela-Castilian. It is maintained even by serious scholars that the batata has been a native of the South Sea Islands since ancient times. I fail to see any tangible evidence for this opinion. There is usually confusion with other species like *Ipomoeà mammosa,* a Convolvulus-like plant with an edible root, but of a distinct botanical character; this is a native of the Moluccas, where it is also cultivated, and this is probably the batata mentioned by Pigafetta on the Moluccas in 1521. The early Spanish records, *e.g.*, those relating to the discovery of the Solomon Islands, mention taro and two kinds of yam, but not the batata, and the Spaniards coming from South America were surely familiar with this plant. Batatas, together with maize, squashes and *Phaseolus pallar*, were introduced from Peru into Tahiti by a Spanish expedition of 1772 under Don Domingo Boenechea, and a few years later Andia y Varela reports that the Tahitians grow two or three varieties of it. The Maori of New Zealand make a clear distinction between their native species and that subsequently introduced from America.

Finally, there is the testimony of the early discoverers of America and the contemporaneous European botanists, above all that of the great Clusius, who visited Spain in 1566, where he observed and described three varieties of batata. He states that it grows spontaneously in the New World, whence it was first brought over to Spain. "We sometimes have them fresh in Belgium," he adds, "but they will not germinate here, the country being too cold." Nicoloso Monardes, physician of Seville in 1572, likewise discusses the batata as a native of America and as widely cultivated and consumed in Spain in his time.

The history of the pineapple is of particular interest because it is so strikingly individualistic, quite in distinction from that of maize, which is so impersonal. A plant of such remarkable characteristics challenged the attention of all observant travelers. There is hardly any other plant the history of which is illuminated by such a flood of interesting documents. More than two hundred documents bearing on the pineapple have been brought together by me, and these have enabled me to trace the steps in its migration with a fair degree of accuracy.

The pineapple, the king of all fruits, as it has frequently been styled, belongs to the order Bromeliae, which consists of twenty-eight genera and 176 species—all natives of the American continent and insular groups, whence their distribution to many parts of the Old World has taken place. It is a particularly interesting point that in consequence of long-continued cultivation the highly cultivated varieties of *Ananassa sativa* have become seedless. This characteristic feature was emphasized as early as 1557 by André Thevet—a sure manifestation of the great antiquity of the cultivation in pre-Columbian America. Seeds are so scarce in the West Indies that there is seldom more than a single one found even in thirty or forty fruits. The plant is therefore usually propagated by means of crowns, slips, suckers and rattoons. The crown furnishes the most vigorous plants and yields the finest products. Plants are raised from seed only for breeding purposes, with a view to obtain novel varieties, but it requires from ten to twelve years to mature a plant from seed. The wild pineapple, on the contrary, is full of seeds, but small, seldom larger than an apple, stringy and rather acidulous in taste. It is only due to the process of cultivation that the fruit has acquired its large size and superior flavor. This was an accomplished fact when the first explorers of the Antilles and South America made its acquaintance.

The documents at our disposal warrant the conclusion that at the time of the conquest the pineapple as a thoroughly cultivated plant occupied the area of Brazil, Guiana, Colombia, parts of Central America and the West Indies. As early as 1519 it is mentioned in Brazil by Pigafetta as "a fruit resembling a pine-cone, extremely sweet and savory, in fact the finest fruit in existence."

André Thevet visited Brazil in 1555–56 and correctly figures and describes the fruit under the Tupi name *nana*. It was then medicinally employed by the natives, who also made a strong wine from it. Thevet's statement, that "the fruit bears no seed whatever and is hence planted by means of small slips, as fruits are grafted in our country," bears out the fact that the cultivation in Brazil must have been many, many centuries old. Jean de Lery, the Huguenot clergyman, who came to Brazil in 1557, is the first to employ the word *ananas* as being derived from the language of "the Savages." Christoval de Acuña, in 1639, found the Indians of the Amazon Valley using pineapples as food. In his "Discoverie of Guiana" (London, 1596), Sir Walter Raleigh speaks of "great abundance of pinas, the princesse of fruits, that grow under the sun, especially those of Guiana."

Although not mentioned by Columbus himself, the pineapple is placed on record by his contemporaries and epigones, who made use of his and his companions' diaries, letters and reports; first of all, by Peter Martyr d'Anghera, who offers three notices of the curious plant in the second and third decades of his *De Orbe Novo*, first published in 1516; this is the earliest record of the pineapple in existence. Martyr describes it as an herb resembling a pinenut, artichoke or acanthus, raised in the gardens of the West-Indian islanders, worthy of a king's table; King Ferdinand of Spain had eaten a fruit shipped from Darien and awarded to it the highest praise. A lengthy, though somewhat verbose and cumbersome description, is inserted in Oviedo's "Historia general y natural de las Indias" (1535). There are numerous old accounts for Cuba, Porto Rico, and other islands, where many varieties were raised at an early date.

In ancient Peru, the pineapple was unknown: it is conspicuously absent in Peruvian archeology, not being found in any graves, nor is it represented on pottery vessels. Joseph Acosta, in his "Historia natural y moral de las Indias," states expressly, "It does not grow in Peru, but is carried there from the Andes, and this fruit is neither good nor ripe." Cieza de Leon mentions it only as growing in Cali, a Spanish settlement of Peru, together with a series of plants introduced from Spain and the West Indies. Finally, there is a formal, though somewhat belated, testimony to the effect that the plant was introduced into Peru from Brazil. G. Piso, in 1658, asserts that, according to a statement of trustworthy old natives of Brazil, the fruit was transmitted from Brazil to Peru. The oral testimony of old men, especially with reference to an event that dates back at least a century, is somewhat subject to suspicion, but there is no doubt of the fact that the pineapple was not imported into Peru until after the conquest. It is not amiss to emphasize this point, in view of the information given in the "Treasury of Botany" that the pineapple first became known to Europeans in Peru, where it is called *nanas;* and most of our dictionaries, even the Oxford New English Dictionary, wrongly define *ananas* as a Peruvian word. In fact, this word does not exist in any Peruvian language, or in Spanish. From the reports of Thevet and de Lery it is perfectly obvious that the word originated in Brazil, where it still occurs in Tupi. The Brazil Portuguese term has conquered all vernaculars of Africa, India, Malaysia, and all European languages, with the exception of modern English, pineapple being modeled after Spanish *piña*, while in seventeenth and eighteenth century English *ananas* was still frequent.

The antiquity of the fruit in Mexico and among the Maya does not seem to be well authenticated: Sahagun apparently does not mention it. Geronimo Benzoni, from Milan, who sojourned in Mexico from 1541 to 1555, gives a brief, though colorless, description of the plant in his "History of the New World" of 1578, without special reference to Mexico. Francisco Hernandez, in his "Rerum Medicarum Hispaniae Thesaurus" (Rome, 1651), writes that he found the pineapple (termed by him *Pinea indica*) in the warm regions of Mexico and Haiti, and furnishes a drawing of it under the Aztec name *matzatli*. This comparatively late evidence, however, does not suffice to regard the pineapple of Mexico as pre-Columbian.

According to W. Popenoe, the pineapple was doubtless cultivated by the ancient Maya, and is still grown in several gardens near Copan. This conclusion, however, is inferred from present-day conditions; it is retrospective and remains to be substantiated by more solid data.

In the West Indies the pineapple occurs solely in the state of cultivation or occasionally as a fugitive, while without any doubt this cultivation is of ancient date and was an accomplished fact at the time of the Spanish conquest. The wild congeners of the plant and many other representatives of the genus, however, are to be found in Brazil and Guiana. It is therefore reasonable to regard Brazil as the mother-country of pineapple cultivation: Brazil was one of the great centers of aboriginal agriculture, from which also emanated the sweet potato, the cassava, the peanut, Capsicum, several species of beans, and others.

The pineapple was introduced to Bermuda from the West Indies under Captain Tucker (1616–19), third governor of the islands.

An early attempt to acclimatize the fruit in Virginia proved a failure. W. Strachey[3] reports in 1614, "The rootes of the delicious Indian pina, sett in a sandy place, thrived, and contynued life, without respect had of yt, untill the cold wynter and the weedes choaked yt; yet is this fruict said to be daintye, nice, and of that nature, that noe art or industry hath be found out hitherto that could preserve yt in any clymate but in the West Indie Islands only." In December, 1621, pines were shipped to Virginia from Bermuda, together with potatoes, sugar-canes and plantains, and began to prosper well, as we learn from a contemporaneous document published by Purchas. I have not yet been able to trace any subsequent documents, and do not know what the outcome of these initial experiments was. Pineapple cultivation in the south of Florida is a recent event, not earlier than 1886.

In English literature the pineapple is first mentioned in 1568 in the "New Found Worlde or Antarctike," which is a translation of André Thevet's "Singularitez de la France antarctique autrement nommee Amerique"; for the second time in 1580 in John Frampton's "Joyfull Newes out of the New Found World," which is based on the Spanish work of Nicoloso Monardes. John Gerard, in his famous "Herball" of 1597, is not yet familiar with the plant; but in the second edition of 1633, prepared by Thomas Johnson, it is described and accompanied by a woodcut of the fruit. John Parkinson, in his "Theatrum Botanicum," published in 1640, is more copious in his description, more accurate in his illustration and gives more information concerning its history. He states that the Anana or West-Indian delicious pine was first brought from Santa Cruz in Brazil, where it grows wild, and was thence introduced to the East and West Indies, being not a native of either; in Brazil, it is called by the natives *nana* and *anana*, but by the Spaniards and Portuguese *piñas*.

In 1657 Oliver Cromwell received four pineapples brought back by an embassy

[3] "Historie of Travaile into Virginia," p. 31. Hakluyt Society.

returning from China. This event is alluded to by John Evelyn in his "Diary," where under date of August 9, 1661, he writes, "I first saw the famous Queen pine brought from Barbadoes and presented to his Majesty Charles II; but the first that were ever seen in England were those sent to Cromwell four years since."

Under date of August 19, 1668, we read the following in Evelyn's "Diary":

> I saw the magnificent entry of the French Ambassador Colbert, received in the Banqueting-house. . . . . Standing by his Majesty at dinner in the presence, there was of that rare fruit called the King-pine, growing in Barbadoes and the West Indies; the first of them I had ever seen. His Majesty, having cut it up, was pleased to give me a piece off his own plate to taste of; but, in my opinion, it falls short of those ravishing varieties of deliciousness described in Captain Ligon's History, and others; but possibly it might, or certainly was, much impaired in coming so far; it has yet a grateful acidity, but tastes more like the quince and melon than of any other fruit he mentions.

It would follow from this notice that the pineapples of King Charles were imported from America.

Lady Mary Wortley Montagu (1689–1762) writes in one of her letters, dated Blankenburg, December 17, 1716:

> I was particularly surprised at the vast number of orange-trees, much larger than any I have ever seen in England, though this climate is certainly colder. But I had more reason to wonder that night at the king's table (in Hanover). There was brought to him from a gentleman of this country, two large baskets full of ripe oranges and lemons of different sorts, many of which were quite new to me; and, what I thought worth all the rest, two ripe ananas, which, to my taste, are a fruit perfectly delicious. You know they are naturally the growth of Brazil, and I could not imagine how they could come there but by enchantment. Upon enquiry, I learnt that they have brought their stoves to such perfection, they lengthen the summer as long as they please, giving to every plant the degree of heat it would receive from the sun in its native soil. The effect is very near the same; I am surprised we do not practise in England so useful an invention.

As to the source from which the two pineapples served at the electoral table of Hanover had come, we receive a bit of information from the philosopher Leibnitz, who wrote about 1714 as follows:

> All the travelers in the world would not have given us by their relations what we are indebted for to a gentleman of this country who cultivates with success the ananas, three leagues from Hanover, almost on the banks of the Weser, and who has found out the method of multiplying them, so that we may perhaps have them one day as plentiful, of our own growth, as the oranges of Portugal, though there will, in all appearance, be some deficiency in the taste.

This remark refers to Otto von Münchhausen, who in his gardens not far from Hameln erected large buildings for the express purpose of raising pineapples, in the beginning of the eighteenth century.

Lady Montagu, however, was not quite correct in her surmise that so useful an invention was not practiced in England. In fact, pineapples were produced in English hothouses several years before they were established in Germany. This was first accomplished in the gardens of Amsterdam by means of plants introduced from Java, Surinam and Curaçao; and from Holland the art spread to England, France and Germany. During the eighteenth century, the culture of pineapples in hothouses was understood by almost every gardener in England. There are a number of very interesting treatises on the subject written by practical English gardeners, such as Speechly, Abercrombie and Baldwin, with full details bearing on the numerous varieties then produced. The Fitzwilliam Museum at Cambridge preserves a landscape by Netcher in which a pineapple is introduced. It is there stated to have been the first that ever fruited in England, and that it was produced in the gardens of Sir Matthews Decker, at Richmond in Surrey (in 1712).

Soon after the discovery of St. Helena in 1502 the Portuguese transplanted there the pineapple, together with many other fruits, vegetables, cereals and cattle, all of which flourished abundantly on the island after a short period. There is no doubt that the Portuguese were active in diffusing the plant along the west and east coasts of Africa and on Madagascar at an early date; through their agency the term *ananaz, nanasi* or *manasi* penetrated into the native languages of Africa. In the account of the Dutch expedition to Guinea in 1602, a lengthy description of the fruit and its usefulness is given; its cultivation on the part of the inhabitants of Guinea then was an accomplished fact. M. Hemmersam, in the account of his voyage to the coast of Guinea (1639–45), writes that "the Moors consume quantities of ananas, as they call this fruit, which is like an artichoke; they also cook it, mixing it with palm-oil which they use for all their food in the place of fat; it belongs to the best fruit of this country and makes an excellent dish when sliced and soaked in Spanish wine, but too much of it entails malady." Etienne de Flacourt, who was governor of Fort-Dauphin on Madagascar from 1648 to 1655, has left interesting observations on pineapple culture in the island, which are inserted in his "Histoire de la grande isle Madagascar" (1661). At the same time, the plant was cultivated on Réunion (Bourbon) and Mauritius, as we learn from Du Quesne and François Leguat. The Mauritius variety is of superior quality, and was also introduced into India. About 1660 the Dutch planted pineapples at the Cape of Good Hope. These had been brought over from Java by Georg Meister, a professional gardener, who himself tells this story in his very interesting diary.

About 1550 or shortly afterwards, the pineapple was introduced into southern India by Garcia da Orta, a Portuguese physician, author of the famous "Coloquios dos simples e drogas da India" (Goa, 1563), who mentions it as an introduction from Brazil; but more than that —it is unmistakably described in the Ain-i Akbari ("History of the Emperor Akbar"), written in 1597 by Abul Fazl Allami. The Emperor Jahangir tells in his "Memoirs," under the year 1616, how a large tray of fruit was brought before him, among these pineapples from the seaports of the Frangi; that is, Franks, Portuguese. "Some plants of this fruit," the emperor goes on to say, "were placed in my private gardens at Agra, and after some time they produced several thousands of that fruit."

Huygen van Linschoten, in 1596, describes ananas as one of the best fruits and of best taste in all India, and adds, "But it is not a proper fruit of India itself, but a strange fruit, for it was first brought by the Portingalles out of Brasille, so that at the first it was sold for a noveltie, at a pardão apiece, and sometimes more, but now there are so many grown in the country that they are very good cheap." Nearly all travelers to India in the seventeenth century describe the fruit and eulogize it. François Pyrard, whose peregrinations in the East extended from 1601 to 1611, even enumerates it among the productions of Nepal. It likewise spread at the same time to Bengal, where Nicolao Manucci reported a more extensive cultivation than anywhere else in India, and where the fruit was also preserved; in the latter part of the seventeenth century it advanced to Assam, Burma and Siam. In Assam, more particularly the Khasia hills, as well as in the forests of Ceylon, it has escaped from cultivation, and has thus become seemingly spontaneous, but even these semi-wild plants are still designated by the natives with the Brazil-Portuguese terms *ananas*, *a*[illegible], etc., which, with slight modifications, are echoed in all languages of India and Malaysia, with the sole exception of Tagalog, which has adopted Spanish *piña*.

Before the close of the sixteenth century, the pineapple had reached Malaka and Java. The Malaka product soon achieved a reputation all over the East. In 1637 Peter Mundy, a prominent English traveler, found pineapples planted, while passing through the strait of Singapore, and even on outlying islands as Pulo Tinggi off the east coast of Johor. Engelbert Kaempfer encountered them in 1690 on his voyage from Batavia to Siam on the island Puli Tumon near the east coast of Malaka.

In China we meet the fruit from the beginning of the seventeenth century. It is not mentioned in Chinese records at an earlier date. It is not correct, however, as has been asserted, that the plant reached China through the Philippines. Chinese authors are reticent as to this point, and it is more probable that the Portuguese brought it from Malaka to Macao, where it is still cultivated and whence it was disseminated into the province of Kwang-tung and the island of Hai-nan. From Hai-nan it was transmitted to Fu-kien and Formosa, where it was noticed as early as 1650 by John Struys. On the other hand, it also appears to have entered Yün-nan Province from Burma. Chinese names for it like "foreign jackfruit," "royal pear" or "phoenix pear" are likewise an index of a recent introduction. Pineapple cultivation and cannery have reached a great economic importance in China and Indo-China, the fruit being consumed by all classes of people in enormous quantity, and the fibers of the leaf being utilized for textiles. As to Japan, the fruit was first introduced into Nagasaki by the Hollanders as late as 1845.

191

关于“烈酒”一词可能源自东方

JOURNAL

OF THE

AMERICAN ORIENTAL SOCIETY

EDITED BY

MAX L. MARGOLIS
Dropsie College

FRANKLIN EDGERTON
Yale University

W. NORMAN BROWN
University of Pennsylvania

VOLUME 49

PUBLISHED BY THE AMERICAN ORIENTAL SOCIETY
ADDRESS, CARE OF
YALE UNIVERSITY PRESS
NEW HAVEN, CONNECTICUT, U. S. A.
1929

On the Possible Oriental Origin of our Word Booze

There is an old Persian-Turkish word *bōza* or *būza* denoting an alcoholic beverage made from millet, barley, or rice, also translated "beer," which is widely distributed over Asia, Europe, and North Africa. Whether in its origin it is Persian or Turkish I am not sure. Johnson-Richardson-Steingass, in their *Persian-English Dictionary,* list it as Persian, but Vambery regards it as a very ancient Turkish word, since it occurs in the Uigur Kudatku Bilik of the eleventh century, and he may be right. According to Radloff's *Wörterbuch der Türk-Dialecte* it occurs in Kazan, Jagatai, and Tobol ("beverage from millet or barley"); and Shaw has registered it in his *Vocabulary of the Turkī Language* as *boza,* "a weak intoxicating liquor made from various grains (at Khokand)." The Mongols have adopted it as a loan-word either from Persian or Turkish in the form *bodzo,* and Kowalewski, in his *Mongol Dictionary,* defines it as "an alcoholic beverage made from barley-meal or milk." The word is well known to the Osmans and is recorded as early as 1674 in the *Epistola de moribus ac institutis Turcarum,* written by Th. Smith of the College Maria Magdalena of Oxford. Speaking of the beverages of the Turks, he goes on to say, "They also have other liquors rather peculiar to them of which I shall only mention Bozza made from millet," etc. Lane (*Manners and Customs of the Modern Egyptians*) mentions *boozeh* or *boozah* as an intoxicating liquor commonly drunk by the boatmen of the Nile and by other persons of the lower orders.

What this kind of booze was is well described by J. L. Burckhardt in his *Travels in Nubia* (London, 1822), p. 201: "Few traders pass through Berber without taking a mistress, if it be only for a fortnight. Drunkenness is the constant companion of this debauchery, and it would seem as if the men in these countries had no other objects in life. The intoxicating liquor which they drink is called *buza.* Strongly leavened bread made from Dhurra [Sorghum] is broken into crumbs, and mixed with water, and the mixture is kept for several hours over a slow fire. Being then removed, water is poured over it, and it is left for two nights to ferment. This liquor, according to its greater or smaller degree of fermentation, takes the names of merin, buza, or om belbel, the mother of nightingales, so called because it makes the drunkard sing. At the beginning of the sitting, some roasted meat, strongly peppered, is generally circulated, but the buza itself (they say) is sufficiently nourishing, and indeed the common sort looks more like soup or porridge than a liquor to be taken at a draught." In another passage he writes, "During the fortnight I remained at Berber I heard of half a dozen quarrels occurring in drinking parties, all of which finished in knife or sword wounds. Nobody goes to a Buza but without taking his sword with him, and the girls are often the first sufferers in the affray." It is evident that in the latter case *buza* has the meaning "drinking-bout."

M. Holderness (*Notes relating to the Manners and Customs of the Crim Tatars,* London, 1821, p. 59), writes, "Another, and I believe the only strong liquor which they are allowed is called booza; it is made either from rice or millet, and with this, it is said, they occasionally get much intoxicated."

From Osmanli our word has migrated into all Slavic languages, also into Rumanian, Hungarian, Albanian, and Neo-Greek. The Turkish origin of the Slavic series was first recognized by F. Miklosich, the eminent Slavic philologist, and has been confirmed by Berneker in his *Slavisches etymologisches Wörterbuch.* Russian *buza* denotes "a beverage made from wheat-flour or oat-meal soaked in water," or according to Pawlowski "a beverage made from buckwheat flour or oatmeal, also cider." In Serbo-Croatian *buza* applies to a drink made from maize flour or the sap of a birch. Czech and Polish have *buza,* Bulgarian *boza.* In Rumanian *boza, bouza,* or *bozan* signifies "a drink made from millet"; and *bozán,* "a kind of beer." These words are classified among the Turkish loans in Rumanian by A. de Cihac, *Dictionnaire d'étymologie dacoromane,* p. 551. In Albanian we meet *bózĕ* as "a beverage made from pea-flour" and *bozadží,* "one engaged in making boza, a booze-maker." In Magyar also, *boza* has been recognized as a Turkish loan-word. The Neo-Greek form is μπόζας or μπουζᾶς, which likewise denotes a millet beer.

Finally the French have adopted this word from Turkish as *bouza* or *bosan* ("Turkish millet beer"), and *buza* is also found in Spanish and Portuguese.

The *Oxford English Dictionary* has also registered the word *buza* in five different spellings (*booza, boza, bosa, bonza, boosa*) as derived from Turkish *boza* with the definition given by Redhouse "a kind of thick white drink made of millet fermented" and with a quotation from Blount's *Glossogr.* (1656): "Boza, a drink in Turky made of seed, much like new mustard, and is very heady."

In view of this situation and the wide distribution of the Turkish word over Europe it occurred to me whether our word *booze* might not be connected with this series. As is well known, many of our culture words have been claimed as European which at closer range have turned out to be of Oriental origin. In English, it seems to me, we must distinguish between the verb *to booze* and the noun *booze* in the sense of strong drink. The verb *to booze* is properly *to bouse,* which is connected by our English lexicographers with German *bausen,* Middle Low German *busen,* akin to *bauschen* ("to bulge, swell up, to revel") and *baus* ("abundance"). This verb, of course, is Germanic, though not traceable to an earlier date than the thirteenth century; but as to the noun *booze* a contamination at least with the Turkish word seems possible.

Whatever the relation of *buza* to *booze* may be, the coincidence itself is suggestive: there is *booze* East and West. There is also the notable semasiological coincidence that both the Oriental *buza* and English *booze* have assumed the double significance: a drink and a drinking bout. The *Oxford Dictionary* defines *booze, boose* 1. a drink, a draught; 2. drinking, a drinking bout.

BERTHOLD LAUFER.

Field Museum, Chicago.

The Feminine Singulars of the Egyptian Demonstrative Pronouns

The following are the forms[1] of one of the Egyptian pronouns meaning *this:*

m. s.—*pn*
f. s.—*tn*
m. p.—*y*[2]*pn*
f. p.—*y*[2]*ptn*

[1] All the forms of the Egyptian pronoun used in this article are taken from section 57, page 29 of *Ägyptische Grammatik* von *Günther Roeder,* zweite Auflage, 1926.

[2] *y* is here used where Roeder uses *j.*

192

中国学生的使命

Chinese Social and Political Science Review

Volume XIII

1929

Published Quarterly by
The Chinese Social & Political Science Association,
Peiping, (Peking) China.

The Chinese Social and Political Science Review

Vol. XIII. July, 1929. No. 3

CONTENTS

Page

1. Unemployment Among Intellectual Workers in China. By L. K. T'ao. 251
2. The Foundations of British Trade in China. By G. E. Hubbard.. 262
3. Mission of Chinese Students. By Dr. Berthold Laufer.. 285
4. Some Thoughts Concerning Economics in the Development of China. By Dr. J.A.L. Waddell 290
5. Some Types of Chinese Historical Thought. By Chang Hsin-hai, Ph. D. (Harvard).............. 321
6. Auditor's Statement of Account for 1928.............. 341
7. Editorial Notes.. 347
8. Public Documents Supplement (separately paged and indexed).. 59

Exchange of Notes Concerning the Revision of Sino-Japanese Commercial Treaty.—Sino-Japanese Settlement of Nanking and Hankow Incidents.—China's Identic Notes to Six Powers Concerning the Abolition of Extraterritoriality.—China's Identic Notes to Six Powers Concerning the Rendition of the Shanghai Provisional Court.—Manifesto of the Second Plenary Conference of the Kuomintang Central Executive Committee, June 18, 1929.

Mission of Chinese Students

By Dr. Berthold Laufer

On February 22, 1929, Washington's Birthday, the American Friends of China, Chicago, gave a reception and tea for the Chinese students of the University of Chicago and Northwesten University at the residence of Mrs. Chauncey B. Borland, 2450 Lake View Avenue. About sixty students and forty members of the Friends of China were present. Mrs. George T. Smith, who lives at the same address, opened her residence to the visitors and gave them an opportunity to view her renowned collection of Chinese jades, porcelains, paintings, and fabrics. After tea Dr. Berthold Laufer addressed the students as follows:

"This afternoon the American Friends of China entertain our young friends from beyond the sea, and we extend to all of you a most hearty warm welcome in our midst. I am not going to deliver official address, shooting oratorical fire-crackers and assuring you of our good will and friendship. You know anyhow that you have all this to the fullest extent without my dilating on this point. So I am not speaking to you as secretary of this organization; but as one who devoted thirty-five years of his life to a study of Chinese civilization and who has derived from it much enjoyment and inspiration, and who is deeply interested in China's welfare, I wish to have a heart-to-heart talk with you and from my point of view to offer you a bit of helpful advice that I hope may benefit you in your future service to your great country and to our country.

"First of all, I have a message of some significance to convey to you. I have just returned from Washington where a conference was held for two days in the interest of the promotion of Chinese studies in this country. A committee was recently appointed for this purpose by the American Council of Learned Societies in Washington. The principal objects of this committee, of which I have the honor to be chairman, are: to disseminate Chinese cultural influence throughout this country, to encourage, foster and promote a serious and profound study of the Chinese language and literature at our universities, to provide scholarships and fellowships for the benefit of those willing to study Chinese seriously, to urge upon American universities the foundation of Chinese departments, to organize a Chinese Research Institute in Washington, to prepare surveys, catalogues, bibliographies, etc., and many other projects have been discussed and recommended at these conferences.

"This new movement will show you that our interest in China is not merely superficial or exclusively political and economic, nor merely platonic and emotional, but is matter-of-fact scientific, solid and intense. We want to buckle down to work and are planning to do things which will really advance the world's knowledge of China.

"But more than that is behind these constructive efforts. We hold the conviction that knowledge of ancient China is part and parcel of a humanistic education, on a par with our old classical curriculum based on the study of ancient Greece and Rome. What was formerly called humanism was narrowly restricted to the civilizations of the Mediterranean region. Our so-called history of the world, as conceived by our old time historians, included merely Greece and Rome and the European nations of medieval and modern times, while the countries of the Indian and Pacific oceans were excluded. At present a new humanism, more comprehensive, more broad-minded, is in the process of formation. In this new humanism the great civilizations of the Orient, to which we are so deeply indebted and from which the foundation of our own culture is largely borrowed, come to the fore and take their place, perhaps even the foremost rank. This is no longer the era of the Mediterranean, but of the Pacific humanism.

"We have come to the conclusion and adopted the platform that a truly humanistic and liberal education is no longer possible without a thorough knowledge of the great achievements of the Chinese in literature and philosophy, in art, science and invention. This is the new gospel of humanism we preach. We advocate the study of sinology—i.e. the scientific study of ancient China —because we are firmly convinced that such a study is of paramount educational and cultural value not only to our country, but to mankind at large, that it has a tendency to broaden our minds, to widen our horizon, to deepen our ideals, to contribute to the progress of a higher learning and to the discovery of a new and beautiful world that is still unknown to us.

"Now, any young friends, do not think for a moment that I am telling you this just to make you feel good or to be merely nice and polite to you. I am telling you this because I want to set you to work, and I am in dead earnest about it. My friends, it is not necessary to live, but it is necesssary to work and to work hard and dig deep and to navigate the sea of knowledge. I want to urge you to take your place in this new movement which I have just outlined, and to contribute your share toward bringing it about. You must not only take from America, but also give America in return. Of course, you have come to our country primarily for the purpose of studying our arts and sciences, to acquaint yourself with our language, our mode of thinking, our social customs, to learn what is best in our civilization and to transplant at a later date these seeds into the fertile soil of the beautiful land of Han. This as it should be, this certainly is your supreme duty. But, on the other hand, I would admonish you, do not become too painfully progressive, remain faithful to the ideals and traditions of your forefathers. This means, of course, that you should by all means keep up the study of your language and literature. This, I know, is more easily suggested than done. From years of hard toil and struggle I have learned what an arduous task it is to master all the intricacies of your literary style. The study of ancient Chinese makes heavy demands on our brain power, and the studies which you are obliged to pursue at the university absorb most of your time and energy. But where there is a will there is a way, and there surely is a way out of this dilemna. The Newberry Library of Chicago offers you a large collection of Chinese books old and new for study.

"But now we come to the most important phase of the question. Sinology is a hard science, 'deep stuff', as our boys say, demanding absolute concentration and a high degree of self-denial. For this reason, we can only hope to produce a small number of sinologists among Americans. Just for this reason we appeal to you, just at this point you can render us the greatest possible assistance. You are our greatest hope, our greatest asset for producing good sinological work. We suggest that some of you or as many of you as desire to enlist in this good cause will devote their energy to the study of sinology entirely and equip yourself with our scientific methods. Later on, when you return to your country, you will be qualified to make first-hand contributions to the scienific knowledge of China that will benefit the world. What we are mostly in need of at present are good and accurate translations of interesting Chinese books (and there are many hundreds of these), well interpreted for the understanding of Western scholars. Some of you should consider to receive a training in phonetics and linguistics to be able to make exact records of Chinese dialects and folklore or contributions to comparative philology. Chinese civilization is as vast as the Pacific, and what we know of it compares in extent to San Francisco Bay. We have not yet crossed the ocean and want to enlist you as mariners and pilots to guide us safely across. Hundreds of problems remain to be studied in this field. There are hundreds of languages spoken by the aboriginal tribes of southern China of which we are woefully ignorant and which are of the utmost importance for the history of the Chinese language. Others of you should enroll in our departments of anthropology, archaeology, or art, and apply the methods evolved by us to fruitful and productive investigations in China. To cite but one example—Harvard University Press has just published a remarkable book by Dr. Li Chi, called "The Formation of the Chinese People." Dr. Li was trained in the Department of Anthropology of Harvard, crowning his career with a Ph. D. He is not only the first Chinese scholar who wrote a book on the racial history of the Chinese from a modern anthropological standpoint, but he is also the first scholar who produced a good and intelligent book on the subject that is bound to be of great service to science. This, I hope, will be encouraging to you and serve as an example worthy of imitation.

"I have been informed by Dr. Kuo, head of the China Institute in New York, who attended our Washington conference, that the National Government has founded an Institute of Scientific Research in Nanking with branches in all provinces, so that there will be good prospects for scientific research in China along the lines I have indicated. My advice to you therefore is briefly this: Honor and love America, but do not forget or neglect the old love by the homely fireside that is waiting for you, continue to love and study China, its glorious, venerable and noble civilization. In this manner you will serve best the interests of your country and the world."

193

麦吉尔大学葛思德中文图书馆致辞

THE GEST CHINESE RESEARCH LIBRARY

at McGILL UNIVERSITY

MONTREAL

by

BERTHOLD LAUFER

THE GEST CHINESE RESEARCH LIBRARY, McGILL UNIVERSITY

FOREWORD

by

GENERAL SIR ARTHUR W. CURRIE, G.C.M.G., K.C.B., LL.D.

Principal of McGill University
Montreal

DR. BERTHOLD LAUFER of the Field Museum of Natural History at Chicago is an erudite student of the Orient. His expeditions to China and Tibet, his love of anthropological research, the breadth and depth of his culture have made him a world-known figure, and when he writes of the Gest Chinese Library, he speaks with authority.

The institution of this remarkable collection coincided with the awakening in Canada of a new interest in China and things Chinese. No change caused by the Great War is more important than that which has affected Canada and China. Canada has acquired new weight in the councils of the nations and in the Empire of which she forms a part; China, freed from the deadening influence which accompanied the Manchu rule, is evolving a new national structure and creating new relationships. Both nations come to the forum of world politics with distinctive viewpoints. They are neighbours, joined rather than separated by the Pacific, and their common interests are yearly increasing.

It is recognized that the influence of education tends to create mutual understanding between nations. The great inheritance of the world's literature and culture is the common property of every country. The man who knows something of the history, the environment and the philosophy of another people tends to look upon that people from a friendly point of view, and in the minds of University students national barriers are breaking down.

It was with these thoughts in mind that we at McGill undertook to develop studies bearing on China; when at the same time we were given the opportunity of adding to our library a collection of Chinese classics, we accepted with enthusiasm, and so in McGill University the books of China have taken their place beside the literature of the western world. We look forward to a time when Chinese as well as western students will make full use of the Gest collection, and we believe that it will prove a real factor in the drawing together of East and West.

THE GEST CHINESE RESEARCH LIBRARY *at* McGILL UNIVERSITY

BY BERTHOLD LAUFER
Field Museum of Natural History, Chicago, Ill.

NEXT in number of volumes to the Chinese Division of the Library of Congress the important collection of Chinese literature made by Mr. Guion M. Gest of New York is the most outstanding and most comprehensive and at the same time outranks others in number of rare works in America. It is justly characterized as a research library, as it enables the student to carry on serious and fruitful investigations in almost any department of Chinese civilization as history, literature, religion, and science. All students interested or actively engaged in Chinese research owe a debt of profound gratitude to Mr. Gest for his unselfish devotion to an ideal which he has pursued for many years of his life. Fascinated by Buddhism since the early days of his boyhood, he has maintained a steadily growing and intelligent interest in the civilizations of the Far East and has created a lasting monument which will be still more profoundly appreciated by future generations as the spiritual bonds connecting us with the Orient grow more intense.

It was my good fortune to spend two days in the Gest Library at Montreal on July 11 and 12 of this year, and I deem it my pleasant duty to record briefly my impressions of what I saw and learned there.

In bringing together a vast collection of Chinese books and depositing it under the name "The Gest Chinese Research Library" at McGill University, Montreal, Mr. Guion M. Gest was actuated by two principal motives: (1) to promote through the study of Chinese literature a better understanding of China and a closer relationship and amity between China and the Western World (it being Mr. Gest's conviction that an *entente cordiale* between nations can be far better accomplished by education, *i.e.*, through a knowledge of Chinese literature and civilization, of which comparatively little is known and translated, than by any other means), and (2) as implied by its name, the object of the Gest Library is to place its books at the disposal of scholars for research work, especially in co-operation with the faculties of McGill and other universities, as well as with sinologists in the United States, Europe, China and Japan. Research work of this character has already been done in medicine, pharmacology, astronomy, etc. A plan is on foot to found a chair of Chinese language and literature at McGill, which was chosen by Mr. Guion M. Gest because of the excellent and profound research work accomplished in the different faculties and departments of this university.

The Gest collection is housed in the attractive library building of McGill University, where it occupies a large room on the second floor. The stacks are of steel, arranged in two stories, the upper one being entirely devoted to the great cyclopedia *Tu shu tsi ch'eng*. The arrangement of the books is so systematic and splendid that any book can be traced at a moment's notice. The reading room is airy and spacious and well equipped. Excellent photographs taken by Mr. Gest himself in the Orient adorn the walls. The floor is laid with Chinese rugs, and Chinese antiquities in a glass cabinet as well as a reproduction in stone of the famous Nestorian tablet lend the room an intimate atmosphere. The library has a special exhibition room where at the time of my visit a most interesting exhibit of Japanese color-prints and Chinese paintings and manuscripts was shown, including a number of very beautiful Tibetan manuscripts in gold and silver writing from Mr. Gest's collection. It may be mentioned also that in the Art Museum of Montreal, Mr. Gest has a very interesting collection illustrating the Lamaist cult of Tibet.

The staff of the Gest Library is formed by Dr. Gerhard R. Lomer, university librarian, and Dr. Robert de Resillac-Roese, who has immediate charge of the cataloguing and pursues his task with a rare zeal and enthusiasm. He is assisted in his work by Miss Nancy Lee Swann, a good Chinese scholar, and by a scholar of Chinese nationality, who at present is Mr. C. B. Kwei. The library's collaborator in China is Mr. I. V. Gillis, who resides in Peking and who has extraordinary ability as a book-hunter. He was formerly naval attaché to the U.S. Legation in Peking.

The Gest Library was informally opened in 1926, on the day of Chinese New Year, February 13th, with assets of 304 large works, consisting of 10,750 volumes. These had been selected by Mr. Ch'en Pao-ch'en, noted statesman and scholar, tutor of the last Manchu emperor, Hsüan T'ung (1908-11). By June 30th, 1929, the collection had increased to a total of 2,054 works, consisting of 50,640 volumes. All these works have been identified, catalogued, doubly card-indexed, labelled, and placed on the shelves in their proper classification sequence. They are all bound in Chinese cases (*t'ao*). For these 2,054 works, 38 catalogues have been typewritten in triplicate and, correspondingly, two sets of index cards, namely: title cards 4,500; authors' cards 4,170; a total of 8,670 cards. Aside from these 50,640 volumes there are 6,000 not yet identified. Another consignment of 5,305 volumes arrived on July 13th of this year at Vancouver, and approximately 15,000 volumes from a famous private library in China are expected the latter part of this summer. Valuable works will be continually added. Beginning from Cat. No. 305 all works were selected by Mr. I. V. Gillis. Excepting a few modern works bought from Peking and Shanghai bookstores, all the works in the collection were formerly in the possession of Manchu princes, well-known statesmen, or bibliophiles.

All books are bound in Chinese style, *i.e.*, cloth cases held together by bone slips (*ku tsien*). The back of every *t'ao* is provided with a label, each label giving the following items: name of library; classification number; accession or running catalogue number; romanization or transliteration of title (in Thomas Wade's system); title in Chinese characters, written with a brush; short contents of work; romanization of author's name; author's name in Chinese; if rare or original edition, thus specified; number of volumes, if more than one.

Works are classified (like those of the Library of Congress, Washington, the Bibliothèque Nationale, Paris, and others) according to the so-called "Four Treasure" system established toward the end of the third century A.D., by Sun Hu, keeper of the Imperial Library of the Wei dynasty, and which was followed by bibliographers and imperial librarians of the subsequent dynasties. In the "Four Treasure" system, all Chinese works are classified under four principal categories, (1) canonical literature (*king*); (2) history (*shi*); (3) philosophers (*tse*); (4) belles-lettres (*tsi*),—designated by the capital letters A, B, C, D.

White index cards are used for original works; red cards for works contained in the *ts'ung shu*, or collections of reprints.

At present the Library has two sets of index cards: title cards and authors' cards. In course of time two sets will be added, namely, title and authors' cards in Chinese, arranged according to the number of strokes in the first character. In addition to this titles of all works in alphabetical order will be written on cards.

On the title cards the following information is given. In the left-hand upper corner: classification number and call number; title of work in Chinese (written with a brush) running vertically parallel to left-hand side of card; romanization of title, running from left to right at top of card; underneath, concise description of contents of work, name or names of authors in Chinese and romanization, of commentators, writers of prefaces, postscripts; date of first publication; date of issue of edition in question; number of chapters; kind of paper; size in millimeters; in left-hand lower corner, accession or running catalogue-number; to the right on same line: number of *t'ao* and volumes.

Authors' cards are written out in the same manner, with contents of work, etc.

The Library is at present well equipped for research work. It is especially strong in dictionaries, historical works, catalogues, encyclopedias, and medicine.

Of rare and old editions the following are deserving of particular mention:

Han wen kung kung k'ao i, collection of works in prose and poetry by Han Yü (A.D. 768-824), compiled by Chu Hi (A.D. 1130-1200), in 8 volumes, issued in the latter part of the 13th century.

T'ung chi, history of China beginning from the mythical emperor Fu Hi down to the end of the T'ang dynasty (A.D. 906), by Cheng Tsiao (A.D. 1108-66), printed in A.D. 1322. Only 50 copies were printed. This edition is rebound in silk covers and interleaved, in 240 volumes.

Chi-ta chung siu Süan-ho Po ku t'u lu, an edition of the Po ku t'u lu, an illustrated catalogue of ancient bronzes, printed in the Chi-ta period (A.D. 1308-11), 30 volumes of 33 x 25 cm. Some pages are lost. Most illustrations are spread over two pages, while in the Ming and Ts'ing editions they are reduced to a single page in octavo. The Library also has a Ming edition of this work printed in 1588.

Nearly 500 *bona fide* Ming editions, printed from wooden blocks. Some of these are:

Tse chi t'ung kien kang mu, Outlines of the Annals of History, compiled by Yin K'i-sin and published A.D. 1473, in 100 volumes.

Tse chi t'ung kien, Annals of History from the beginning of the Fourth Century B.C. down to the end of the Five Dynasties (A.D. 960) by Se-ma Kwang, published A.D. 1544, in 120 volumes.

Li tai kün ch'en kien, Biographies of emperors and high ministers of state, by the emperor King T'ai, published in A.D. 1453, in 32 volumes.

Wen hien t'ung k'ao, Cyclopedia, dealing with political economy, government offices, ceremonies, etc., by Ma Twan-lin, published in A.D. 1521, in 80 volumes.

Tung i pao kien, a general compilation of medicine, of Korean origin, by Hü Tsun, printed in A.D. 1577, in 25 large volumes.

Shi wu ki yüan, Cyclopedia bearing on the origin of things, selected by Kao Ch'eng of the Sung period, compiled by Yen King, published by Li Kwo in A.D. 1472, in 12 volumes.

An-yang tsi, Posthumous collection of poetry and prose works by Han K'i (A.D. 1008-75), published in A.D. 1514, in 16 volumes.

T'ai-po tsi, Collected poems of Li Po (A.D. 705-762), edited by Yang Tse-kien, published in A.D. 1543 (the Kwo Family edition).

Of palace or imperial editions, printed on special paper in only 50 copies (volumes in the imperial yellow color), the Library has 48 works. Among these are:

Shu king t'u shwo, Text and explanation of the Shu king, "Book of History," compiled by order of the Empress Dowager, richly illustrated; issued in A.D. 1905, 16 volumes.

Ta Ts'ing hui tien, Laws and Statutes of the Ts'ing Dynasty, compiled by order of the emperor Kwang Sü, richly illustrated, Ed. of 1887, in 500 volumes.

Ta Ts'ing shi ch'ao sheng sün, Imperial edicts and proclamations of the Ts'ing emperors, from T'ien Ming (1616-27) down to the emperor T'ung Chi (1862-75), printed during the reign of Kwang Sü, in 608 volumes.

K'in ting ts'i sheng fang lio, Chronicles of the suppression of the various rebels during the T'ai-p'ing Rebellion (1851-65), compiled under supervision of Prince Kung, printed in A.D. 1896, in 1,156 volumes.

The original edition of the great cyclopedia *T'u shu tsi ch'eng*, compiled between 1686 and 1726, printed in 1726, consisting of 5,020 volumes, bound in 502 *t'ao*; the first large Chinese work printed with movable copper type (the only other complete cópy outside of China in British Museum).

Two copies of the *Wu Ying tien chü chen pan ts'ung shu*, a collection of 139 works selected by the emperor K'ien Lung, reprinted during 1773-94, the first large Chinese work printed with movable wooden type, in 811 and 600 volumes.

Shi san king, Rubbings from the stone tablets in the Hall of Classics connected with the Confucian Temple, Peking, of the 13 canonical books, the text of which, consisting of more than 800,000 characters, was written during twelve years and completed in A.D. 1740 by Tsiang Heng (1672-1743) and by order of Emperor K'ien Lung cut in stone, with imperial writings; in 208 volumes.

The *Cheng t'ung Tao tsang*, Collection of the works contained in the Taoist Canon, lithographically reprinted on extra white paper, published by the Commercial Press, Shanghai, 1923-25, in 1,200 volumes.

The original manuscript of the *P'ei wen yün fu*, Concordance of terms and phrases, compiled in 1704-11, by a board of 76 scholars under personal supervision of the emperor K'ang Hi, printed first in A.D. 1724 in 160 volumes (also in the possession of the Gest Library), in 105 volumes.

In view of the fact that first editions of older works are no longer obtainable, the so-called *ts'ung-shu* are of fundamental importance for any Chinese library. These are collections of reprints of old books now out of print or not easily accessible separately. The greater part of earlier literature can now be found in the *ts'ung-shu* exclusively. The Gest Library is fortunate in owning fifty-five such *ts'ung-shu*.

One of the greatest treasures of Mr. Gest is an extensive collection of Sutras from a Tripitaka edition which were obtained in a remote part of China. At the time of my sojourn in Montreal (July 11 and 12, 1929) the consignment had just reached Vancouver, B.C., and I had no occasion to see it; but Mr. Gest was good enough to show me photostats of a number of pages. This collection consists of

698 volumes printed under the Sung in A.D. 1246.
1,635 volumes printed under the Yuan (14th century), mostly in A.D. 1306.
876 volumes printed under the Ming (16th century).
2,114 volumes in manuscript, dated A.D. 1600.

Total 5,323 volumes.

This collection will undoubtedly form the most superb Buddhistic library anywhere in existence.

194

1928年北美考古田野作业

American Anthropologist

NEW SERIES

ORGAN OF THE AMERICAN ANTHROPOLOGICAL ASSOCIATION, THE ANTHROPOLOGICAL SOCIETY OF WASHINGTON, AND THE AMERICAN ETHNOLOGICAL SOCIETY OF NEW YORK

ROBERT H. LOWIE, *Editor*, Berkeley, California
FRANK G. SPECK and E. W. GIFFORD, *Associate Editors*

VOLUME 31

MENASHA, WISCONSIN, U. S. A.

PUBLISHED FOR

THE AMERICAN ANTHROPOLOGICAL ASSOCIATION

1929

two localities in the Okanagan Indian area; one on the west side of Okanagan lake near Kelowna, and another on the right side of the Penticton-Keremeos road.

Mr. W. J. Wintemberg spent the field season of 1928 in making an archaeological reconnaissance in three areas; the Richelieu river valley in southwestern Quebec; the Canadian Labrador; and New Brunswick. He discovered a few widely scattered evidences of Algonkian and Iroquoian occupation in the Richelieu valley. Along the north coast of the gulf of St. Lawrence extensive workshop sites, where scrapers, celts, knives, and projectile points were manufactured, were discovered near Blanc Sablon and Bradore, near the Newfoundland-Canadian boundary. Near the latter place, a few fragments of pottery were found, one of them bearing an Algonkian type of decoration apparently produced with a rocking stamp. This is the northernmost recorded occurrence of aboriginal pottery in eastern North America. A few human bones obtained from a grave under an overhanging rock near Blanc Sablon were covered with red paint and are probably remains of Beothuk Indians. Cylindrical beads made from the columella of large ocean shells were found in the same grave. A few evidences of the former presence of Eskimo were discovered, consisting of a number of post-European walled stone graves on an island near St. Augustine, and a rubbed slate point for a man's knife from the mainland near Harrington harbor. Most interesting was the discovery of some Iroquois pottery in two small shell-heaps near Kegashka, about two hundred miles west of Blanc Sablon, and about one hundred and fifty miles east of the Gaspe coast where Cartier met Iroquois in 1534. A few small workshop sites were found near Natashquan, and a small camp and workshop site was examined near Seven Islands; all probably Algonkian. The work in New Brunswick was confined to visiting a site near Fredericton and an extensive camp and workshop site at Redbank on a branch of the Mirimichi river, where the site of a grave, in which the artifacts were covered with red paint, was examined.

Harlan I. Smith,
National Museum of Canada

The archaeological work carried on by Dr. William Duncan Strong, when he accompanied the Rawson-Macmillan Subartic Expedition under the auspices of the Field Museum of Natural History, as anthropologist, included the two summer seasons of

1927 and 1928 spent along the coast and interior of northeastern Labrador, and a brief reconnaissance trip around Frobisher bay in Baffinland during the former season. In July, 1927, a visit was paid to the reputed Norse site on Sculpin island near Nain. On a rocky shallow bay with a northern exposure are located about a dozen round and rectangular stone enclosures with various stone walls or windbreaks. Most of these enclosures had central stone squares or circles within which the soil was very black and greasy. These appeared to be depositories for blubber and occur in the many old tupik rings encountered here and elsewhere along the coast. None of those examined had any traces of charcoal, nor did they seem to have been used as fireplaces. This opinion differs from that expressed by Cadzow (*Indian Notes, Heye Foundation*, 100, January, 1928), who had examined the site shortly before. The stonework is crude and uneven and rather inferior to that of the Eskimo cairn burials in the vicinity. Near the stone enclosures were two characteristic old tupik rings of unevenly distributed stones. As in the case of several of the stone enclosures, their doorways were indicated rather by an absence of stones than any attempt at stone masonry. The fact that this stonework does not resemble that to be seen in photographs of Norse ruins in Greenland, that no typical Norse implements have been reported from the site, that similar circular stone enclosures occur in association with Eskimo cairn burials and summer camp sites at Black island near Hopedale, and that almost identical house types of Eskimo origin were found in Frobisher bay, leads one to conclude that the Sculpin island site represents an Eskimo spring or autumn whaling camp, possibly of the Thule culture, as Mathiassen suggests (AMER. ANTHROP., n. s., 3.: 578).

A hasty reconnaissance of Frobisher bay (August 9-25, 1927) revealed several interesting Eskimo sites. On Fletcher's island occur eight beacons five feet high, two with horizontal slabs suggesting a cross, and associated with them two empty cairns of a burial type. At Koojesse inlet are many stone meat caches, circular tupik rings, and four stone enclosures with two-to three-foot walls similar to those on Sculpin island. On Bishop's island are three well made stone slab houses, meat caches, kayak rests, and a few tupik rings. Digging in the shallow soil of the houses revealed a harpoon point of Thule type and a few other Eskimo artifacts. On Kodlunarn island in Countess of Warwick sound the remains of Sir Martin Frobisher's old settlement of 1577 were examined. The considerable traces of houses, ship's pits, and Caucasian artifacts encountered left no doubt as to the correctness of Hall's conclusions in 1865. At Brewster's point nine old stone and sod iglus of large size were superficially examined. This place, which at present supports a small Hudson's Bay Company post and a village of Nugumiut Eskimo, is a very promising archaeological site.

During the summer of 1928 a canoe trip was made some fifty miles into the interior of Labrador to examine an old camp site discovered by the Naskapi Indians on a lake at the head of Hunt's river (Jack Lane's bay). This camp seems to be that of Stone Age Eskimo and if so carries their occupation of this region back a considerable period. The only other interior site where stone tools had been found by the Indians is a day's trip by canoe up the Adlatuk river (head of Hopedale bay), but this was not visited. Since the Indians knew of no Stone Age sites in the Indian House Lake region, although they manifest considerable interest in such things, and as the known interior sites seem to be Eskimoan, it can be safely stated that the Barren Ground and Davis Inlet Naskapi Indian bands are late-comers to a region long occupied by the Eskimo. The ethnological results of the winter's work with these Indians strongly confirm this opinion first expressed by Lucien Turner in 1894.

The cultural remains revealed by the summer's excavation on the coast during 1928 indicate two main periods of Eskimo occupation of the region between Port Manvers and Hopedale. The first is a Stone Age culture found in small exposed camp sites marked by well chipped chalcedony, quartz, or flint points and blades; heavy ground stone pot fragments, adze blades, gouges, stone ulus, and a marked absence of bone or ivory artifacts. Some small fragments of fossilized bone were found at these sites, but no worked implements. An old native quarry of colorless chalcedony, its lower exposures covered by two feet of moss and soil, was discovered at the head of Jack Lane's bay. Hammerstones and characteristic stone implements were found in the bare wind-eroded exposures near-by. This quarry marks the only occurrence of the mineral known in the region and the site shows evidence of extensive work. The character of the stone ulus, adze blades, and of one steatite charm indicates that these people were Eskimoan, though there still remains a faint possibility that this old coastal culture antedates both Eskimo and Naskapi in the region. Since most of the camp sites had been exposed by wind action and were unassociated with later remains, their antiquity is largely assumed from the character of the artifacts themselves.

The second culture is much later and may be distinguished as that of the Early Mission Period. It is characterized by large rectangular stone, sod, and whalebone iglus (similar to those at Brewster point), well made stone grave and gift cairns on the high ridges behind the villages, and equally well made stone box traps with sliding stone-slab doors associated with the graves. The articles from the iglus and gift caches show much greater use of bone and ivory as well as advanced work in steatite. Practically all of the village sites examined at Hopedale, Spirit island, September harbor, and Nukasujuktok island, revealed considerable evidence of early Caucasian contact. A superficial examination of the Eskimoan artifacts from these sites suggests many Thule culture characteristics, but this must await more detailed comparison.

In conclusion it may be stated that a long Eskimoan occupation of northeastern Labrador is indicated by the old stone culture which contrasts strongly with the bone and ivory working cultures of later times. Between the former and the Early Mission Period there was probably a long interval of which we as yet know nothing. The linking up of these two cultures and the determination of their relation to the old Indian cultures of Newfoundland and the adjacent mainland rests with the future

Berthold Laufer,
Field Museum of Natural History

Alabama. Archaeological researches in Alabama during the past year have been conducted by the members of the Alabama Anthropological Society in Walker, Lawrence, Blount, Montgomery, Elmore, Lowndes, Jackson, Lee, Russell, and Autauga counties. Further finds of urn burials have been made as far north as 32° 28′. This is the most northern reported occurrence of this trait. Natural weathering of the soil has exposed interesting burials at the Thirty Acre Field Mound in Montgomery county, in the territory investigated by Moore in 1899.

Peter A. Brannon,
Alabama Anthropological Society

Alaska. Mr. Henry B. Collins, Jr., conducted archaeological field work on the St. Lawrence and Punuk islands, and on the Seward peninsula in northwestern Alaska. Skeletal material was collected at Golofnin bay and on Sledge island, in Norton sound. The work on the St. Lawrence and Punuk islands revealed extensive pre-Russian

195

乔治·尤摩弗帕勒斯的中国青铜器

NUMBER CCCXV VOLUME LIV JUNE 1929

THE BURLINGTON MAGAZINE

for Connoisseurs Illustrated & Published Monthly

 CONTENTS 

A REDISCOVERED BLAKE—BY LAURENCE BINYON

THE RIDDLE OF THE MAITRE DE FLEMALLE—BY EMILE RENDERS

ENGLISH PRIMITIVES: THE ASCOLI COPE AND LONDON ARTISTS—BY W. R. LETHABY

AMERICAN FURNITURE PROBLEMS—BY HERBERT CESCINSKY

THE *VENUS AND ADONIS* TAPESTRIES AFTER ALBANI—BY H. C. MARILLIER

ENGLISH INFLUENCES ON PARISIAN PAINTING OF ABOUT 1300—BY ROBERT FREYHAN

THE EUMORFOPOULOS CHINESE BRONZES—BY BERTHOLD LAUFER

LIMOGES ENAMELS IN THE BIRMINGHAM GALLERY—BY W. B. HONEY

SHORTER NOTICES: LUCAS VAN LEYDEN'S *NATIVITY* IN THE LOUVRE (KURT ERDMANN); THE NATIONAL ART-COLLECTIONS FUND

THE LITERATURE OF ART

FORTHCOMING SALES—BY A. C. R. CARTER

LONDON: THE BURLINGTON MAGAZINE LTD., 16A ST. JAMES'S STREET, S.W.1.
PARIS: BRENTANO'S, 37 AVENUE DE L'OPERA. FLORENCE: B. SEEBER, 20 VIA TORNABUONI;
NEW YORK: BRENTANO'S INCOR., 1 WEST 47TH STREET; E. WEYHE, 794 LEXINGTON AVENUE.
CHICAGO. WASHINGTON: BRENTANO'S INCOR. AMSTERDAM: J. G. ROBBERS, SINGEL 151-153

PRICE HALF A CROWN NET; ANNUAL SUBSCRIPTION (INCLUDING INDEXES) THIRTY-TWO SHILLINGS.
PRICE IN THE UNITED STATES ONE DOLLAR NET. ANNUAL SUBSCRIPTION NINE DOLLARS NET.

Bronze wine vessel of the *ku* class. Date doubtful. Height, 28.7 cm. (Mr. George Eumorfopoulos)

The Eumorfopoulos Chinese Bronzes

relationship that the illustration bears to it and the high quality of the work all subscribe to such a contention). The representation spread —in changed but recognizable form[8]—in the fourteenth century all over England, Germany, France and Italy; it was taken over by mural decorators in the latter half of the century and eventually developed into the Dance of Death. (Earliest known version 1382 in Minden.)

It owes its origin to the happy union of a simple compiler who freed the attractive legend of all learned and rhetorical dross with a real artist—the Master of the Arundel MS. 83, II.

[8] In the fourteenth century considerable changes are introduced: The Dead stand in front of sarcophagi, the Living are on horseback, and between the two groups stands St. Makarius as interpreter. The latest example of a direct dependence upon the Arundel composition is the diary (hour book) of Bonne de Luxembourg, wife of King Johann—this is dated and covers the period 1332 (marriage) to 1349 (death). (Reproduction in the " Catalogue des livres précieux, manuscrits et imprimés faisant partie de la Bibliothèque de M. Ambroise Firmin Didot," Vol. II, Paris, 1882, pl. 2.) F 320 a: " Ci après commence une moult merveilleuse et horrible exemplaire que len dit des III vis et des III mors." Makarius does not yet appear between the two groups, the Living are on horseback and very unlike our version, though isolated motives are still clearly recognizable. The three Dead, however, and particularly the third, are more closely related to Arundel than to Arsénal! This is naturally only in respect of the movement, the style is that of the mature Jean Pucelle. The text is a mixture of Baudouin (introduction) and dialogue from after Nicholes de Marginal. An Italian example of the same time: Florence, Bibl. Nazionale, Cod. Magliab. II, b. (Reproduction in: Bartoli, I manoscritti della Biblioteca Nazionale di Firenze, I, Firenze, 1897) has much in common with our composition, though here St. Makarius is already present.

THE EUMORFOPOULOS CHINESE BRONZES
BY BERTHOLD LAUFER

HE appearance of the first volume of a new catalogue of the renowned George Eumorfopoulos Collection is an artistic and scientific event of the first order.[1] We have every reason to extend to Mr. Eumorfopoulos and his faithful collaborator, Mr. W. Perceval Yetts, sincere and hearty congratulations for rendering accessible to students of Chinese archæology this very important collection of ancient Chinese bronzes and presenting it in so admirable a volume, which is a work of art in itself. One, who like myself, has spent many years on collecting bronzes in China and many more years on studying them, knows how to appreciate the immense labour performed by both the collector and the cataloguer, who have shunned no effort to make the results of their work attractive, not only to art-students, but also to scholars who here find for the first time a well-laid foundation for a serious study and discussion of Chinese achievements in bronze.

The volume before us is modestly styled a catalogue, but it is far more than that, as Mr. Yetts has availed himself of this opportunity to make this collection the subject of a fundamental piece of research-work which for a long time will serve as a guide into the labyrinth of Chinese bronzes. He has contributed an elaborate study on the archaic inscriptions with which many bronzes are covered, and there are very important ones in the Eumorfopoulos Collection; he has written a most interesting and valuable essay on the technique of bronze-casting, and has rendered a useful service to all of us by conducting a penetrating enquiry into the classification of the types, names and uses of ancient bronze vessels. All these studies, profound and erudite as they are, will be gratefully appreciated and will open new vistas and new paths in the jungle. In his sympathetic foreword Mr. Eumorfopoulos treats us to a document of human interest, telling us of his experiences and thrills as a collector of bronzes, glass and jade, and no one will deny that he has acquitted himself of his task as a collector and patron of art in the most royal manner and has contributed a large share not only to the progress of our knowledge, but also to the intensity of our joy, enthusiasm and inspiration. As the readers of this magazine are chiefly interested in art, I shall confine myself to some observations on the bronzes themselves.

The catalogue contains seventy-five plates representing a total of 172 illustrations, but this number is not identical with that of the bronzes in the collection of which there are 128, as of many specimens two and even three views are given, each being provided with a separate number. In a case like this, I believe, it is preferable to assign but one number to each object and to distinguish the various views by suffixing letters to this single number. Of a total of 128, a respectable figure for a private collection, there are sixty-eight bronze vessels inclusive of lamps, while the remainder are bronze implements and weapons or relics of China's bronze age. One third, i.e., twenty-five, of the plates are coloured, and are really magnificent works of art, the most superb coloured reproductions of Chinese bronzes which to my knowledge have ever been executed. All the plates, plain and coloured ones, have been made with utmost care, and even exhibit the finest details of patina and

[1] " The George Eumorfopoulos Collection: Catalogue of the Chinese and Corean Bronzes, Sculpture, Jades, Jewellery and Miscellaneous Objects." By W. Perceval Yetts. Volume I Bronzes: Ritual and Other Vessels, Weapons, etc. Ernest Benn, Ltd., Bouverie House, London. xii + 90 pages, large folio, + 75 plates. Price £12 12s.

ornamentation. It is no exaggeration to say that these reproductions are so exact and faithful that they are the best possible substitutes for the originals, and permit one to study these intelligently. While I always decline to express an opinion on the date or value of a Chinese bronze merely through the medium of a photograph, I do not hesitate to say that I can clearly visualize and evaluate the Eumorfopoulos bronzes from the plates of this admirable folio, and I think no higher praise could be bestowed on the photographers and artists for the distinguished quality of their work.

Our knowledge of Chinese bronze is still in the initial stages; while much has been done, there is much more that remains to be done. Chinese and Japanese contributions to the subject, on the whole, are praiseworthy, but are rather epigraphical than archæological. Some progress has been made in the classification and chronology of bronzes. Four essential periods can now be easily differentiated: (1) the archaic period or the genuine bronze age, comprising the times of antiquity (Shang, Chou and Ts'in dynasties); (2) the age of transition (Han and Six dynasties); (3) the medieval period (T'ang, Sung and Yüan dynasties); and (4) the modern period (Ming and Ts'ing dynasties). There is no bronze that could not unhesitatingly be assigned to one of these four large divisions; and for a practical museum-man who deals with the general public and is obliged to popularize the subject this division is obvious, efficient and sufficient. It is equally obvious, but a factor often overlooked even by specialists, that Chinese dynastic periods do not necessarily synchronize with cultural or artistic periods; the latter merge into one another and may overlap from dynasty to dynasty. Recourse to the name of a dynasty, in this case, is merely a symbol which may stand for an idea, a type, a form, a definite artistic motif or expression. There are, for instance, certain ideas, many shapes of bronze and pottery vases, many artistic motifs created in the age of the Han and typical of this epoch; and these belong to the culture of the Han whether they fall within the years of the dynasty or a half century after the collapse. In some cases a doubt is possible as to whether a bronze vessel may be Shang or Chou, as our knowledge of what constitutes Shang culture is still somewhat limited; and while there is no doubt of what a Chou bronze and its characteristics are, it is not always certain whether it may be early, middle, or late Chou. Han bronzes are unmistakable both for their technique, forms and decorations, and so are Sung and Ming bronzes. Some degree of uncertainty still prevails as to T'ang bronzes; and even Chinese connoisseurs, to evade the issue, habitually take refuge in the phrase "T'ang-Sung" (corresponding to my "medieval"), but I am convinced that the time is not distant when we shall arrive at making a net distinction between T'ang and Sung bronzes. A single collection, of course, is not conducive to furnishing the entire material for the solution of all chronological problems; we need large series of objects, and only a comparative study of the extant types in all prominent collections will ultimately yield safe results. There are great private collections of bronzes in China which have not yet been published and which remain to be heard from in the future.

In his chronological attributions of the Eumorfopoulos bronzes, Mr. Yetts is very cautious. Thirty pieces are labelled "date doubtful"; twenty-four are qualified by a "probably" preceding Chou or Han; thirty-four bear the prefix "perhaps" before the name of the dynasty; two are dated "perhaps Han or earlier," others "Chou or later," "Han or later," or "perhaps Han or later." Only two pieces (Nos. 45 and 47) are given a clean bill of health with a straight date "Han." This restraint in itself is laudable and worthy of a scholar, but it seems to me that in the present stage of our knowledge we need not be quite so reserved and conservative. I am less timid, and perhaps due to my long familiarity with bronzes have cast aside the "safety first" method and instead have adopted an aggressive policy; while I may err, I have at least the courage of my conviction, nor can we hope to make progress without the hazard of committing errors, and I am convinced that the majority of the Eumorfopoulos bronzes can be dated positively.

The bronzes of "date doubtful" fall into two distinct classes. First, these are bronze vessels, some of them Chou like Nos. 2 and 3, one (No. 8) being Shang, most of them T'ang like Nos. 46-47, 60, 73, 80, 86 (same specimen in Field Museum), 89, and one (No. 62) being Sung. Second, these are bronze implements which partially for lack of definite characteristics cannot perhaps be safely attributed to a certain dynasty, but all of which must date in the Bronze Age in which an early period of spontaneous activity and a later reproductive and retrospective period can be distinguished.

The "perhaps" and "probably" prefixed to Chou and Han may safely be cancelled in the following pieces:—1, 11, 16, 29, 30, 38, 39, 40, 64, 65, 66 (omit also "or later," likewise in 10, 13, 36, 68, and 96), 71, 72, 74-77 (the bear feet not pointed out in the text are characteristically Han), 90, 93-95, 102-105, 114-121, 166.

Not all bronze vessels in museums and private collections which at present are labelled "Chou" on traditional grounds or on purely emotional impressions are really Chou. There is such a pseudo-Chou (fortunately only one) in the Eumorfopoulos Collection. It is impossible for me to accept the date "probably Chou" for the wine-vessel 10 in Plate 7 [PLATE], nor can I agree with its description as "a splendid example of archaic design and craftsmanship," my objection being restricted to the attribute "archaic." I have seen too many of this class (and they are all exactly alike in shape, ornamentation, and technique) to believe that they are archaic. They are archaistic. They represent a highly and fully developed, volitional, rational, and self-conscious art with studied and well-devised proportions and sharp contours; and all bronzes of this class, in my opinion, are products of the T'ang. They are certainly beautiful and the Eumorfopoulos specimen is a gem, but they decidedly lack the spontaneity, directness, and spiritual grandeur of the archaic epoch. No. 13 in Plate 10, a vessel in the shape of a sacrificial ox, assuredly is Sung.

On the other hand, I am ahead of Mr. Yetts in dating a number of pieces. The covered tripod vessel *ting* in Plate 3, in my estimation, is not "perhaps Han," as stated, but, on the contrary, is a very archaic product of the early Chou. The decorative bands of knotted rope designs (somewhat like our hawser-bend) enclosed in rectangles are foreign to Han art, and point to a rather primitive stage of culture and decorative art, having their origin in textiles. This specimen is of documentary value: in their origin these *ting* were simply cooking-vessels of ordinary daily use fashioned from crude pottery; they were set over a fire, supported by a three-legged stand which was first detached, like our camera tripod, and became subsequently attached and fitted to the vessel. When the process of cooking was finished, the pot had to be removed from the fire, and was handled by means of coarse hempen fabrics thrown across it; or, as suggested by the *ting* under consideration, was covered with a sort of network which in the ceremonial bronzes assumed the role of a pattern. The rope design on the loop-handles suggests that the original handles in the clay cooking-pots were plainly of twine. The rings atop the cover of the bronze cauldrons likewise served for the passage of a cord from which the cover was lifted. Note also that the twisted rope band laid around the body of the vessel. For similar reasons I am disposed to think that the globular cauldron 5 in Plate 4 is not "probably Former Han," but is middle or late Chou; the spiral composition which decorates the surfaces of lid and body is not Han, but typically Chou. On the other hand, this example is less archaic than the preceding one, inasmuch as the heavy, obtrusive loop-handles are replaced with elegant rings like those on the cover, and as a greater refinement of form is displayed in the feet.

The marvellous wine-vessel 8 (Plate 5) is a brilliant and venerable example of Shang art; whether actually made under the Shang dynasty or under the early Chou does not matter. Nothing unfortunately is said about the vigorous decoration which would merit a detailed study. The wine-vessel *ku* (No. 9, Plate 6) I would confidently assign to the Shang against "probably Chou," as suggested.

One of the monumental and most extraordinary vessels in the collection is a cup decorated with four stylized elephants arranged in two confronted pairs (Plate 13) and provided with an inscription of great historical importance. This is early Chou, as likewise are the wine-vessel in shape of a double ram (Plates 8-9) and another in shape of an owl (Plate 11). Mr. Eumorfopoulos, in his foreword, remarks justly with reference to the elephant bronze, "One feels that here the Chinese bronze worker is at his best and that behind this bronze there must be a centuries-old tradition of bronze-casting." Mr. Yetts contributes some very interesting observations on the significance of the owl in ancient Chinese lore, though these are far from exhausting the subject to which I hope to devote a monograph some day. Here a few remarks may suffice. While the owl was regarded in ancient China as a bird of evil voice and omen and as unfilial (it was even supposed to kill its mother), there is a song in the *Shi king* in which owls settle on the trees about the Confucian college of the state of Lu in Shantung. These owls, it is said, "eat the fruits of our mulberry trees and salute us with fine notes"; that is to say, the good influence spreading from this college makes even an owl sing delightfully, and the owl in this song alludes to the wild tribes on the river Hwai who should be transformed into civilized people. It is therefore not impossible that the owl represented in a bronze vessel has an allegorical significance and hints at a civilizing influence or a note of wisdom.

The bronze beaker in Plate 32, without any doubt Chou, is equally beautiful for its pleasing, well-balanced proportions, noble simplicity, and purity of form and design. The bronze basin in Plate 47, with its two extraordinary panels of cicada designs, is the most beautiful of its type I have ever seen.

No. 73 is characterized as "a late bronze which does not conform to a classical type; date doubtful." The design of lotus petals brought out in high relief on this bowl stamps it plainly as a T'ang production. For the same reason, the wine-vessel *hu*, No. 33, decorated with beautiful inlaid designs, cannot be Han; for the lid is surmounted by a four-petalled lotus, and the six leaves surrounding the edge of the lid are likewise traceable to the lotus pattern. We must concede that the T'ang era, aside from bronze mirrors which were long ago determined, was also productive in bronze vases of the highest quality; and those in the Eumorfopoulos Collection form one of its most attractive features.

The supernatural bird carrying a wine-vessel on its back (Plate 12) is surely post-Han, as stated, but may with certainty be dated in the T'ang period. The author speaks merely of a stand, but the essential point is that this stand assumes the shape of a drum and as a matter of fact is a drum with a most interesting ornamentation reminiscent of Han style. This is the prototype or starting-point for the well-known combination of the cock and the drum in Japanese art.

No. 57 is dated "probably post-Han," with the correct remark that "certain features of the design suggest memories of an archaic prototype, yet other features date the vessel as a late product." I think we shall not err in regarding this vessel as a splendid example of the T'ang founder's art.

If the elephant in No. 87, as intimated, was used as a ewer for pouring water on an ink-slab, the hollow process on the elephant's trunk is not difficult to explain: it is a tube for holding a writing-brush. This object is dated "perhaps Han or later." In my estimation it belongs to the T'ang.

The beautiful paintings in the *lien* illustrated in Plate 54 merit attentive and close study. Besides this one I have seen only one other painted Han bronze. With the Han painted lacquers discovered in Korea and numerous pieces of painted pottery we now have a fairly good foundation for a study of the pictorial art of the Han.

Under No. 97 (Plate 59) is described and figured "an ornament to top of a staff, the middle part of the base being socketed, and across the socket runs a pin; design repeated on other side; date doubtful." Exactly the same object was obtained by me in China some twenty years ago, and there never was a doubt in my mind that it is a product of the Han period. The essential point is that the design consisting of rows of circles with central dot or rows of dots are laid out in concentric zones, the bands being set off by circles. This principle of concentric ornamentation is characteristically Han: it is successfully applied to metal mirrors and drums as well as to the discs of roofing-tiles and the lids of pottery vessels. The exact purpose for which this object served is not yet known; it must have been a sort of ceremonial emblem carried by means of a staff which was inserted into the socket. Numbers, as is well known, are always symbolic and significant in ancient Chinese lore. In the Eumorfopoulos emblem there are one circle in the centre, a band of twelve circles in the inner zone, and a band of twenty-four circles in the outer zone; further, there are four and three concentric rings of dots and five plain circles,—all this being suggestive of a kind of numerical symbolism.

The Field Museum also secured from me a specimen of the collapsible pocket-lamp, as shown in Plate 58, save that it is plain and minus the designs which decorate the Eumorfopoulos example. The inscription on this lamp (Figs. 37 and 43), *i tse sun* (a well-known typical Han formula), is translated, "May [the owner of this lamp] have his due of sons and grandsons." It means, however, "May it [the] lamp still benefit my sons and grandsons," or "May it last so long that it will be of use to my descendants."

The two fragmentary stands in Plate 68 of the Han period, surmounted by figures of bears, are correctly reconstructed on the basis of Fig. 44. There are two very fine complete specimens of this type in the collections of the Field Museum, which were excavated from a Han grave near Ho-nan Fu under my own eyes in 1909, together with two bronze lamps forming the top of similar standards. There is a possibility that, aside from serving as holders of a lamp or brazier, these standards may also have served for holding a scale for weighing.

In explanation of the object in Plate 56 the comment is made, "The three Manchu catalogues call this type an ice container, and two assign it to the Chou period; but surely it is a brazier, as described in the earliest Sung catalogue, and it can hardly date before the Han." In fact there is no contradiction between these, but seemingly conflicting views; both are correct. These braziers were used for holding ice in the summer with the object of cooling the air in the apartments, and were filled with burning charcoal in the winter for heating purposes. This is known from actual specimens still made in the K'ien-lung era and used in the imperial palaces with this double function. The cowering men of Central Asia as caryatids of this type of vessel have likewise persisted down to that period. The Eumorfopoulos specimen appears to belong to the era of the Six Dynasties, as the censer in shape of a rooster in the following plate does likewise.

These cursory notes, which I would have gladly increased if more time had been granted to me by the Editor, suffice to indicate what a wealth of novel, significant and superb material is amassed in this excellent book which is a boon to collectors, artists and students alike.

LIMOGES ENAMELS IN THE BIRMINGHAM GALLERY
BY W. B. HONEY

HE city of Birmingham is fortunate in possessing, in the collection bequeathed to the Birmingham and Midland Institute by Sir Francis Scott, a representative little series of Limoges enamels, both early and late. Included in it are the four pieces here reproduced [PLATES I, A, B, and II, A, B], which not only show fine quality as works of art, but touch the history of the painted enamel at some of its most difficult points.

The early period of Limoges enamel-painting has been exhaustively described and classified by M. J. J. Marquet de Vasselot,[1] and there can be no doubt that the beautiful piece figured in PLATE I, A belongs to the group associated by him with the " Louis XII Triptych " at South Kensington.[2] When writing his book, M. Marquet de Vasselot gave his opinion (based on a bad photograph, the whereabouts of the enamel being at the time unknown to him) that the plaque here illustrated was painted by a less gifted pupil of the Master of the Triptych : but a comparison with one of the panels of the latter, here figured in PLATE I, C, can leave little doubt that the Birmingham plaque is actually by the same hand. The treatment of the draperies, and the heads with heavy-lidded eyes show a precisely similar manner. The superb colour characteristic of the class is admirably represented by this piece. The striking composition depicts the group variously known as " The Three Maries," " The Lineage of St. Anne" and "The Holy Kith and Kin," showing St. Anne with her children by her three husbands, Joachim, Cleophas and Salomas, and her seven grandchildren. In the middle is St. Anne with the Virgin and Child on her lap; to the left is Mary-Cleophas, wife of Alpheus, with her children St. James the Less, Joseph the Just, St. Simon and St. Jude; to the right, Mary-Salome, wife of Zebedee, with St. James the Greater and St. John the Evangelist, and their cousin St. John the Baptist, who I am told[3] is usually brought into the group though not actually a grandchild of St. Anne. The theme is one that appears only in later medieval art. M. Mâle, in his "L'Art Religieux de la Fin du Moyen Age en France " (pp. 227-229) points out some examples in stained glass and engravings dating from about 1500, but in none of them does the composition in the least resemble that of this plaque. As in the case of most of these enamels, the subject was doubtless copied from a print or book-illustration, though the actual engravings that were used for the earliest examples have seldom been identified. The occurrence of some of the motives in rather different form in prints of Martin Schongauer was pointed out by M. Marquet de Vasselot, but apart from this no exact correspondence has been traced in the case of any enamels of the class (formerly grouped loosely under the name Pénicaud School) to which this specimen belongs. The publication of this striking enamel may, I hope, bring to light the early French or (more probably) German book from which the subjects were copied.

That the prints were copied with some exactness is shown by PLATES I, B and D. Part of the same composition, with details modified and suppressed, was used on a plaque in the Salting Collection by the same hand, which is that of the painter for long known as " Kip," from the occurrence of these letters (variously arranged) on much of his work. The late Mr. H. P. Mitchell, in a long article devoted to this painter, published in the BURLINGTON MAGAZINE,[4] gave good reason for believing the enameller using these initials to have been a goldsmith named Jean Poillevé, and the small size of the present as of most other examples of his work tends to support the belief that his enamelled work was made for the decoration of coffers and the like. Poillevé's style—essentially Gothic in handling, in spite of the Renaissance compositions employed—is well represented by this charming piece, while his use of the anonymous Italian print[5] here published [PLATE I, D] is a further proof of the wide dissemination

[1] *Les Emaux Limousins de la fin du XVe Siècle et de la première partie du XVIe.* (Paris, 1921.)

[2] No. 552—1877 : figured by M. Marquet de Vasselot, *op cit.* No. 127 (Pl. XLIX).

[3] Mr. R. P. Bedford has kindly helped me with the iconography of this subject, which is worked out in detail in his *St. James the Less : a study in Christian Iconography* (1911).

[4] Vol. XIV (1908-9), pp. 278 to 290. The Salting plaque is figured in Pl. I, c. Another enamel with the same composition was sold with the Marczell von Nemes collection at Amsterdam in November, 1928.

[5] British Museum, Catalogue B.I. 11 : Plates. Text : B.I. 10 (2).

196

毡的早期历史

American Anthropologist

NEW SERIES

ORGAN OF THE AMERICAN ANTHROPOLOGICAL ASSOCIATION, THE ANTHROPOLOGICAL SOCIETY OF WASHINGTON, AND THE AMERICAN ETHNOLOGICAL SOCIETY OF NEW YORK

ROBERT H. LOWIE, *Editor*, Berkeley, California
FRANK G. SPECK and E. W. GIFFORD, *Associate Editors*

VOLUME 32

MENASHA, WISCONSIN, U. S. A.

PUBLISHED FOR

THE AMERICAN ANTHROPOLOGICAL ASSOCIATION

1930

American Anthropologist

NEW SERIES

Vol. 32 JANUARY-MARCH, 1930 No. 1

THE EARLY HISTORY OF FELT

By BERTHOLD LAUFER

THE art of making felt by rolling, beating, and pressing animal hair or flocks of wool into a compact mass of even consistency is assuredly older than the art of spinning and weaving. In point of time, felted stuffs followed immediately, or originated contemporaneously with, the custom of using animal skins or furs as garments. Felting was practised in times of great antiquity both in Asia and Europe, but it was restricted to these two continents. It is noteworthy that it has always been absent in Africa. Even in ancient Egypt where sheep were reared and their wool woven into cloth felt was unknown. It did not exist either in aboriginal America. The ancient Peruvians, although they had domesticated the llama and alpaca, did not conceive the notion of felt.

There are ancient records extant that give references to felt in Chinese, Greek, and Latin literatures. We must not imagine, however, that for this reason the Chinese, Greeks, and Romans were the first nations to have made use of felt. The Greeks lived in proximity to the roving Scythians of southern Russia; and the vast steppes stretching east of the Ural and the Caspian sea across Russian and Chinese Turkestan into southern Siberia and Mongolia were, from earliest times, the playground of ever moving tribes, restless like the waves of the oceans, of Iranian, Turkish, Mongol, and Tungusian nationalities. These tribesmen of nomadic habits subsisted on the wealth of their flocks consisting of cattle, camels, sheep, goats, and horses. The making of felt naturally presupposes the existence of wool-furnishing domestic animals like sheep, goat, and camel. While it is true that felt can be made and has been made from the hair of wild animals, the supply of such hair is not plentiful enough to establish the industry on a large scale. It is therefore clear that solely peoples who possess a large stock of herds of wool-bearing sheep and camels could call into life a flourishing felt industry. This reason alone, however, would hardly be sufficient to ascribe the invention of felt to the nomadic population of Asia and to disclaim it for the Chinese and the Greeks; the two latter nations also had domesticated sheep, and the Greeks manufactured sheep-wool into garments. The ancient Chinese, although they had sheep, never utilized its wool for clothing. The Chinese, as well as the Greeks and Romans, assuredly understood the preparation of felt, and the Chinese still prepare it, but the manufacture of this article had a limited importance among them. Eliminate felt from Chinese, Greek, and Roman civilizations, and they would still remain what they are, not being in the least affected by this minus. Eliminate the same element from the life of the nomadic populations, and they would cease to exist, they would never have come into existence. With these peoples felt is a fundamental of culture, an absolutely essential feature and necessity of life, while with the highly civilized nations like the Chinese, Indians, Greeks, and Romans it is a side issue, an incident, an element of occasional and minor importance.

The use of felt, therefore, has reached its maximum intensity and its climax among the nomadic tribes of Asia, and this is the principal reason why we are compelled to attribute the invention of felt, both the initiative and the perfection of the process, to the Asiatic nomads. This means, of course, that the Chinese, the Indians, and the Greeks learned the art from the latter, while the Romans adopted matter and word from their masters, the Greeks. Another interesting point of difference is that to the civilized nations felt was simply a utilitarian product which they adopted because it was useful and practical, whereas among the nomads it was associated with religious and ceremonial practices. It was part and parcel of their life, inseparable from their inward thoughts. Which of the many hundreds of tribes of inner Asia was the original inventor of felt, it is impossible to ferret out in the present state of our knowledge. The beginnings of the art are lost in the dawn of human civilization. Neither the ancient Scythian nor the ancient Turkish tribes had any system of writing so that no records of their earliest history are preserved in their own languages; for all that we know about them we are indebted to the records of the Chinese and the Greeks. Archaeology, to some extent, comes to our rescue, for some ancient remains of felt have been discovered in Central Asia. More than that, the ancient mode of life of the Turkish, Mongol, and Tibetan tribes is still preserved in full vigor: they still manufacture felt as their ancestors did thousands of years ago, they still utilize it for exactly the same purposes. By combining historical, ethnological, and archaeological methods the early history of felt can be reconstructed with a fair degree of accuracy and completeness.

In the light of the preceding remarks it is clear that no credence can be attached to the European legend according to which the invention of felt is ascribed to Saint Clement, who while on a pilgrimage placed carded wool in his shoes to protect his feet, the constant pressure and moisture changing this wool into felt.

No detail of the early process of felting is preserved by any Chinese or Greek author, but there can be no doubt that in principle the ancient process was identical with that prevailing in Asia at the present time. This primitive process is practically the same everywhere. The principal instrument used is a large mat. The wool is spread out on this mat layer upon layer until the desired thickness is secured, the wool for the upper layers being generally of a better quality or finer texture than that in the interior and lower layers. Grease or oil mixed in water serves as size. The mat is rolled up under firm pressure with the feet (some people use the back of the forearm in this process), then it is unrolled and rerolled from the opposite end. This manipulation of rolling forward and backward occupies a considerable time; revolving is continued for four or five hours, when the fibers become firmly and closely intertwisted. The felt is now taken up, washed with soap and water, dried, and again stretched on the mat and dried in the sun. Colored patches of felt or wool are arranged on it in India and Turkestan, and the whole is then again subjected to the rolling process for several hours, when the material is completed and fit for use. In India the finer kinds of felt are trimmed with a mowing-knife, which greatly improves the appearance and brings out the distinctness of the colors.

Felt in China

In the earliest documents of the Chinese, the Book of Songs (*Shi king*) and the Book of History (*Shu king*) no mention is made of felt. It appears in Chinese records toward the end of the Chou dynasty (fourth to third century B.C.), and felt rugs seem to have been used at that time as mattresses to sleep upon. At the outset it is improbable that the Chinese could be regarded as the inventors of felt. They raised sheep, but never utilized their wool for any fabrics. Hemp and other fibrous plants, as well as silk, furnished the staple for clothing. Woolen materials have always been alien to Chinese civilization. There was no cattle-breeding on a large scale, and consumption of milk and any dairy products was unknown. The Chinese were (and still are) essentially a nation of agriculturists. From early times, the north of China was in close contact with central and northern Asia teeming with a vast pastoral population, for the greater part of Turkish and Tungusian nationality. These ever restless hordes perpetually poured over the Chinese frontiers and raided and pillaged the villages of the farmers. The most dreaded of these predatory foes were the Hiung-nu, as they are styled in the Chinese annals, who have been identified with the Huns. From about 1400 B.C. the Chinese were constantly engaged with them in a life and death struggle. The Chinese armies in the beginning were usually the losers as they opposed their infantry to the mobile cavalry and mounted archers of their enemies. The Hiung-nu, a Turkish tribe, subsisted on cattle, fed upon flesh and milk, and used leather obtained from the skins of their domestic animals as clothing and armor; in addition to leather garments they wore coats or overcoats of felt and lived in tents covered with the same material. It is very probable that the Chinese made their first acquaintance with felt during their long military and diplomatic intercourse with the Hiung-nu which lasted for many centuries. In 307 B.C. Wu-ling, king of the principality Chao, adopted the clothing and the tactics of shooting with the bow on horseback from the nomadic tribes. Chinese garments were spacious, loose, and flowing, and a serious obstacle to riding and shooting, while the costume of the nomads was tight-fitting and equipped with tall boots. There is no doubt that on the occasion of this reform movement in dress also articles of felt and perhaps the manufacture of felt itself were adopted by the Chinese. The country inhabited by the nomads is known to them under the name "the land of felt."

Under the Han dynasty (201 B.C.–A.D. 220) felt was well established in China and used in the form of mats. The Emperor Wen (179–152 B.C.) of this dynasty wore a felt cap on his hunting expeditions. The felt of the nomads is alluded to by the philosopher Huai-nan-tse, who lived in the second century B.C.; his statement implies that in his time felt was still unknown south of the Yangtse region.

At the end of the third century A.D. the use of felt was still regarded as something foreign and barbaric, for it is on record that in the period T'ai-k'ang (A.D. 280–290) when fillets and girdles of felt were introduced as a novel fashion, the people ridiculed this custom and said, "China apparently has been conquered by the nomads (Hu), for felt is a product of the nomads, and now with felt fillets and girdles we adopt their styles."

In A.D. 532 Yüan Siu was placed upon the throne as tenth emperor of the Northern Wei dynasty by Kao Huan, who sent four hundred horsemen to meet him. The future emperor betook himself into a felt tent to don imperial regalia. He was then escorted to the east gate of the palace, and according to an ancient custom of the Toba, one of the northern nomad tribes from which the Wei dynasty issued, he was lifted by seven men on a piece of black felt; and while seated on it, he bowed toward the west, imploring Heaven. This was an old usage of the nomads of central Asia, and we shall encounter it again among Turks and Mongols.

A certain Liu Ling-chʻu, who lived in the fifth century A.D., is said to have cut human figures out of felt for magical purposes. This idea was doubtless borrowed from the nomads, for it was an ancient Turkish and Mongol custom, more of which will be said below, to fashion religious images from felt and to keep them in leather cases.

A felt cap is referred to in the *Tsʻan luan lu*, a diary kept by Fan Chʻeng-ta during his journey from the capital to Kwei-lin in Kwang-si, on his appointment to that prefecture in A.D. 1172.

Not only in the north, but also in the west and southwest were the Chinese surrounded by felt-using nations. The vast area occupied at present by the provinces of Se-chʻwan and Yün-nan was anciently populated by many different aboriginal groups of tribes partially related to the Tibetans, partially to the Siamese (Tai family), and partially of independent stock, prior to the advent of the Chinese. The latter, in the course of several centuries, penetrated those regions, subdued the very warlike aborigines, and colonized the country. Many of the tribes were annihilated, others were pushed back into inhospitable high mountains, still others migrated into Siam and Burma, others survive to this day. The earliest reference to felt in this territory is made in the Annals of the Han Dynasty with reference to a tribe inhabiting Se-chʻwan, called the Jan-mang, who were essentially sheep-breeders and manufactured felt as well as various kinds of woolen stuffs; the Chinese annalist records as a remarkable fact that they understood the art of treating the diseases of sheep.

The present province of Yün-nan was formerly occupied by the powerful kingdom of Nan-chao from which at a later date the present-day Siamese issued. The men of the Nan-chao tribes of Yün-nan wore one-piece blankets of felt in the ninth century (according to the *Man shu*, written by Fan Chʻo about A.D. 860). The same author also relates the curious fact that many men in the country Pʻiao wore white felt. Now Pʻiao was situated 75 days' journey south of Yung-chʻang in Yün-nan and corresponds to Pyū, name of the prominent tribe at Prome, the ancient capital of Burma. Whether felt was at that time manufactured in Burma is not known; it seems more likely that it was imported there from Yün-nan.

An important document bearing on felt is contained in the *Ling wai tai ta*, written by Chou Kʻü-fei in A.D. 1148 (ch. 6, p. 12). This work gives a geographical description of the two southern provinces, Kwang-tung and

Kwang-si, as well as many valuable notes on the ethnography of the native inhabitants, their customs, products, and manufactures. The author emphasizes the wealth of sheep in the land of the Southwestern Man as they are called by the Chinese, and says that they produce felt and woolen cloth in great quantity.

From their chieftains downward to the lowest man there is not one who would not throw over his shoulders a piece of felt. The sole difference between the two classes is that the chieftains wear an embroidered shirt on their skin and don the felt over it, while the common people wear the felt directly over their skin. The felt of northern China is thick and solid; in the south felt pieces are made to a length of over thirty feet and to a width of from sixteen to seventeen feet. These are doubled along their width, and the two ends are sewed together, so that they are from eight to nine feet wide. They take a piece of felt lengthwise and wrap it around their body, fastening it with a belt around their loins. The women follow the same practice. During the daytime they are thus wrapped up; at night they sleep in their felt blankets; whether it rains or the sun shines, whether it is cold or warm, these are never separated from their bodies. In their upper part these blankets are decorated with designs like walnuts. Those which are long and big and yet light in weight are held in the highest esteem, and those manufactured in the country of Ta-li (in Yün-nan) are regarded the best.

What this Chinese author noted some eight hundred years ago still holds good for the majority of aboriginal tribes in Yün-nan and southern China. Most of these, particularly the Lolo and Moso, still wear a blanket or a sort of sleeveless coat made of a single piece of white felt as a protection against chill and rain both in winter and summer. Many authors relate with amazement that they never part with this outfit, even in intensely hot weather.

In 1863 S. Wells Williams (*The Chinese Commercial Guide*, 5th ed., Hongkong, 1863, p. 119) wrote,

Felt caps are worn by the poor throughout the whole country. They are of various shapes and different degrees of fineness; some are made hollow so that when pulled out, they resemble a double cone. The felt cuttings are collected from the manufacture of druggets, caps, soles of shoes, and leggings, to be boiled down and felted over again.

Felt is still manufactured in China into caps, rain-hats, coats, stockings, shoes, shoe-soles, tablecloths, rugs, and carpet-bags. In Suchow the industry is still very much alive. Boys are fond of felt caps, especially when trimmed with colored silk and provided with ear-muffs of fur. The fishermen on the Great Lake (T'ai Hu) wear large, broad-brimmed felt hats plain or trimmed with black satin (specimens collected by me in Field Museum, Chicago).

The method employed by the Chinese in preparing felt is the same as that used by the Tibetans, Mongols, and Turks, with a single exception: the first step they take is to loosen the wool by means of a large bow by tightening the string and jerking it off in rapid motion. This process is derived from that of treating cotton, and the bow in either case is identical. The layers of wool are heaped up on a bamboo mat and carefully moistened with water sprayed from the mouth in the same manner as our Chinese laundrymen moisten linen. Then the wool is rolled up in the mat which is rolled to and fro, and pressed by means of the feet.

Felt in Tibet

In ancient times felt and hide formed the common material for the clothing of the Tibetans, according to the Chinese Annals of the T'ang Dynasty (A.D. 618-906). Felt was also used in Tibet for plates. Even the kings of Tibet were clad in garments of felt; when Srong-btsan sgam-po, the first king of Tibet known in history, married a Chinese princess in A.D. 641, he adopted, in order to please his refined consort, the cultured manners and customs of China and discarded his felt and fur robes which had to give way to Chinese silk and brocade. The Chinese Annals inform us also that the men of rank in Tibet lived in large felt tents called *fu-lu* (Tibetan *sbra*); this kind of tent served for military purposes, and there were big ones capable of holding several hundred men; they formed a military camp. The pastoral population of Eastern Tibet, however, has always lived in square tents covered with a black cloth densely woven from yak-hair. In this respect and in its quadrangular shape the Tibetan tent contrasts with the Mongol circular felt tents and represents a dwelling-type of its own. This tent of yak-hair stuff goes back to a venerable age, for it is referred to as early as the sixth century in the Annals of the Sui Dynasty with reference to the Tang-hiang (Tangut), a Tibetan tribe living in the vicinity of the Kukunor. The same people, however, as emphasized by the Chinese annalist, held felt in highest esteem and looked upon it as the finest ornament.

In central Tibet all men, even the Dalai Lama, wear a high-crowned, red-fringed felt hat; the women wear a red felt hat in the summer. The felt made by them is praised by a Chinese author of the eighteenth century; it is also worked up, he adds, into boots. In fact, the women of Tibet mostly wear high felt boots. These are of the same shape as the leather boots usually worn by men and reaching up to the knees. These felt boots are trimmed with colored patches, the lower part white, then red and green. Like the leather boots they are lined with woolen cloth, while the soles are always of leather. The Tibetan boot is devoid of a heel. The Tibetan nomads wear high conical felt hats with a large brim turned downward.

The most interesting object made of felt by the Tibetans is a poncho which consists of a long rectangular strip of felt with a hole in the center to put the head through, and which is used on horseback in rainy weather. Most Tibetans spend the whole day in the saddle. When traveling in Tibet for more than a year, I always carried such a felt poncho with me and found it immensely useful; it was a perfectly safe protection in the most violent rain and snowstorms and completely envelops the horse as well as the rider. Similar rain ponchos are used in Asia Minor.

W. W. Rockhill describes the production of felt in Tibet as follows:

Its mode of manufacture is extremely simple. The wool, having been first picked over, is spread out a handful at a time on a large piece of felt on the ground, each handful overlapping the preceding one in such a way that a piece of uniform thickness and of whatever size is desired is made. This is rolled up tightly and with much pounding of the closed fist and then unrolled, and this work is kept up for an hour or more; then the roll is soaked in water and the work of rolling, unrolling, kneading, and beating with the closed fist goes on for another hour or two. I was told that a piece of felt had to be kneaded at least 1,000 times before it was ready for use. After the roll has been left to dry for a while it is opened, and by pulling it slightly in different directions the surface is made smooth, and the edges are trimmed with a knife. Sometimes it is bleached. Altogether, Tibetan and Mongol felt is vastly inferior to that made by the Chinese.

Felt in India

Felt appears to have been known in India in ancient times. Nearchus, who accompanied Alexander the Great on his expedition to India and as admiral of his fleet in 325 B.C. discovered a sea route between the Indus and the Euphrates, reports that the inhabitants of India understood the art of felting wool (Strabo XV. 1, 67). It is on record in the Chinese Annals of the T'ang Dynasty that in the beginning of the period T'ien-pao (A. D. 742-756) tribute gifts were dispatched to the imperial court by the king of the island of Ceylon, and among these presents pieces of white felt figured conspicuously. In this connection it is worthy of mention also that according to an old Chinese account of Java two kinds of felt were obtainable on the island—one dyed a color like granite and another dyed a deep crimson.

John Fryer, who traveled in India and Persia from 1672 to 1681, writes that at Surat the horses were covered warmly with a kind of felt or flock-work, two or three double. Both woven and felted blankets (*kambala*) were made in northern India.

In India felt is at present manufactured in Ladak, Jeypore, Rajputana, Hyderabad and other places, felts being used for blankets, carpets, cushions, bedding, cloaks, and leggings. Colored wool is often used with great effect in producing patterns on the surface of the material. The best sort of felt consists entirely of sheep's wool, or is a mixture of wool with goat's and camel's hair picked and cleaned.

Felt Among Iranians and Turks

The Chinese Buddhist pilgrim Fa Hien started in A.D. 399 on his memorable long journey to India overland from China by way of Central Asia, of which he has left a fascinating account. Passing through the kingdom of Shen-shen south of and not far from Lake Lob (Lob-nor), he made this entry in his diary:

> The clothes of the common people are coarse, and like those worn in our land of Han (China), some wearing felt and others coarse serge or cloth of hair; this was the only difference seen among them.

This is the earliest account of the use of felt in a region of what is now Chinese Turkestan. Turkestan means "land of the Turks." At the time of Fa Hien's visit, however, Turkestan was not yet conquered by Turks, who were then confined to southern Mongolia, but was densely populated by Iranian tribes, members of the Indo-European family, who had a highly flourishing civilization. The Iranian stock at that time covered an immense territory, stretching from the confines of China on the west through the plains of Chinese and Russian Turkestan far into the steppes of southern Russia; for the Scythians so called by the Greek historians are members of the same group, and all of them are close relatives of the Persians. All the tribes belonging to this great Iranian family were active and energetic producers of felt, and it may even very well be the case that they were the initiators of the technique. Certain it is that woven rugs and carpets were first produced in their midst, and as in my estimation carpet-weaving sprang up after and as a consequence of felted rugs, it stands to reason that it was Iranians who invented the manufacture of felt.

Herodotus (IV, 46) describes the Scythians as living on carts which were the only houses they possessed. Rawlinson comments justly that their wagons carried a tent consisting of a light framework of wood covered with felt or matting, which could be readily transferred from the wheels to the ground. Hesiod, the Greek poet, says that Phineus was carried by the Harpies "to the land of the milk-fed nations whose houses are wagons." Aeschylus (Fettered Prometheus 709) sings of the "wandering Scyths who dwell in latticed huts high-poised on easy wheels." The Scythians also were in the constant habit of wearing felt caps or hats.

The fact of an Iranian felt industry is signally confirmed by the combined testimony of Chinese and Greek observers. Fa Hien has just been called to the witness-stand. According to the Chinese Annals of the T'ang Dynasty, the king of Sogdiana, who resided at Samarkand, was in the habit of wearing a felt hat adorned with gold and precious stones.

The Persian Magi, the priests of Zoroaster, wore high turbans of felt, reaching down on each side so as to cover the lips and sides of the cheeks (Strabo XV. 3, 15). The Lycians who accompanied Xerxes, king of Persia, on his expedition to Greece, were clothed with felt caps surrounded by plumes (Herodotus VII,92). The Persian soldiers in Xerxes' army wore light and flexible caps of felt which were called tiaras. The Medes and Bactrians were equipped with the same kind of headgear as the Persians. The Armenians were also styled "wearers of felt." Strabo characterizes the Persian cap as "a felt in the shape of a tower," adding that these caps were necessary in Media on account of the cold climate. The king of Persia was distinguished by a stiff felt hat which stood erect, whereas his subjects wore their tiaras folded and bent forward (Xenophon, Anabasis II. 5, 23). Hence in *The Birds* of Aristophanes, the father of comedy, the cock is ludicrously compared to the Great King, his erect comb being called his "Persian cap" (*kyrbasia*). The Athenians no doubt considered this form of the tiara as an expression of pride and arrogance. Xenophon alludes to felted quilts manufactured in Media and spread out as couches or rugs on the ground to sit upon. The Medes also availed themselves of bags and sacks made of felt, and the Persians used felt for the trappings of their horses.

In Anglo-Indian a rug felt is styled *numda* or *numna*. This word is derived from Hindustani *namda* and Persian *namad*. These felt rugs to this day form a special product of the home industry of Khotan whence large consignments are annually exported to Ladak and Kashmir. Sir Aurel Stein (*Sand-buried Ruins of Khotan*, p. 402) has discovered the earliest mention of these felt rugs under the name *namadis* in a Kharoshthi document found in the ruins of Khotan and dated in the ninth year of King Jitroghavarshman, which relates a transaction by a certain Buddhagosha concerning some household goods pawned perhaps or taken over on mortgage. The articles are enumerated in detail, and their value is indicated. Besides sheep, vessels, wool-weaving appliances, and some other implements, this list contains also the felt rugs *namadis*.

Still more fortunate, Sir Aurel succeeded in wresting from ancient refuse heaps and buried temple-ruins of Chinese Turkestan numerous remains of old felts, which are described in his monumental work *Serindia*. These, in all probability, are the earliest felt remains now in existence that have survived the ravages of time; they are preserved in the British Museum. They should be carefully examined and analyzed some day by a felt expert. Such a study may throw an unexpected light on the early technique of felt making and its historical associations.

Of felt pieces and fragments discovered by Sir Aurel Stein in Chinese Turkestan may particularly be mentioned a felt pad of kidney shape covered with buff silk, a conical headgear in carefully gored yellow felt shaped like a Phrygian cap, shoe-soles, a fragment with a wave-scroll pattern in thin crimson felt sewed on; fragments of felt dyed yellow, red, and scarlet; small pieces of yellow felt painted on a tempera surface with floral and geometrical designs in a variety of colors, and many others. With reference to his discovery of crimson felt it may not be amiss to call to mind the purple or scarlet felt used for draping the funeral pile of Hephaestion when this friend of Alexander the Great died at Ecbatana in 324 B.C. and was interred at Babylon with splendid obsequies by order of his master.

In his work *Ruins of Desert Cathay* Sir Aurel writes,

> Kök-yar is famous throughout Turkestan for its excellent felts, and a good deal of the manifest ease prevailing in these homesteads was no doubt derived from the profits of this flourishing industry.

In the ruined fort of Miran he found a well-preserved felt pouch which might have formed part of a soldier's equipment (plate 138, fig. 27). Kök-yar is also renowned for its felt socks called *paipak*, and Karghalik is the great market for them. In another passage he says,

> Clean mud walls and gaily-colored Khotan felts (*kirgiz*) make even a bare little room look cheerful and homely on a winter evening.

Another archaeological discovery of importance was made two generations ago by W. Radloff in graves of southern Siberia which belong to the Iron age. From these he brought to light a felt boot or sock, the sole of which was wrought from a very fine kind of felt. This was the product of some ancient Turkish tribe. Pointed caps appear frequently on stone monuments or on bronze plaques of southern Siberia, and these were doubtless made of felt.

Reference was made above to the ancient Hiung-nu or Huns as having dwelt under felt tents, and this type of habitation has been characteristic of most Turkish tribes in Asia through all ages. In the sixth century of our era a new Turkish nation inhabiting what is at present southern Mongolia came to the attention of the Chinese, and was called by them Tu-küe which transcribes the very word "Turk" and which represents the first appearance of this name in history. These Tu-küe, in the same manner as their predecessors, clothed themselves in hide and wool and lived in felt tents. The Kozlov expedition in northern Mongolia, the results of which were published in 1925, found a felt carpet bordered with embroidered silk beneath the coffin in the main tomb excavated. This splendid specimen may be attributed to Hiung-nu workmanship, and is believed to date from the first century before our era. Thick felt soles embroidered with silk or thin thread were also brought to light from the same group of graves. For illustrations see *Burlington Magazine*, April, 1926.

The Kirgiz, another ancient Turkish tribe, according to the Chinese Annals, wore white felt caps, with the exception of their chief, who in the winter wore a sable hat and in the summer a pointed metal helmet with a turned-up tip. They joined pieces of felt together to make tents; the chiefs lived in small tents.

The Shi-wei, an ancient tribe of Manchuria (now extinct), although they lived in huts covered with coarse mats, had felt tents in the Turkish manner placed on carts; these were obviously used for traveling. In lieu of felt, the Chinese Annals say, they placed a package of grass under the saddles of their horses.

In electing their chieftains the Turkish tribes were accustomed to lift them on a white felt rug, not on a carpet. In ceremonial ritual the oldest customs of a tribe are purely preserved and rigidly adhered to, and it is plainly manifested by this practice that the use of felt rugs preceded that of woven rugs among the Turks. It is an interesting fact also that in the Turkish epic poems which clearly mirror a true picture of their ancient primitive life the art of weaving is never mentioned, whereas sewing, embroidering, and felting are referred to as the sole pastime and handicraft of women.

The manufacture of felt covers is the most important home industry of the Kirgiz-Kaizak in Russian Turkestan, and is almost exclusively the business of women. Felts are used by them for covering their tents (yurts), as rugs, door curtains, saddle covers, pouches, bottle cases, mittens, and mattresses. Their sale forms a significant source of income for them; for the Russians also, especially the Cossacks, and the sedentary town-population of Turkestan like the Sarts, make ample use of felt material, e. g. for window shutters, mattresses, and particularly for packing merchandise to be transported by caravan. Owing to the preponderance of felt used in a variety of ways in their equipment, the Cossacks have received from the regular troops the nickname "felt troops." The inhabitants of the towns of

Russian Turkestan also produce felt, but this article is less durable and inferior in quality to the Kirgiz felt. The town products are cheaper and even finer, softer, and smoother than the unpretentious Kirgiz felt, but the latter is ten times as strong as that of the Sarts. This point is of great interest, for it confirms my opinion that felt was originally an invention of pastoral, not sedentary peoples. The latter have merely imitated the former, and while their product is more elegant and refined in appearance, it does not rival the original in solidity and durability. The Kirgiz make white and black felt; the former is regarded as the better one. Besides felt covers the women also make felt hats from white wool for the men. The Turkmens produce from felt slings for the use of boys in killing birds.

For the making of felt the summer wool of sheep is preferred, especially the first wool of the lambs born in the spring. Oil-cake serves as size, and is mixed with the water which is sprinkled over the wool spread over a reed mat. It is first beaten with rods until the mass reaches the same level. The wool is usually arranged in two layers, a lower one of brown cheaper wool and an upper one of white wool. The mat is then rolled up as tightly as possible and tied with cords. This package is rolled to and fro over the ground, pulled along with a rope by some experienced old people, and pushed with the feet by a number of girls following it. The cords are tightened from time to time. Finally the mat is removed, the wool is rolled up again and rolled and rerolled for several hours, while water is continually sprinkled on it. The woolen layers are then spread out, dried at the sun, and the felt is ready, supple and smooth like cloth. Patterns are cut out of colored felt, laid on the felt rug and beaten into it.

Among the Turkish tribes of Central Asia the white wool is first separated from the dark one. The layers are spread out on horse skins and are beaten. They are then sprinkled with water and rolled between two reed mats until the mass is solid. First it is rolled with the hands, then continued with the feet, while six or eight women with arms akimbo shove the roll along in equal pace not unlike the movements of a dance, and songs are chanted at the same time. Patterns, if desired, are laid out in dyed wool.

Franz von Schwarz, formerly astronomer of the Tashkent Observatory, in his book *Turkestan* (1900), makes the following interesting observation:

Among the natives of Russian Turkestan the belief prevails that scorpions, phalanges, tarantulas, karakurts and snakes cannot move on felt mattresses and that consequently one is safe from their attacks by sleeping on felt covers. In how far this opinion is founded on fact I cannot say with certainty; but this much I know that I myself during my travels when as a rule I used felt covers as a padding for my camp-bed, was never attacked by scorpions, etc., even in places which teemed with this vermin.

Felt among the Mongols

Marco Polo (book I, ch. 52), the Venetian traveler of the thirteenth century, writes that

the houses of the Mongols are circular and are made of wands covered with felts. These are carried along with them whithersoever they go; for the wands are so strongly bound together, and likewise so well combined, that the frame can be made very light. They also have wagons covered with black felt so efficaciously that no rain can get in. These are drawn by oxen and camels, and the women and children travel in them.

In the same manner Plano Carpini, in 1246, describes the Mongol houses as

round and artificially made like tents, of rods and twigs interwoven, having a round hole in the middle of the roof for the admission of light and the passage of smoke, the whole being covered with felt, of which likewise the doors are made.

Ibn Batuta, the eminent Arabic traveler of the fourteenth century, when he betook himself to Sarai, was conveyed in a four-wheeled wagon on which he says was placed a sort of pavilion of wands laced together with narrow thongs; it was very light, covered with felt or cloth, and equipped with latticed windows, so that the traveler inside could look out without being seen; he could change his position at pleasure, sleeping or eating, reading or writing during the journey.

Some of the tents were collapsible, others were massive and stationary. On this point we are informed by Carpini as follows:

Some of the huts are speedily taken to pieces and put up again; such are packed on the beasts. Others cannot be taken to pieces, but are carried bodily on the wagons. To carry the smaller tents on a wagon a single ox may serve; for the larger ones three oxen or four, or even more, according to size.

The carts that were used to transport the valuables of the Mongols were covered with felt soaked in tallow or ewe's milk, to make them waterproof. The stilts of these carts were rectangular, in the form of a large trunk.

White felt played a significant role among the Mongols during the coronation ceremony. The king was placed on a mat of white felt which was spread on the ground. In A.D. 1206 Temuchin was crowned emperor at an assembly of the princes of Mongolia when he assumed the title Chingiz Khan. On this occasion he was seated upon a rug of white felt and was reminded of the importance of the duties to which he was called. An orator who spoke in the name of the nation addressed the new lord thus:

Direct thy eyes on the felt on which thou sitteth. If thou wilt well govern thy kingdom, thou wilt rule gloriously, and the whole world will submit to thy sway; but if thou wilt do the reverse, thou wilt be unhappy and be outcast and become so indigent that thou wilt not even have a piece of felt on which to sit.

This was not merely intended as a moral exhortation, but the ceremony was imbued with a deeper significance. Among the Mongols, even of the present time, white felt is a material endowed with a sacred character. Placing a person on a white felt rug means expressing to him good wishes for his welfare. For this reason a bride is seated on a white felt during the marriage ceremony, or people at the point of starting on a long journey receive this honor. An animal selected for a sacrifice to the gods is slaughtered on a white felt. The women therefore, in speaking of felt, carefully avoid the common word for it (*ishighei*), which is a term of respect, but substitute for it the words *dzulakhai* or *tolok*. It is on record also that the felt rug which served for the inauguration of Chingiz, dignified by the fortune of the world conqueror, was long preserved by his successors as a palladium and sacred relic.

Timur or Tamerlan (1336-1405), the formidable conqueror, is credited with the invention of a kind of felt hat for the use of his troops when he invaded Persia. These headgears guarded his soldiers more efficiently from the sun and rain than turbans, and distinguished them from their enemies.

Of all facts connected with the history of felt the most singular is that the images of their gods were fashioned by the Mongols from this material. Plano Carpini, who in the year 1246 went as ambassador to the Great Khan of the Mongols, informs us:

They have certain idols made of felt in the image of a man, and these they place on either side of the door of their dwelling; and above these they place things made of felt in the shape of teats, and these they believe to be the guardians of their flocks, and that they insure them increase of milk and colts. Whenever they begin to eat or drink, they first offer these idols a portion of their food or drink.

Friar Rubruk, who also made the wearisome journey to Mongolia, has this story to tell:

And over the head of the master is always an image of felt, like a doll or statuette, which they call the brother of the master; another similar one is above the head of the mistress, which they call the brother of the mistress, and they are attached to the wall; and higher up between the two of them is a little lank one, who is, as it were, the guardian of the whole dwelling.

Marco Polo, with reference to the god of the "Tartars," says,

They have a certain god of theirs called Natigay, and they say he is the god of the earth, who watches over their children, cattle, and crops. They show him great worship and honor, and every man has a figure of him in his house, made of felt and cloth; and they also make in the same manner images of his wife and children. The wife they put on the left hand and the children in front. And when they eat, they take the fat of the meat and grease the god's mouth withal, as well as the mouths of his wife and children.

Friar Odoric of Pordenone, who visited northern China between 1322 and 1328, speaks of the Minor Friars as exorcising devils among the Mongols and throwing into the fire their idols which are made of felt, while all the people of the country round assemble to see their neighbors' gods burnt.

Felt gods formerly existed among the Turks also. Captain John Smith, the same who wrote *The General History of Virginia, New England and the Summer Isles*, has given a vivid description of the life of the Tartars of southern Russia in his *True Travels, Adventures, and Observations in Europe, Asia, Africa, and America, from 1593 to 1629* (London, 1630). He describes the houses of the princes as

very artificially wrought, both the foundation, sides, and roof of wickers, ascending round to the top like a dove-coat; this they cover with white felt or white earth tempered with the powder of bones, that it may shine the whiter, sometimes with black felt, curiously painted with vines, trees, birds, and beasts.

His most interesting contribution is the description of the felt gods as follows:

Having taken their houses from the carts, they place the master alwayes towards the north; over whose head is alwayes an image like a puppet, made of felt, which they call his brother; the women on his left hand, and over the chiefe mistris her head, such another brother; and betweene them a little one, which is the keeper of the house; at the good wives beds-feet is a kids skinne, stuffed with wooll, and neere it a puppet looking towards the maids; next the doore another, with a dried cowes udder, for the women that milke the kine, because only the men milke mares; every morning, those images in their orders they besprinkle with that they drinke, bee it cossmos [kumis] or whatsoever, but all the white mares milke is reserved for the prince. Then without the doore, thrice to the south, every one bowing his knee in honour of the fire; then the like to the east, in honour of the aire; then to the west, in honour of the water; and lastly to the north, in behalfe of the dead.

The Mongols, in making felt, wet and beat sheep's wool with sticks, then press it, and tie the rough strips of wool to grazing horses who drag them across the smooth grass surface of the plain and thus complete them.

Felt among Greeks and Romans

The earliest Greek allusion to felt (Greek *pīlos*) occurs in Homer's Iliad (X, 265), where it is said that Odysseus wore a hide helmet lined with felt.

Felt was used by the Greeks for cuirasses and garments, especially rain cloaks; chiefly, however, for tight-fitting caps of a conical shape to be pulled over one's ears to ward off cold or rain (Greek *pilidion*, Latin *pilleolum*). Such a cap was generally worn by artisans and sailors, and appears in artistic representations as their characteristic outfit. Hephaestus and Daedalus wear it as craftsmen; Charon and Odysseus, as seafarers. Brimmed hats also were made of felt. It is a curious coincidence that the Greek fishermen were equipped with a felt cap as their fellow-workers in China still are. In the description of a fisherman's apparatus Philippus mentions "the felt cap encompassing his head and protecting it from wet."

Boots and socks were likewise made of felt, and there is an instance on record that it was used in lieu of armor by Caesar's soldiers when they were much annoyed by Pompey's archers and in need of arrow-proof jerkins (*Bellum civile* III, 44). Thucydides refers to a similar expedient to protect the body from arrows. Even in besieging and defending cities felt was used, together with hides and sackcloth, to cover the wooden towers and military engines.

The ancients used chiefly sheep wool for making felt, more rarely the hair of goat, camel, hare, and beaver. It seems that felt was sometimes used to cover the bodies of animals. According to Aristotle, the Greeks clothed their sheep with soft wool either with skins or with pieces of felt, and the wool turned gray in consequence.

The Romans received the use of felt together with its name from the Greeks (Latin *pileus, pilleus, pileum* or *pilleum*); this word, in particular, denotes the tight-fitting felt cap worn by the Romans at meals, theatrical performances, and festivals. It is a curious fact that the felt cap was among the Romans a symbol of liberty; when a slave obtained his freedom, he had his head shaved and wore the skull-cap of undyed felt. On the other hand, slaves when they were sold by their master, wore this cap as a sign that the seller would not offer any guaranty for them. The phrase *ad pileum vocare* ("to call to the felt cap") had the meaning "to call the slaves to freedom, to provoke them to rebellion through promises of freedom." At the death of Nero in A.D. 68 the common people roamed about in the streets of Rome as an expression of their joy. Suetonius, in his Life of Nero, speaks on this occasion of the "felted mob" (*plebs pileata*). In allusion to this custom the figure of Liberty on the coins of Antoninus Pius (A.D. 138-161) holds the cap in her right hand.

Pliny (VIII, 73) writes that

wool is compressed also for making felt, which when soaked in vinegar is capable of even resisting iron; and what is still more, after having gone through the last process, wool will even resist fire.

Papadopoulo-Vretos, in 1845, made this communication to the Academy of Inscriptions and Letters of Paris:

I have macerated unbleached flax in vinegar saturated with salt, and after compression have obtained a felt, with a power of resistance quite comparable with that of the famous armor of Conrad of Montferrat; seeing that neither the point of a sword, nor even balls discharged from fire-arms, were able to penetrate it.

The felting process was denoted by the verb *cogere* ("to bring together, to pile up"). A felter was called a *coactor*, *coactiliarius*, or *coactor lanarius* ("wool felter"); his art was designated *ars coactiliaria;* felt products were styled *coacta*. In an edict of the Emperor Diocletianus (A.D. 285-305) is mentioned a horse-cover of felt under the term *centunclum equestre coactile*.

The question may be raised whether the Romans transmitted the knowledge of felt to the Celtic and Germanic peoples, or in other words whether the use of felt in mediaeval and modern Europe is a heritage of classical civilization. The Germanic languages have a word for felt in common: German *filz*, Dutch *vild*, Danish-Swedish *filt*, Anglo-Saxon *felt*. This word is connected by linguists with Old Slavic *plusti*. It is noteworthy that the word for felt in the Romanic languages is not based, as might be expected, on Latin *pileus*, but on the Germanic word: Italian and Portuguese *feltro*, Spanish *fieltro*, French *feutre* (Italian *feltrare*, French *feutrer*, "to felt"), hence mediaeval Latin *filtrum*. It is therefore probable that the Romanic nations received the knowledge of felt not from the ancient Romans, but from Germanic tribes early in the middle ages. The latter may have acquired the art from their eastern neighbors, the Slavs; and the Slavs derived their knowledge from Scytho-Siberian-Turkish peoples. The Russian word for felt, *woilok*, is a loan-word based on Turkish *oilik* ("that which serves as a cover"); the same word appears in Polish as *wojlok*.

FIELD MUSEUM,
CHICAGO, ILLINOIS

197

一份中文-希伯来文手卷：中国犹太人史的新史料

The American Journal of SEMITIC LANGUAGES AND LITERATURES

FOUNDED BY WILLIAM RAINEY HARPER

VOLUME XLVI

October, 1929—July, 1930

THE UNIVERSITY OF CHICAGO PRESS
CHICAGO, ILLINOIS

A CHINESE-HEBREW MANUSCRIPT, A NEW SOURCE FOR THE HISTORY OF THE CHINESE JEWS[1]

By Berthold Laufer
Field Museum of Natural History, Chicago, Illinois

In 1927 when the American Oriental Society held its annual meeting at Cincinnati and enjoyed the hospitality of the Hebrew Union College, President Morgenstern very kindly showed me a collection of Hebrew manuscripts originating from the Chinese Jews of K^cai-fung fu in Honan and preserved in the library of the college. In looking these manuscripts over I was particularly attracted by a booklet of seventy-six small pages, because most of these were inscribed with Hebrew and Chinese characters alternating. The mere fact that it was the only Chinese-Hebrew manuscript I had ever laid my hands on and presumably the only one in existence proved a magnetic attraction in itself. I was permitted to take this manuscript along to Chicago where I had a photostat made of it. It turned out to contain a register of the Jewish congregation of K^cai-fung fu drawn up between the years 1660 and 1670, giving first the names of male individuals, then those of women, both in Hebrew and Chinese. Although practically a dry list of names, this unique manuscript is one of great historical interest. Before proceeding to offer some remarks on its contents and significance, a brief outline of the history of the Chinese Jews is presented, as I cannot expect that everyone is familiar with the subject and especially in view of the fact that many fantastic notions are still current about it in our cyclopedias and among the public in general.

There are very few well-authenticated dates and facts to be gleaned from the history of the Chinese Jews. Of the inner life of this small community we are almost ignorant. The principal sources for our information are three Chinese inscriptions of considerable length on stone tablets written by Jews themselves and formerly erected in the synagogue of K^cai-fung, which ceased to exist between 1840 and 1850. These inscriptions are dated 1489, 1512, and 1663, which means that they are of recent date, belonging to the time of the two last dynasties,

[1] Read at the meeting of the American Oriental Society at Cambridge, April 3, 1929.

the Ming and the Tsᶜing, so that their chronological data with reference to events prior to the Ming period must be viewed with critical eyes. In 1903, while in China, I obtained rubbings of these three inscriptions, and as I had occasion to meet at that time several Chinese Jews, I became much interested in their history and vicissitudes, and laid the results of my investigations before the International Congress for the History of Religions held at Basel, Switzerland, in September, 1904. This article was subsequently published in *Globus* (1905), and its results have generally been accepted in scientific circles.

Besides the lapidary inscriptions there were twenty-three horizontal inscriptions on wooden tablets hung in the synagogue and containing only brief maxims or devices, but interesting for the names and dates of Chinese Jewish officials who dedicated them to the temple. For the seventeenth and eighteenth centuries we have several reports anent the Jews from Jesuit missionaries beginning from Matteo Ricci, the first European who in 1605 had an interview with a Chinese Jew in Peking. The Jesuit relations contain a great deal of interesting information, but must also be taken with criticism. Two Chinese Protestants were delegated to Kᶜai-fung in 1850 by the London Society for Promoting Christianity among the Jews, and their report was published at Shanghai in 1851 by George Smith, Lord Bishop of Victoria, Hongkong. This, as well as the later accounts of several travelers, is merely of secondary or limited importance, as the Jewish community then was in a deplorable state of disintegration and had forgotten almost all its traditions; the little knowledge they were then able to offer is all traceable to their inscriptions.

At the outset we are confronted by two singular phenomena:

1. The Chinese, with their immense wealth of historical documents, leave us entirely in the lurch as regards the Jews, while they give us many notices of Nestorians, Manicheans, Zoroastrians, Mohammedans, and even Catholics. All that has thus far been discovered are three brief references to Jews in the *Yüan shi*, the annals of the Yüan or Mongol dynasty: Under the year 1329 the Jews are mentioned on the occasion of the reinforcement of a law concerning a levy of taxes on dissenters (chap. xxxiii); in 1340 the levirate was interdicted to Mohammedans and Jews (the levirate was an abomination in the eyes of the Chinese, and under the Manchu dynasty was prohibited

on pain of death); in 1354, in consequence of several insurrections, rich Mohammedans and Jews were summoned to Peking and called upon to render services in the army. A few more references occur in the *Yüan tien chang*, "The Statutes of the Yüan Dynasty." For the rest there is complete silence in the Chinese camp, which it is difficult to explain in view of the fact that the Jewish inscriptions refer, for instance, to a Sung emperor permitting the Jews to settle at Kʿai-fung, to Yung-lo's consent to rebuild the synagogue, and to other important events in their history which we should expect or should like to see confirmed in the Chinese annals—also considering the fact that many Jews filled high offices in the army, civil administration, and as physicians.

2. Another peculiar deficiency is that the Chinese Jews unfortunately failed to produce any literature, while there is a considerable literary output on the part of Mohammedans both in Chinese and Arabic. The Jewish inscription of 1663 mentions two tracts—one written by Chao Ying-chʿeng on "The History of the Holy Scriptures" (*Sheng king ki pien*), and another treatise by his younger brother Ying-tou entitled *Ming tao sü* ("Introduction to the Understanding of the Doctrine"), in ten sections, a sort of apology of the Jewish religion. Neither of these tracts has survived. The Jesuit Gabriel Brotier informs us that they printed in Chinese only a single very small book on their religion which they presented to the mandarins when menaced by a persecution, and this may be identical with the tract of Chao Ying-tou.

Two facts are conspicuous in the history of the Chinese Jews: they hailed from Persia and India and reached China by way of the sea. The historical portion of the earliest inscription of 1489 points to India (Tʿien-chu) as the country from which the Jews had started on their way to China—seventy families, bringing cotton goods of the Western countries as tribute to the court of the Sung and settling at Pien-liang (the older name for Kʿai-fung). No date for this event is fixed, nor is the name of the Sung emperor given. All that can be safely asserted is that the first settlement of Jews in the Sung capital took place between the years 960 and 1126 when the city was conquered by the Jurchi and the capital was removed to Hang-chou. The first date on record is the year 1163 as that when the construction of the synagogue was commenced. The gift of cotton goods points directly to India, as the cotton plant was not yet cultivated in China under the Sung, and Indian cotton fabrics were highly appreciated there; it thus stands to reason that it was the cotton trade in the interest of which the Jews came to China. In the second inscription of 1512 the origin of the first ancestor, Adam, is traced to India, and in the third inscription of 1663 it is stated that "the Jewish religion took its origin in India." The official designation of the Chinese Jews was "religion of India," and this name has persisted until recent times and was the only one known to the Chinese Jews whom I had occasion to interrogate in 1903. The Indian Jews had emigrated from Persia, and Persian influence is plainly evident among the Chinese Jews. Like the Persian Jews, they divided the Pentateuch into fifty-three sections (instead of fifty-four), the Masoretic fifty-second and fifty-third sections being combined into one, which was recited during the week of the Feast of Tabernacles. Like the Persian Jews, they counted twenty-seven letters of the Hebrew alphabet (instead of the standard of twenty-two) by rating the final kaph, mem, nun, pe, and tsad as separate letters. All directions as to the recitation of prayers were given in Persian, and according to Dr. E. N. Adler,[1] a Judeo-Persian translation is added to some hymns in a prayer-book for the Passover service. The most interesting point is that the Chinese Jews designated the rabbi by the Persian word *ustād* ("teacher," "master"), used in the same sense by the Persian Jews; thus, our earliest inscription speaks of a Lie-wei Wu-se-ta, "Rabbi Levi." What should be stressed in particular is that not a trace of Pehlevi or Middle Persian has been found among the Chinese Jews, but that the Iranian element in their midst is strictly New Persian which, as generally assumed, developed from about the tenth century, so that their immigration into China could hardly have taken place before that period. The language spoken by them at that time was most probably New Persian, which was the *lingua franca* all over the Far East during the Middle Ages. The best example to illustrate this point is the name of the Jews, as it is on record in the annals of the Yüan dynasty to which I alluded; this name is a very exact phonetic transcription of N.Pers. *Djuhūd* or *Djahūd* with initial palatal sonant, while in Middle Persian the word is *Yahūt*, correspond-

[1] *Jewish Quarterly Review*, X, 624.

ing to Heb. *Yehūdi* and Arab. *Yahūd*. The change of initial *y* into *j* is peculiar to New Persian. For the Chinese it was just as easy to transcribe *ya* as *dja* or *dju*, but the fact that they transcribed *Djuhūd* goes to show that they heard the New Persian form and that they could not have learned the name of the Jews before the tenth century.

In the course of a few generations the small band of Jews became almost completely sinicized, adopting the Chinese language, attire, manners, and customs and eagerly absorbing Chinese literature and Confucian ethics. In matters of phonetics they adapted themselves to Chinese to such a degree that in Chinese fashion they dropped the liquid *r*, replacing it by *l*, and forgot how to articulate the sonants; thus, they pronounced *Thaula* for *Thora*, *Tavite* for *David*, *Etunoi* for *Adonai*, *I-se-lo-ye* for *Israel*, *Ie-le mei-hung* for *Jeremiah*, etc. In other words, they applied Chinese phonetics to the pronunciation of Hebrew.

Another point to be emphasized is that the Jewish technical terminology, as revealed in their inscriptions, is much dependent on that of the Chinese Mohammedans. From these the Jews adopted, e.g., the term *Mollah* (transcribed in Chinese *man-la*) and the name of the synagogue, *Tsᶜing-chen se*, which the older translators rendered literally, but wrongly, "the pure and true temple." Even Dr. Martin, in 1906, translated it "the Temple of the Pure and True." *Tsᶜing-chen*, however, is the technical Islamic term for "Allah," and *Tsᶜing-chen se* is simply a mosque; for the Jews it signified "temple of God" or simply "synagogue." The synagogue of Kᶜai-fung was built after the model of a mosque. In company of Arabic and Persian Mohammedans the Jews must have made their first appearance in China, for the various stages of their migration can be traced with a fair degree of exactness; we meet them in the same ports of southern China as the Arabs and Persians: at Zaitūn (the Arabic name for Tsᶜüan-chou fu in Fu-kien Province), Ning-po, Hang-chou, Nanking, Yang-chou, finally advancing into the metropolis of the Northern Sung, Kᶜai-fung, and in the fourteenth century also in Peking. It is not necessary to assume that there was but a single stream of their immigration into China; more probably they poured in gradually, in small detachments, but they always entered China from India over the maritime route at the southern ports, not, as was formerly believed without reason, over the land route by way of Central Asia. The first immigration may be assigned to the ninth or tenth century.

The Chinese-Hebrew manuscript here in question came from Kᶜai-fung fu to Shanghai as far back as 1851, and was briefly noticed in the *North-China Herald* of Shanghai[1] together with several copies of the Pentateuch and rituals. It was defined there as "a genealogical table of the principal Jewish families of Kᶜai-fung." The Chinese characters are very crudely written, apparently with a stylus and by several inexperienced scribes. The Chinese Jews used Chinese paper, several sheets of which were pasted together, but they did not use Chinese writing-brushes or ink for sacred purposes; they availed themselves of a bamboo stylus and annually made sufficient ink at the Feast of Tabernacles for the ensuing year.

The register contains first the names of 453 men distributed over 7 clans indicated by the Chinese family names Ai, Li, Chang, Kao, Chao, Kin, and Shi, and presumably including about 200 individual families. The inscription of 1663 also speaks of about 200 families with reference to the year 1642. These 7 clan names are traceable to the oldest inscription of the year 1489 as being among those who first settled at Kᶜai-fung under the Sung; this inscription mentions 70 families and enumerates 16 clan names; accordingly, 9 names of this inscription do not appear in our register, although several of these are recorded in the names of women. The strongest clan is that of Li represented by 109 individuals, followed by the Kao with 76, the Chao with 74, the Chang with 73, the Ai with 56, the Kin with 42, and the Shi with 23 names (total, 453). In order to arrive at a satisfactory date of the register, I drew up a careful list of all the names with biographical data, which occur in the three stone and the twenty-three wooden tablets, with the result that half-a-dozen names listed in the most recent inscription of 1663 recur also in our register, so that the latter must be coeval with the date of this inscription or must have been prepared shortly afterward, say, roughly, during the decade of 1660–70. Moreover, the 7 clan names of the register are contained on the reverse of the inscription tablet of 1663 as the names of those who contributed funds for the reconstruction of the synagogue which had been destroyed by an inundation of the Yellow River in 1642.[2] The

[1] August 16, 1851; reproduced in *Chinese Repository*, XX (1851), 465.

[2] Tobar, *Inscriptions juives*, p. 83.

register consequently is a thoroughly authentic document. There is no relation between the Chinese and Hebrew names. Ben Israel, Ben Josef, Ben Aron, Ben Mosheh, Ben Jehosha, and Abraham Ben Israel are among the most frequent Hebrew names.

In the section devoted to the women, a total of 259 names is listed, in most cases only the name of the family from which the woman originated; in some cases, however, her personal name is added. It appears that many of these women were Mohammedans or of pure Chinese stock; in one case there is even a woman from the orthodox clan Kᶜung (Confucius) and another née Mong (Mencius). In the Jesuit relations it is asserted that the Jews, while they freely intermarried with Gentiles, did not allow their daughters to contract a marriage with one outside their religion. In several cases it is indicated in the register that "Mme So-and-So" is the wife of "Mr. So-and-So" or the mother of "So-and-So." The most frequent Hebrew names of women are "Daughter of Adam" and "Daughter of Israel." Each section winds up with a prayer in Hebrew. The writer expresses the wish that the men whose names are inscribed in the register may be united with the seven ancient righteous sages—Abraham, Isaac, Jacob, Moses, Aaron, Elijah, and Elisha—and meet with them under the tree of life in the gardens of Eden. A similar prayer is devoted to the women. Names and number of children are unfortunately not given, so that the register has but little value to the student of vital statistics. The total number of individuals recorded is 712. Assuming that there were several hundred children and that there were a number of Jewish farmers scattered over the villages in the environment of the city and not officially registered by the synagogue of Kᶜai-fung, we may arrive at an estimate of about a thousand souls. This result is in accord with a contemporaneous report of the Portuguese Jesuit, Pater Gozani, who visited Kᶜai-fung in 1704 on direct instructions from Rome and who writes that the number of Jewish families (he means, of course, clans) was then reduced to 7 and that the total population amounted to about 1,000. By 1850 the number of Jews in Kᶜai-fung had diminished to about 200 individuals, but the 7 clan names were still recorded by the Protestant delegates.

The last statistical information I was able to obtain came in a letter of Li King-sheng, a Chinese Jew then about fifty-two years old who died in 1903, addressed to the Shanghai Society for the Rescue of

Chinese Jews and dated April 5, 1901. Li wrote that at that time there were about 50 families in existence of the names Kao, Li, Chao, Shi, Kin, and Chang, numbering about 250 souls. None of them, he said, could write or read Hebrew; none observed the Mosaic Law. The Sabbath was not kept. They were scattered about all over the city, some employed in government offices as junior assistants, others keeping small shops, and the sole distinction between them and the other Chinese being that they did not worship idols and did abstain from pork.

I have referred above to an interview of the Jesuit Matteo Ricci with a Jew in 1605. Pelliot[1] has devoted a special notice to this Jew. This Jew of whom Ricci gives only his family name Ai had come to Peking to obtain an official post. Ricci reports that this man, who was about sixty years old, told him that because he had followed the career of one of the Chinese litterati he had been expelled from the synagogue by the archpriest who is their chief, and had almost been excommunicated, and that he would have easily abandoned his religion if he had been able to obtain the Doctor's degree as the Musulmans do, who if successful in obtaining the Doctor's degree no longer have fear of their Mollahs and abandon their religion. Now Pelliot has identified this interlocutor of Ricci with a certain Ai Tᶜien whose name he traced in the Chinese Gazetteer of Kᶜai-fung fu as having obtained the degree of licentiate in 1573 and as having reached the position of district magistrate (*chi-hien*). The fact that the name of a Jewish official is traceable in a local gazetteer is interesting in itself and also encouraging in raising hopes to find more Jewish names in Chinese records. But Pelliot's identification of this Ai Tᶜien with the Mr. Ai of Ricci is not conclusive, for he has overlooked a very important fact, and this is that the said Ai Tᶜien is the author and donor of an orthodox Jewish inscription tablet to the synagogue of Kᶜai-fung,[2] and in this document signs himself as a disciple of the Jewish religion. Ricci asserts that this Jew, according to his story, had from childhood studied only Chinese and had never learned the Hebrew letters; but in his inscription this alleged sinicized Jew proclaims, "We recite the 53 sections of our sacred books and instruct our families in the knowledge of the 27

[1] "Le Juif Ngai, informateur du P. Mathieu Ricci, *Tᶜoung Pao*, XX (1920–21), 32–39.

[2] Tobar, *op. cit.*, p. 28, No. XV.

letters of our alphabet." Moreover, this alleged heretic Ai Tᶜien had a son, Ai Ying-kwei, who on his part had five sons, all named in the last inscription of 1663 as having taken an active part in the rebuilding of the synagogue. One of Ai Tᶜien's grandsons even had his grandfather's inscription tablet restored and re-engraved. All these data go to prove incontrovertibly that Ai Tᶜien was not a renegade, as Ricci's story makes him out, but on the contrary was a good and faithful Jew. There is but one alternative: either Pelliot's identification of Ricci's Ai with Ai Tᶜien is untenable, or if it be correct, Ricci's story cannot be true—or Ricci, despite his excellent knowledge of Chinese, may have misunderstood his informant, or the Jew Ai must have had some reason for mystifying Ricci with a yarn.

I have mentioned this incident not from a desire to antagonize Ricci, for whom I have a keen admiration, but as an example to show that a study of the lives and genealogy of the Chinese Jews is of real historical interest. For this reason I am planning to publish this register *in extenso*, giving the Chinese names in one column with the corresponding Hebrew names in the next column. The importance of this document rests on the fact that it supplies us with an arsenal of weapons, the names of 453 men definitely identified as Jews, and that these names offer us an opportunity of looking for further information in regard to them in Chinese records, especially in the local gazetteers which contain chapters giving lists of the graduates of the districts and officials who served in them.

At the same time I am also planning to publish a new translation of the Jewish inscriptions with an analytic commentary. Despite all that has been written on the Chinese Jews the real work remains to be done. There is not one complete or reliable English translation of their fundamental inscriptions. The only critical edition of the inscriptions we owe to the Jesuit Jerome Tobar,[1] whose translation in general is good, but suffers from many defects in details and lacks interpretation. The whole Jewish terminology remains to be studied at close range.

[1] *Inscriptions juives de Kᶜai-fong-fou, Variétés sinologiques*, No. 17, 1900.

198

新增玉器收藏

Field Museum of Natural History

Founded by Marshall Field, 1893

Roosevelt Road and Lake Michigan, Chicago

FIELD MUSEUM NEWS

STEPHEN C. SIMMS, *Director of the Museum*.....*Editor*

CONTRIBUTING EDITORS

BERTHOLD LAUFER..........*Curator of Anthropology*
B. E. DAHLGREN...........*Acting Curator of Botany*
O. C. FARRINGTON................*Curator of Geology*
WILFRED H. OSGOOD..............*Curator of Zoology*

H. B. HARTE............*Managing Editor*

Field Museum is open every day of the year as follows:

November, December, January 9 A.M. to 4:30 P.M.
February, March, April, October 9 A.M. to 5:00 P.M.
May, June, July, August, September 9 A.M. to 6:00 P.M.

Admission is free to Members on all days. Other adults are admitted free on Thursdays, Saturdays and Sundays; non-members pay 25 cents on other days. Children are admitted free on all days. Students and faculty members of educational institutions are admitted free any day upon presentation of credentials.

The Library of the Museum, containing some 92,000 volumes on natural history subjects is open for reference daily except Sunday.

Traveling exhibits are circulated in the schools of Chicago by the Museum's Department of the N. W. Harris Public School Extension.

Lecturers for school classrooms and assemblies, and special entertainments and lecture tours for children at the Museum, are provided by the James Nelson and Anna Louise Raymond Foundation for Public School and Children's Lectures.

Announcements of courses of free illustrated lectures on science and travel for the public, and special lectures for Members of the Museum, will appear in FIELD MUSEUM NEWS.

In the Museum is a cafeteria where luncheon is served for visitors. Other rooms are provided for those bringing their lunches.

Members should inform Museum promptly of changes of address.

A VACATION SUGGESTION

Assuming what is probable, in view of the attendance records of the past several years, considerably more than one million persons will visit Field Museum this year, and thereby, it is hoped and confidently expected, profit both in the matters of enjoyment and adding to their knowledge.

Another million, more or less, will think about visiting the Museum—"some day when I have time." Perhaps you are one of the many who realize fully the advantages of visiting the Museum, and who frequently make a mental note to do so, whenever some Museum activity comes to notice through the press or otherwise, and then defer your visit until finally months pass by and it does not materialize.

Of course, as many people say to themselves, "The Museum will always be there, so why hurry?" But too often this attitude results in failure to make the desired visit at all.

If you are one of those who has failed to make that long-planned Museum visit due to actual lack of time, your opportunity is now at hand. Vacation time is here. A mere fraction of your vacation—say a half-day—spent at the Museum will, it seems sure, amply repay you in pleasure and in intellectual stimulation. Even if you have made a visit recently, come again—additions and improvements are constantly being made, and it is safe to say that no person, in a single visit or even a series of visits, has more than begun to cover all there is here to interest him. Participate in some of the guide-lecture tours, a schedule of which is published elsewhere in FIELD MUSEUM NEWS. Members of the Museum, of course, are admitted free on all days, and other persons on Thursdays, Saturdays and Sundays. Remind your friends. Bring or send the children, too—they are always admitted free. The Museum can help to solve your problem on those vacation days when you don't know just what to do with the youngsters.

Museum Librarian Resigns

Miss Elsie Lippincott, Librarian of Field Museum, has resigned after thirty-three years of service on the staff of the institution. Her resignation, tendered on account of ill health, was accepted with regret, and became effective on June 30.

Miss Lippincott joined the staff of the Museum in 1897 as Assistant Librarian, and was appointed Librarian in charge of the Museum's Library in 1900. Her physician had advised her retirement for some time past, but she delayed the step as long as possible because of her strong attachment to and deep interest in her work, and her long association with so many members of the Museum staff.

Miss Lippincott's administration of the Library was marked by extreme ability and loyalty, and in addition she brought to her work a kindly and helpful personality which was of much assistance to members of the Museum staff and to visitors from outside whenever they had occasion to consult the books on the library shelves.

Mrs. Emily M. Wilcoxson, Assistant Librarian since 1905, has been appointed Librarian to succeed Miss Lippincott.

Gifts to the Museum

Following is a list of some of the principal gifts received by the Museum during the last month:

From John G. Shedd Aquarium—14 tropical fishes; from James J. Mooney—one plated lizard; from Stanley Field—a stone meteorite weighing 745 pounds; from Frank Vondrasek—30 specimens of minerals, concretions, flint arrowheads and spearheads; from R. T. Crane, Jr.—one decorated white jade ax, one inscribed jade slab from a jade book and three archaic jade carvings of deer, dragon and ox, China; from Moise Dreyfus—a Navaho blanket; from C. D. Mell—77 herbarium specimens, Mexico; from E. E. Sherff—23 herbarium specimens, Hawaii.

BEQUESTS AND ENDOWMENTS

Bequests to Field Museum of Natural History may be made in securities, money, books or collections. They may, if desired, take the form of a memorial to a person or cause, to be named by the giver. For those desirous of making bequests, the following form is suggested:

FORM OF BEQUEST

I do hereby give and bequeath to Field Museum of Natural History of the City of Chicago, State of Illinois,

....

....

Cash contributions made within the taxable year to Field Museum not exceeding 15 per cent of the taxpayer's net income are allowable as deductions in computing net income under Article 251 of Regulation 69 relating to income tax under the Revenue Act of 1926.

Endowments may be made to the Museum with the provision that an annuity be paid to the patron for life. These annuities are tax-free and are guaranteed against fluctuation in amount.

ADDITIONS TO JADE COLLECTIONS

By BERTHOLD LAUFER

Curator, Department of Anthropology

Five important objects of Chinese jade were recently presented to Field Museum by Richard T. Crane, Jr., a Trustee of the institution, and will soon be placed on exhibition in the Jade Room now in course of installation. One of these is a ceremonial battle-ax carved from a grayish white jade and beautifully ornamented on both sides with conventionalized monster heads of archaic style, which are symbolic of attack. Four animal heads jut out from the edges, and the notches between them are emblematic of "the teeth of war."

Jade Battle-ax

The period and significance of this unique weapon are revealed by an inscription of eight characters in ancient style, four on the obverse and four on the reverse. This inscription reads, "Made by order of the Great Sung dynasty and bestowed upon the President of the Board of War." The Sung emperors, who reigned from A.D. 960 to 1279, maintained ateliers in which pottery, bronzes and jades were manufactured for use in the palaces of the court or to be presented by the sovereign to deserving officials. The jade ax in question is a product of the imperial studios and was conferred by the emperor on the minister of war as a badge of office and emblem of power.

Jade slabs were used in ancient China as writing material, and documents carved in such slabs were united into books. The Museum owns such a jade book consisting of ten slabs, thirty pounds in weight, and inscribed by the emperor K'ang-hi of the Manchu dynasty. Through a lucky chance, a slab from a jade book containing the handwriting of his grandson, the emperor K'ien-lung, has now come into the Museum's possession, so that the two greatest sovereigns of the Manchu dynasty are now represented here with facsimiles of their compositions in jade.

The jade slab presented by Mr. Crane is engraved with a pair of rampant five-clawed dragons soaring in clouds and striving for a flaming pearl, the ocean and an island emerging from the waves below. The center is occupied by the title of the book, which reads, "A Dissertation on Talents and Virtues with Reference to the Counsels of Kao Yao—an Imperial Essay." Kao Yao was minister of justice to the ancient emperor Shun, and is still regarded as the model for all administration of law. His wise counsels form a chapter of the Shu king, the oldest historical book of China.

In earliest times carvings of jade were buried with the dead in the belief that this stone, regarded as the most precious jewel and as embodying the quintessence of nature, would have the tendency to preserve the body from decay and to promote its resurrection. Small figures of animals delicately carved from jade were attached to the shroud. Three very fine and rare examples of this type of the early archaic period—an elk, an ox-head, and a fish-monster—are included in the gift of Mr. Crane.

199

中国的铃、鼓和镜

NUMBER CCCXXXI VOLUME LVII OCTOBER 1930

THE BURLINGTON MAGAZINE

for Connoisseurs

Illustrated & Published Monthly

CONTENTS

A NEWLY DISCOVERED TONDO BY BOTTICELLI—BY GIUSEPPE FIOCCO
THE DUBLIN APOCALYPSE—BY JAMES WARDROP
MAESTRO PAOLO VENEZIANO—BY EVELYN SANDBERG VAVALA
CHINESE BELLS, DRUMS AND MIRRORS—BY BERTHOLD LAUFER
A SILVER CHASSE-RELIQUARY—BY W. W. WATTS
THE IRANIAN ELEMENT IN PERSIAN ART—BY SAVA POPOVITCH
THE LITERATURE OF ART
SHORTER NOTICES: THE NEW MUSEUM BUILDINGS IN BERLIN (D.L.); MR. JAMES L. CAW; SIR PHILIP SASSOON
LETTERS: "ART AND SEX" (M. J. NICOLSON); "THE NEW REMBRANDT" (EDITH A. STANDEN)

LONDON: THE BURLINGTON MAGAZINE LTD., 16A ST. JAMES'S STREET, S.W.1.
PARIS: GALIGNANI'S, 224 RUE DE RIVOLI. FLORENCE: B. SEEBER, 20, VIA TORNABUONI.
NEW YORK: BRENTANO'S INCOR., 1 WEST 47TH STREET; E. WEYHE, 794 LEXINGTON AVENUE.
CHICAGO. WASHINGTON: BRENTANO'S INCOR. AMSTERDAM: J. G. ROBBERS, SINGEL, 151-153.

PRICE HALF A CROWN NET; ANNUAL SUBSCRIPTION (INCLUDING INDEXES) THIRTY-TWO SHILLINGS painted brocades of Venice, and the Oriental love of jewellery—see the large round brooches of the Madonna.

Human proportions and types are also Byzantine; Byzantine with a slight, as yet a very slight, humanization of the Oriental model. Sad-visaged ancients with goat-beards and vast, domed, corrugated foreheads, stiff-legged, haggard Baptists with rough hair and vehement gestures, who clench their emaciated hands on open scrolls, stand as guardians to the hieratic Madonnas, in whom alone there is seen some approach to human sentiment, but who, like their attendant angels, go back as closely as did the mystic creations of a Cimabue or a Duccio a hundred or fifty years earlier to the Byzantine model, with its delicate pathos, its reserved sadness, its tiny pinched features and bent head, its long prehensile hands. The Gothic current, destined to meet the Byzantine on almost equal terms in the art of Lorenzo Veneziano, and to overcome it and obliterate in Giambono and Jacobello del Fiore, is here only faintly incipient. We find it, at most, in the elongated proportion—too long even for correct Byzantinism, in the swing of the draperies and occasionally in a slight movement of the entire figure, as for example in the Madonna of Sir Joseph Duveen, the saints of the Worcester polyptych and the unforgettably beautiful angels of the Bologna polyptych.

CHINESE BELLS, DRUMS AND MIRRORS

BY BERTHOLD LAUFER

THIS monumental volume,[1] illustrated by seventy-five superb plates, demonstrates that Mr. Eumorfopoulos, in acquiring his bronzes, was not merely guided by æsthetic motives or solely by the desire to accumulate beautiful objects, but primarily aimed at having a collection thoroughly representative of all types of ancient bronze objects found in China and selected with discriminating taste. In view of that, the work is a most important contribution to Chinese archælogy, and Dr. W. Perceval Yetts, the editor of this six-volume series, has performed an admirable task, for which his fellow-workers will be grateful to him. In the text, which precedes the plates, the author endeavours to establish facts based on a serious study of Chinese records and palæography, and steers clear of theories and speculations which are now the fashion among students of Chinese art without knowledge of the Chinese language and civilization. The writer has learned a great deal from this book, which is packed with new and solid information.

There are two magnificent bells in the Eumorfopoulos Collection. The ancient bronze bells have been but little studied by our scholars, chiefly because of lack of material. The large ones are among the rarest objects in our Oriental collections, being difficult to obtain, as most of them are covered with more or less lengthy inscriptions of historical value and are therefore so highly prized by the Chinese that they are reluctant to part with them. Aside from a few remarks of Hirth (*T'oung Pao*, 1896, pp. 496-498) and some stray notes in the literature on Chinese musical instruments, nothing worth reading has been written about bells. Dr. Yetts is the first to present us with a detailed and instructive investigation of this difficult subject, and to describe minutely the different types of bells and their construction. The decipherment, translation, and interpretation of the lengthy and intricate inscription on the Eumorfopoulos bell, which Dr. Yetts assigns to a period between 561 and 340 B.C., constitute a most scholarly and painstaking piece of research.

Dr. Yetts's dissertation on the drums used in ancient China is an interesting and valuable scientific contribution to this little explored subject. The various types of drums are figured and discussed by him, and he winds up with a learned and enlightening review of our present knowledge of the famous bronze drums, enlivening it with many novel and refreshing points. His inference that " an actual prototype of the mysterious bronze drums did exist in ancient China " is worthy of serious consideration, and his observations on the heron motive and the connexion of the heron with drums are ingenious and helpful. With the general conclusions reached by the author (pp. 28-29) I am entirely in accord and, like himself, feel that only systematic excavations in southern China and Indo-China will enable us to advance our present knowledge and to establish the desired historical chronology of these drums instead of relying on alleged history elicited from a stylistic classification that must remain more or less subjective and hazardous.

The subject of bronze drums has engaged my attention for almost thirty years, and I do

[1] The George Eumorfopoulos Collection Catalogue of the Chinese and Corean Bronzes, Sculpture, Jades, Jewellery and Miscellaneous Objects. By W. Perceval Yetts. Vol. II. Bronzes: Bells, Drums, Mirrors, etc. VIII + 99 pp. with 44 figs. + 75 pl. (25 in colour). (Benn.) £12 12s.

not flatter myself that I have solved the whole series of complex problems which confront us here. I obtained four bronze drums at Suchow in 1901, six at Ch'êng-tu, capital of Szechwan, in 1908, and subsequently two of enormous size with decorations of frogs from Kwang-si Province and two from the Karen tribe of Upper Burma. I have also collected a large number of important Chinese texts bearing on the subject and unnoticed by De Groot and Hirth. In the preface of his book Dr. Yetts states that there is a voluminous literature concerning bronze drums, but that it is not in English. I beg to point out that one of the very first contributions to this fascinating subject is in English and is due to T. de Lacouperie (*On Antique and Sacred Bronze Drums of Non-China,* in *The Babylonian and Oriental Record,* VII, 1894, pp. 193-204, 217-220). I do not think much of the work of F. Heger, which is merely a good collection of materials, but which lacks intelligence and vision. But the notice by A. Fischer (*Ueber die Herkunft der Shan-trommeln,* in *Zeitschrift für Ethnologie,* 1903, p. 668) is important, because it contains a description of the modern manufacture of bronze drums in eastern Karenni. During the last hundred years this industry has also flourished in Shantung, as I infer from statements made in local gazetteers of this province, so that the dating of bronze drums remains somewhat problematical. The view that most of them go back to the Han period is untenable; and Dr. Yetts sounds a timely warning as to dates on bronze drums which in my opinion are all subject to suspicion. Neither Mongolia nor New Guinea are to be included in the area in which bronze drums are found, nor do I believe that their designs can in any way be correlated with those encountered on shamans' drums of Mongolia and Siberia. No bronze drums have ever been found in northern China; they occur only in Szechwan and the southern provinces of the country, in the Shan States of Upper Burma and sporadically in Indo-China, Java, and some other islands of the Archipelago, whither they were probably brought in the course of trade. The theory that they originated in Camboja is due to W. Foy, not to Heger, who merely copied his predecessor. It is entirely unconvincing and lacks both historical and archæological evidence.

The catalogue illustrates a fine series of metal mirrors. "To some they make a strong æsthetic appeal; for no other artistic product of China shows such a high standard of nearly uniform excellence in craftsmanship and decorative invention. To others they are human documents which, with their inscriptions and designs, convey precious clues to bygone customs and beliefs," Dr. Yetts comments. The mirrors are arranged in approximately chronological order, but this arrangement is said to be tentative, as all efforts to catalogue mirrors must be in our present state of knowledge. Dated mirrors are numerous, yet a large number of types still remain unrepresented by dated specimens, and opinion concerning these depends on conjecture and clues derived from excavated examples. The Eumorfopoulos collection contains no mirror inscribed with the actual year when it was made, but there is one that bears an allusion to the period A.D. 601-604. In the study of mirrors, Dr. Yetts has arrived at many novel and remarkable conclusions. Mirrors of the class B 10 to 12, which formerly were assigned to the early part of the Six Dynasties and which were thus dated by myself, are now attributed to the Former Han period on the evidence of excavations in Korea. There is an elaborate discussion of the Four Supernatural Animals symbolizing the Four Quadrants of the vault of heaven, with an interesting digression into astronomical problems. There is no evidence to prove that the Four Animals existed as a symbolic group earlier than about the third century B.C., and the author is inclined to think that with the advance of research they may be found less "Chinese" than is generally imagined.

I do not think that the name of the goddess Si Wang Mu is to be taken as the transcription of a foreign name (p. 39), but that it must be taken in its literal sense: "Royal Mother of the West." In all discussions and controversies concerning this name and its significance one important fact has been overlooked, and this is that in the account of Ta Ts'in (the Roman or Hellenistic Orient), inserted in the *Wei lio* and *Hou Han shu,* Si Wang Mu appears as a goddess of Western Asia (*cf.* Chavannes, *T'oung Pao,* 1905, p. 556, and 1907, p. 185; and Hirth, *China and the Roman Orient,* pp. 43, 51, 77, 292). This, of course, is the well-known Great Goddess, the Ishtar of Babylonia, worshipped in the Near East down to Hellenistic times as a deity of fertility, in the same manner as Si Wang Mu in China. The notion that the mirrors decorated with designs of grapes, birds, and fabulous animals go back to Han times was first expounded by Wang Fu in his *Po ku t'u lu,* who argues naïvely that General Chang Kien, because he introduced the grapevine into China must have also brought along the artistic motive of grape bunches. On my first visit to Si-an fu in 1902, the date of this class of mirror was the first subject I discussed with native antiquarians, and the consensus of opinion there was that these mirrors are productions of the sixth and seventh centuries. I presume that the impetus to the design was received from Sasanian art, but that the Chinese adapted it freely to their own decorative style. I am entirely in accord with Dr. Yetts in regarding as Chinese the technique of gold, silver, and mother-of-pearl ornaments embedded in lacquer and in explaining the glassy black lacquer-like surface of some mirrors not as an accidental patina, but as an artifact for which siliceous matter in all probability was used.

Bishop White, who resides at K'ai-fung, Ho-nan, recently sent me a number of rubbings of mirrors found in his province, which he says date from the Ts'in period. They correspond in style of design to B8 of the Eumorfopoulos collection [PLATE]. Dr. Yetts discusses the theory which assigns this type to the third century B.C. and to a local school of craftsmanship in the Huai Valley. Though he regards it as the earliest mirror in the collection, he reserves conclusions as to time and place until the results of excavation elsewhere in China are better known.

The catalogue contains a number of Korean mirrors. Such must have frequently been traded from Korea to China, for in a group of more than two hundred mirrors gathered by me in China for the Field Museum, Mr. Umehara identified about half a dozen as Korean. A pair of dragons may be a favourite design for Korean mirrors (B65), but one of the same design was found by me in 1903 at Tsi-nan, Shan-tung, bearing a date-mark of the Yüan period.

The subject on the mirror B69, in my opinion, represents the imaginary journey of the Emperor Ming Huang to the palace of the moon, whither he was conducted by the magician Lo Kung-yüan, who threw his staff into the air; the staff changed into a dazzling bridge over which the travellers passed with safety (*cf.* Giles, *Chinese Biogr. Dict.*, No. 1389). On the mirror there is represented the bridge which the Emperor and the magician are about to cross; beyond are the palace of the moon and three fairies in the middle distance; the drug-pounding hare and the toad as lunar emblems are pointed out by Dr. Yetts himself.

The author has traced two intriguing Chinese-Greek parallels. Both Greeks and Chinese saw in Orion an armed man—a coincidence the more noteworthy as little in the configuration of its stars suggests such a conception; and Hellenic influence may be sought in the disposition of the stars in Scorpio, suggestive of a tailed reptile, and this conception may have helped to establish the dragon in an astronomical setting which accorded thoroughly with native tradition. As to the theory mentioned by Dr. Yetts, that the concepion of a dragon was originally based on the alligator which haunts the Yangtse, this must be restricted to a certain type of dragon; there are many varieties of the dragon, and not all of these are traceable to the alligator. A special investigation of this problem is urgently needed.

The remainder of the work deals with buckles, plaques, horse's bits, coins, tallies, seals, stands, pommels, and Korean bronze vessels.

In regard to the hooks B227-228, which Dr. Yetts defines as "finials for staffs used in the furniture of tombs to support canopies or other hangings," the most reasonable explanation seems to be that they were used in hanging on to a wall or taking down wooden tablets used for writing; they have a striking family resemblance to the modern metal hooks still used for hanging pictures or taking them down. I have many such hooks of gilt bronze and also of jade.

Dr. Yetts promises to treat in the fourth volume "that difficult question of the animal art conveniently termed Scythian." We anticipate this study with the greatest interest. The bibliography is very complete and conscientious.

A SILVER CHÂSSE-RELIQUARY BY W. W. WATTS

RARELY does a work of art of this character [PLATES I and II] pass into the possession of a private collector; it seems more reasonable to expect to meet with it in the treasure of some ancient church or cathedral. Here there is no clue to its original possessor, and whatever relic it may have contained has been carefully removed.

Its provenance may, however, be determined by the study of its workmanship and decoration. The form follows the lines of a well-known type. It is an oblong chest standing on a wide base, with a heavy overhanging pent roof which may have originally been furnished with a ridge cresting. The first noticeable feature is its great weight and solidity; it is throughout of oak about an inch and a half in thickness, the obvious determination being to make robbery difficult if not impossible. This is overlaid with thin sheet silver. The relic was probably the arm-bone of a saint, and undoubtedly invested with considerable sanctity. Intended to be placed on or above an altar and

Chinese bronze mirror; third to first century, B.C. Actual size. (Mr. George Eumorfopoulos)

Chinese Bells, Drums and Mirrors